# The IMA Volumes in Mathematics and its Applications

## Volume 70

*Series Editors*
Avner Friedman Willard Miller, Jr.

# Institute for Mathematics and its Applications
## IMA

The **Institute for Mathematics and its Applications** was established by a grant from the National Science Foundation to the University of Minnesota in 1982. The IMA seeks to encourage the development and study of fresh mathematical concepts and questions of concern to the other sciences by bringing together mathematicians and scientists from diverse fields in an atmosphere that will stimulate discussion and collaboration.

The IMA Volumes are intended to involve the broader scientific community in this process.

Avner Friedman, Director
Willard Miller, Jr., Associate Director

* * * * * * * * * *

### IMA ANNUAL PROGRAMS

| | |
|---|---|
| 1982–1983 | **Statistical and Continuum Approaches to Phase Transition** |
| 1983–1984 | **Mathematical Models for the Economics of Decentralized Resource Allocation** |
| 1984–1985 | **Continuum Physics and Partial Differential Equations** |
| 1985–1986 | **Stochastic Differential Equations and Their Applications** |
| 1986–1987 | **Scientific Computation** |
| 1987–1988 | **Applied Combinatorics** |
| 1988–1989 | **Nonlinear Waves** |
| 1989–1990 | **Dynamical Systems and Their Applications** |
| 1990–1991 | **Phase Transitions and Free Boundaries** |
| 1991–1992 | **Applied Linear Algebra** |
| 1992–1993 | **Control Theory and its Applications** |
| 1993–1994 | **Emerging Applications of Probability** |
| 1994–1995 | **Waves and Scattering** |
| 1995–1996 | **Mathematical Methods in Material Science** |

### IMA SUMMER PROGRAMS

| | |
|---|---|
| 1987 | **Robotics** |
| 1988 | **Signal Processing** |
| 1989 | **Robustness, Diagnostics, Computing and Graphics in Statistics** |
| 1990 | **Radar and Sonar (June 18 - June 29)** |
| | **New Directions in Time Series Analysis (July 2 - July 27)** |
| 1991 | **Semiconductors** |
| 1992 | **Environmental Studies: Mathematical, Computational, and Statistical Analysis** |
| 1993 | **Modeling, Mesh Generation, and Adaptive Numerical Methods for Partial Differential Equations** |
| 1994 | **Molecular Biology** |

* * * * * * * * * *

### SPRINGER LECTURE NOTES FROM THE IMA:

*The Mathematics and Physics of Disordered Media*

Editors: Barry Hughes and Barry Ninham
(Lecture Notes in Math., Volume 1035, 1983)

*Orienting Polymers*

Editor: J.L. Ericksen
(Lecture Notes in Math., Volume 1063, 1984)

*New Perspectives in Thermodynamics*

Editor: James Serrin
(Springer-Verlag, 1986)

*Models of Economic Dynamics*

Editor: Hugo Sonnenschein
(Lecture Notes in Econ., Volume 264, 1986)

John E. Lagnese David L. Russell
Luther W. White
Editors

# Control and Optimal Design of Distributed Parameter Systems

With 19 Illustrations

Springer-Verlag
New York Berlin Heidelberg London Paris
Tokyo Hong Kong Barcelona Budapest

John E. Lagnese
Department of Mathematics
Georgetown University
Washington, DC 20057 USA

David L. Russell
Department of Mathematics
Virginia Polytechnic and
State University
Blacksburg, VA 24061-0123 USA

Luther W. White
Department of Mathematics
University of Oklahoma
601 Elm Avenue, Room 423
Norman, OK 73019-0315 USA



Mathematics Subject Classifications (1991): 93C20s, 93B05, 93E10, 93E12

Library of Congress Cataloging-in-Publication Data
Control and optimal design of distributed parameter systems / John E. Lagnese, David L. Russell, Luther W. White, editors.
p. cm. — (The IMA volumes in mathematics and its applications ; v. 70)
Papers based on the proceedings of a 1992 workshop.
Includes bibliographical references.
ISBN 0-387-94490-7 (acid-free)
1. Control theory—Congresses. 2. Distributed parameter systems—Congresses. 3. Optimal designs (Statistics)—Congresses.
I. Lagnese, J. II. Russell, David L. III. White, Luther William.
IV. Series.
QA402.3.C628 1995
620.1′01′178—dc20 95-2606

Printed on acid-free paper.



Production managed by Hal Henglein; manufacturing supervised by Jeffrey Taub.
Camera-ready copy prepared by the IMA.
Printed and bound by Braun-Brumfield, Ann Arbor, MI.
Printed in the United States of America.

9 8 7 6 5 4 3 2 1

ISBN 0-387-94490-7 Springer-Verlag New York Berlin Heidelberg

# The IMA Volumes
## in Mathematics and its Applications

### Current Volumes:

Volume 1: Homogenization and Effective Moduli of Materials and Media
Editors: Jerry Ericksen, David Kinderlehrer, Robert Kohn, and J.-L. Lions

Volume 2: Oscillation Theory, Computation, and Methods of Compensated Compactness
Editors: Constantine Dafermos, Jerry Ericksen, David Kinderlehrer, and Marshall Slemrod

Volume 3: Metastability and Incompletely Posed Problems
Editors: Stuart Antman, Jerry Ericksen, David Kinderlehrer, and Ingo Muller

Volume 4: Dynamical Problems in Continuum Physics
Editors: Jerry Bona, Constantine Dafermos, Jerry Ericksen, and David Kinderlehrer

Volume 5: Theory and Applications of Liquid Crystals
Editors: Jerry Ericksen and David Kinderlehrer

Volume 6: Amorphous Polymers and Non-Newtonian Fluids
Editors: Constantine Dafermos, Jerry Ericksen, and David Kinderlehrer

Volume 7: Random Media
Editor: George Papanicolaou

Volume 8: Percolation Theory and Ergodic Theory of Infinite Particle Systems
Editor: Harry Kesten

Volume 9: Hydrodynamic Behavior and Interacting Particle Systems
Editor: George Papanicolaou

Volume 10: Stochastic Differential Systems, Stochastic Control Theory, and Applications
Editors: Wendell Fleming and Pierre-Louis Lions

Volume 11: Numerical Simulation in Oil Recovery
Editor: Mary Fanett Wheeler

Volume 12: Computational Fluid Dynamics and Reacting Gas Flows
Editors: Bjorn Engquist, M. Luskin, and Andrew Majda

Volume 13: Numerical Algorithms for Parallel Computer Architectures
Editor: Martin H. Schultz

Volume 14: Mathematical Aspects of Scientific Software
Editor: J.R. Rice

Volume 15: Mathematical Frontiers in Computational Chemical Physics
Editor: D. Truhlar

Volume 16: Mathematics in Industrial Problems
by Avner Friedman

Volume 17: Applications of Combinatorics and Graph Theory to the Biological and Social Sciences
Editor: Fred Roberts

Volume 18: $q$-Series and Partitions
Editor: Dennis Stanton

Volume 19: Invariant Theory and Tableaux
Editor: Dennis Stanton

Volume 20: Coding Theory and Design Theory Part I: Coding Theory
Editor: Dijen Ray-Chaudhuri

Volume 21: Coding Theory and Design Theory Part II: Design Theory
Editor: Dijen Ray-Chaudhuri

Volume 22: Signal Processing: Part I - Signal Processing Theory
Editors: L. Auslander, F.A. Grünbaum, J.W. Helton, T. Kailath, P. Khargonekar, and S. Mitter

Volume 23: Signal Processing: Part II - Control Theory and Applications of Signal Processing
Editors: L. Auslander, F.A. Grünbaum, J.W. Helton, T. Kailath, P. Khargonekar, and S. Mitter

Volume 24: Mathematics in Industrial Problems, Part 2
by Avner Friedman

Volume 25: Solitons in Physics, Mathematics, and Nonlinear Optics
Editors: Peter J. Olver and David H. Sattinger

Volume 26: Two Phase Flows and Waves
Editors: Daniel D. Joseph and David G. Schaeffer

Volume 27: Nonlinear Evolution Equations that Change Type
Editors: Barbara Lee Keyfitz and Michael Shearer

Volume 28: Computer Aided Proofs in Analysis
Editors: Kenneth Meyer and Dieter Schmidt

Volume 29: Multidimensional Hyperbolic Problems and Computations
Editors: Andrew Majda and Jim Glimm

Volume 30: Microlocal Analysis and Nonlinear Waves
Editors: Michael Beals, R. Melrose, and J. Rauch

Volume 31: Mathematics in Industrial Problems, Part 3
by Avner Friedman

Volume 32: Radar and Sonar, Part I
by Richard Blahut, Willard Miller, Jr., and Calvin Wilcox

Volume 33: Directions in Robust Statistics and Diagnostics: Part I
Editors: Werner A. Stahel and Sanford Weisberg

Volume 34: Directions in Robust Statistics and Diagnostics: Part II
Editors: Werner A. Stahel and Sanford Weisberg

Volume 35: Dynamical Issues in Combustion Theory
Editors: P. Fife, A. Liñán, and F.A. Williams

Volume 36: Computing and Graphics in Statistics
Editors: Andreas Buja and Paul Tukey

Volume 37: Patterns and Dynamics in Reactive Media
Editors: Harry Swinney, Gus Aris, and Don Aronson

Volume 38: Mathematics in Industrial Problems, Part 4
by Avner Friedman

Volume 39: Radar and Sonar, Part II
Editors: F. Alberto Grünbaum, Marvin Bernfeld, and Richard E. Blahut

Volume 40: Nonlinear Phenomena in Atmospheric and Oceanic Sciences
Editors: George F. Carnevale and Raymond T. Pierrehumbert

Volume 41: Chaotic Processes in the Geological Sciences
Editor: David A. Yuen

Volume 42: Partial Differential Equations with Minimal Smoothness and Applications
Editors: B. Dahlberg, E. Fabes, R. Fefferman, D. Jerison, C. Kenig, and J. Pipher

Volume 43: On the Evolution of Phase Boundaries
Editors: Morton E. Gurtin and Geoffrey B. McFadden

Volume 44: Twist Mappings and Their Applications
Editors: Richard McGehee and Kenneth R. Meyer

Volume 45: New Directions in Time Series Analysis, Part I
Editors: David Brillinger, Peter Caines, John Geweke, Emanuel Parzen, Murray Rosenblatt, and Murad S. Taqqu

Volume 46: New Directions in Time Series Analysis, Part II
Editors: David Brillinger, Peter Caines, John Geweke, Emanuel Parzen, Murray Rosenblatt, and Murad S. Taqqu

Volume 47: Degenerate Diffusions
Editors: Wei-Ming Ni, L.A. Peletier, and J.-L. Vazquez

Volume 48: Linear Algebra, Markov Chains, and Queueing Models
Editors: Carl D. Meyer and Robert J. Plemmons

Volume 49: Mathematics in Industrial Problems, Part 5
by Avner Friedman

Volume 50: Combinatorial and Graph-Theoretic Problems in Linear Algebra
Editors: Richard A. Brualdi, Shmuel Friedland, and Victor Klee

Volume 51: Statistical Thermodynamics and Differential Geometry of Microstructured Materials
Editors: H. Ted Davis and Johannes C.C. Nitsche

Volume 52: Shock Induced Transitions and Phase Structures in General Media
Editors: J.E. Dunn, Roger Fosdick, and Marshall Slemrod

Volume 53: Variational and Free Boundary Problems
Editors: Avner Friedman and Joel Spruck

Volume 54: Microstructure and Phase Transitions
Editors: David Kinderlehrer, Richard James, Mitchell Luskin, and Jerry L. Ericksen

Volume 55: Turbulence in Fluid Flows: A Dynamical Systems Approach
Editors: George R. Sell, Ciprian Foias, and Roger Temam

Volume 56: Graph Theory and Sparse Matrix Computation
Editors: Alan George, John R. Gilbert, and Joseph W.H. Liu

Volume 57: Mathematics in Industrial Problems, Part 6
by Avner Friedman

Volume 58: Semiconductors, Part I
Editors: W.M. Coughran, Jr., Julian Cole, Peter Lloyd, and Jacob White

Volume 59: Semiconductors, Part II
Editors: W.M. Coughran, Jr., Julian Cole, Peter Lloyd, and Jacob White

Volume 60: Recent Advances in Iterative Methods
Editors: Gene Golub, Anne Greenbaum, and Mitchell Luskin

Volume 61: Free Boundaries in Viscous Flows
Editors: Robert A. Brown and Stephen H. Davis

Volume 62: Linear Algebra for Control Theory
Editors: Paul Van Dooren and Bostwick Wyman

Volume 63: Hamiltonian Dynamical Systems: History, Theory, and Applications
Editors: H.S. Dumas, K.R. Meyer, and D.S. Schmidt

Volume 64: Systems and Control Theory for Power Systems
Editors: Joe H. Chow, Petar V. Kokotovic, and Robert J. Thomas

Volume 65: Mathematical Finance
Editors: Mark H.A. Davis, Darrell Duffie, Wendell H. Fleming, and Steven E. Shreve

Volume 66: Robust Control Theory
Editors: Bruce A. Francis and Pramod P. Khargonekar

Volume 67: Mathematics in Industrial Problems, Part 7
by Avner Friedman

Volume 68: Flow Control
Editor: Max D. Gunzburger

Volume 69: Linear Algebra for Signal Processing
Editors: Adam Bojanczyk and George Cybenko

Volume 70: Control and Optimal Design of Distributed Parameter Systems
Editors: John E. Lagnese, David L. Russell, and Luther W. White

**Forthcoming Volumes**:

1992 Summer Program: *Environmental Studies*

1992–1993: *Control Theory*

Robotics

Nonsmooth Analysis & Geometric Methods in Deterministic Optimal Control

Adaptive Control, Filtering and Signal Processing

Discrete Event Systems, Manufacturing, Systems, and Communication Networks

1993 Summer Program: *Modeling, Mesh Generation, and Adaptive Numerical Methods for Partial Differential Equations*

1993-1994: *Emerging Applications of Probability*

Discrete Probability and Algorithms

Random Discrete Structures

Mathematical Population Genetics

Stochastic Networks

Stochastic Problems for Nonlinear Partial Differential Equations

Image Models (and their Speech Model Cousins)

Stochastic Models in Geosystems

Classical and Modern Branching Processes

1994 Summer Program: *Molecular Biology*

# FOREWORD

This IMA Volume in Mathematics and its Applications

## CONTROL AND OPTIMAL DESIGN OF DISTRIBUTED PARAMETER SYSTEMS

is based on the proceedings of a workshop that was an integral part of the 1992–93 IMA program on "Control Theory and its Applications." The workshop dealt with current control, stability, and optimal control problems for nonlinear partial differential equations and with questions of optimal design as they relate to partial differential equations, variational inequalities, etc., and it provided an assessment of work on control and stability of nonlinear partial differential equations. We thank John E. Lagnese, David L. Russell, and Luther W. White for organizing the workshop and for editing the proceedings. We also take this opportunity to thank the National Science Foundation, whose financial support made the workshop possible.

Avner Friedman

Willard Miller, Jr.

## PREFACE

As part of the IMA Year of Emphasis on Control Theory, a *Workshop on Control and Optimal Design of Distributed Parameter Systems* was convened in Minneapolis during November, 1992. Coordinators of the meeting were J.E. Lagnese (Georgetown University), D.L. Russell (Virginia Polytechnic Institute and State University), and L.W. White (University of Oklahoma, Norman). The workshop followed an extensive period of preliminary planning, at the outset of which a decision was made to concentrate on the control theory of systems governed by nonlinear partial differential equations and questions of identification and optimal design for general (linear and nonlinear) PDE systems, the latter to include material relating to the modelling of advanced materials. It was also decided that emphasis would be placed on the uniqueness and significance of contributions rather than on any attempt at completeness of coverage, so that ample time and space could be allotted to the contributors both in the workshop itself and in these proceedings.

The choice of subject matter and the timing of the workshop complemented each other quite agreeably. To a considerable extent it can be asserted that the control theory of linear PDEs had developed and matured during the twenty-five year period from 1965 to 1990, not in the sense that all questions had been settled but in the sense that systems of the major PDE types and questions naturally arising within those broad categories had by 1990 been extensively studied by some of the most able researchers in applied PDE theory with impressive results; results, indeed, of such depth and scope as we should hardly have dared hope for at the outset of our work in the 1960s.

By the time our workshop was convened in 1992, a number of researchers had begun serious studies in control and stabilization of nonlinear PDE systems, which is not to say that such questions had been wholly neglected before that; there were, indeed, some noteworthy contributions already in the 1970s which are precluded from listing here by conviction that any merit that a preface may have lies foremost in its brevity. We are fortunate to have included in this volume several papers representative of current studies in the area.

*Optimal design* of systems governed by PDEs is a relatively new area of study, now particularly relevant because of interest in optimization of fluid flow in domains of variable configuration, advanced and composite materials studies, and "smart" materials, which include possibilities for built in sensing and control actuation. Mathematically, optimal design has much in common with the more established area of *system, or parameter, identification*, which has enjoyed considerable popularity because of its applications to hydrology and petrology, with much of the theory required for optimal design purposes already in place as a consequence of earlier identification studies. Because of this connection, papers treating both optimal design and system identification were included in the workshop and appear now in these proceedings.

We are confident that the papers presented here will find a large audience and hopeful that they may stimulate new and more far-reaching studies in these areas of increasing significance and applicability. In concluding this preface we, as the workshop organizers, would like to express our gratitude to IMA as an institution, to its directorate, and to its most able and helpful staff for their invaluable cooperation in bringing this program to fruition.

John E. Lagnese
Georgetown University

David L. Russell
Virginia Tech

Luther W. White
University of Oklahoma, Norman

July, 1994

## CONTENTS

# A SHAPE OPTIMIZATION PROBLEM IN INVERSE ACOUSTICS*

GIOVANNI CROSTA†

**Abstract.** Given the incident field, a plane wave or a superposition thereof, and the scattered far field, the problem addressed herewith consists of determining the shape of a sound soft obstacle, which is axially symmetric, star shaped with respect to the origin and smooth. The unknowns are the parameters, by which shape is represented as the linear combination of e.g., trigonometric functions. Some methods of solution proposed in the past consist in transforming the problem into the constrained minimization of an objective function, which consisted of the boundary defect and a penalty term. The boundary defect is a squared $L^2$ norm of the incident plus approximate scattered field at the obstacle surface. The penalty term usually compares the computed and measured scattered far fields. The method presented herewith retains the minimization approach: it however replaces the constraint contained in the penalty term by a functional relationship between the coefficients, which appear in approximate representations of the far and resp., boundary scattered field. Said relationship is expressed by the approximate back propagation (ABP) operator, a product of some matrices, the entries of which depend on the shape parameters because they are inner products of suitable basis functions on the obstacle surface. As a consequence, a boundary defect alone has to be minimized with respect to the shape parameters. Some properties of the ABP operator are presented: they depend on those of the basis functions (spherical wave functions, their real parts and the normal derivatives of both at the obstacle surface) and are related e.g., to the properties of the **T** matrix method, used in the solution of forward scattering problems. Two classes of numerical problems are considered herewith: *i*) shape identification from the boundary coefficients of the scattered field and *ii*) shape identification from the far field coefficients. In either case the performance of the minimization algorithm agrees with an error estimate based on applying the projection theorem to the boundary defect. All examples have been selected to comply with a known uniqueness result, which would apply if the scattered field were exactly known: uniqueness is however lost because of approximation. The role of supplementary information in improving numerical reconstruction is demonstrated. Some preliminary results are also given for shape identification from the approximate scattering amplitude.

**Key words.** Inverse acoustic scattering, shape identification, exterior boundary value problems.

**AMS(MOS) subject classifications.** 35J05, 35R30, 76Q05

---

* The early stages of this work were funded by the Italian Ministry of University and Scientific Research (MURST 60%, 1987-89) and by the Department of Mathematical Sciences, University of Delaware (1988-89). Related travel support by the *Frequent Flyer Program* of Delta Air Lines, Inc., Atlanta (GA) and by the former *FT WorldPass Program* of Pan American World Airways, Inc., New York (NY) is gratefully acknowledged. This research has also been supported in part by the Institute of Mathematics and its Applications (1992) with funds provided by the National Science Foundation. The author is deeply indebted to T. S. Angell and R.E. Kleinman (Newark, DE) for the introduction to this class of problems. A.G. Ramm (Manhattan, KS) is thanked for advice about some key references

† Universita' degli Studi di Milano, Dipartimento di Scienze dell' Informazione, 39, via Comelico, I-20135 Milan, Italy; e_mail: crosta@imiucca.csi.unimi.it

**1. Introduction.** This is an attempt towards answering the following question (FIG. 1.1)

*"Can one hear the shape of an obstacle?"*

and at providing some hints towards answering the related question *"Can one design an obstacle, which generates the desired radiation pattern?"*

It is shown herewith that the answers are *"yes, provided .."* some conditions are fulfilled.

Assume that the obstacle, $\Omega$, is axially symmetric and star shaped w.r. to the origin, has a sufficiently smooth boundary, $\Gamma$, described by the function $r(\theta)$, where $\theta$ is colatitude, and is impenetrable e.g., sound soft. Let $r(\theta)$ or a function thereof be the linear combination of suitable base functions weighted by some coefficients, the *shape parameters*. Let the obstacle sit in a homogeneous, unbounded medium. The incident acoustic field, $u$, is chosen to be a plane wave or a superposition thereof s.t., wavelength and obstacle diametre are comparable. This gives rise to resonant scattering. The scattered field, $v$, is the solution to an exterior Dirichlet BV problem for Helmholtz equation, subject to a radiation condition.

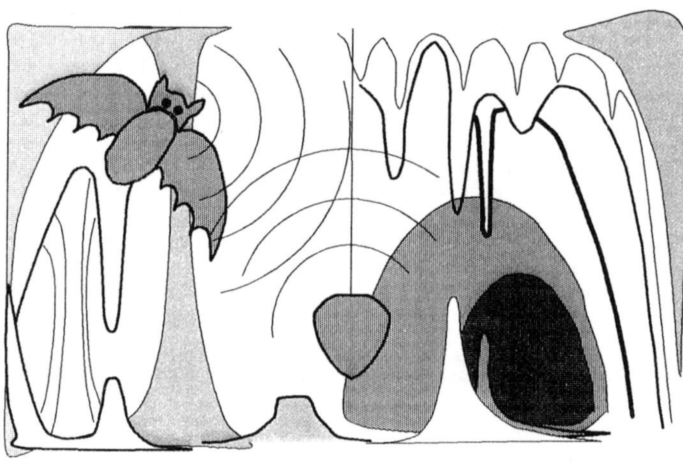

FIG. 1.1. *Can one hear the shape of an obstacle?*

It is well known that on the (smooth) surface $\Sigma$ of any domain, which contains the sphere circumscribed to the obstacle, the field $v$ can be represented by the series

$$v = \sum_{\lambda \in \Lambda} v_\lambda f_\lambda \tag{1.1}$$

where $\Lambda$ is a countable set of multi-indices, the functions $\{v_\lambda | \lambda \in \Lambda\}$ are outgoing spherical waves. The (infinite dimensional) vector of the *far field coefficients* $\mathbf{f} := [f_\lambda]$ depends on the obstacle shape and on $u$.

Given $u$ and some of the far field coefficients e.g., $\{f_\lambda | \in \Lambda(L)\}$, where $L$ is the prescribed order of approximation and $\Lambda(L) \subset \Lambda$ a finite set, the inverse problem addressed below consists of determining the shape parameters. Several methods of solution have been proposed before: some of them have led to constructive algorithms, which yield an approximate solution. Since no comprehensive existence result is available, the identification of shape parameters is often transformed into a constrained minimization problem, where the objective function consists of one or more terms. Traditionally, the first term is the *boundary defect*

$$B^{(L)} := \|v^{(L)} + u\|^2_{L^2(\Gamma)}, \tag{1.2}$$

where $v^{(L)}$ is the approximate scattered field on the boundary, more properly denoted by $v^{(L)}|_\Gamma$. Existing identification methods differ by the implemented relationship between $v^{(L)}|_\Gamma$ and information about the far field. Actually, an attempt at representing either $v|_\Gamma$ or $v^{(L)}|_\Gamma$ by (1.1) or, resp., by the corresponding linear combination, need not succeed in general, because the set $\{v_\lambda\}$ is not a Schauder basis in $L^2(\Gamma)$. Nonetheless, as suggested by Barantsev and Kozachek [2] (see also Ch. 4 of [18]), it makes sense to *approximate* the surface scattered field by the sum

$$v^{(L)} = \sum_{\lambda \in \Lambda(L)} v_\lambda c_\lambda^{(L)}, \tag{1.3}$$

where $\mathbf{c}^{(L)} := [c_\lambda^{(L)}], \lambda \in \Lambda(L)$ is the vector of *boundary coefficients.* The latter depend on $L$ and are determined such that, they minimize the boundary defect $B^{(L)}$.

Early identification methods relied on a penalty term in an objective function to turn the far field coefficients into boundary coefficients by a purely numerical procedure. On the other hand it is intuitive that a functional relationship must exist between the vectors $\mathbf{c}^{(L)}$ and $\mathbf{f}$. One of the key features of the method presented herewith is the introduction of an approximate *back propagation* (ABP) operator, $\mathbf{W}^{(L)}$, by which boundary coefficients of order $L$ are obtained from far field coefficients. As a consequence, the objective function to be minimized consists of *one term*, the boundary defect (1.2). The penalty term is thus eliminated; only the constraints on the smoothness of $r(\theta)$ and on its extreme values are retained.

SECTION 2 contains some preparatory material about the direct problem, the inverse one and the complete families to be used in approximation schemes.

SECTION 3 deals with the boundary coefficients and applications. First, the determination of $\mathbf{c}^{(L)}$ is considered when $u$ is known and the obstacle is given by means of the reference vector of shape parameters $\vec{\psi}_{\text{ref}} \in \mathbb{R}^{I+1}$. Existence, uniqueness, convergence and numerical computation of $\mathbf{c}^{(L)}$ are covered. In particular the minimum value of $B^{(L)}(\vec{\psi}_{\text{ref}})$ is obtained from the projection theorem. Next, these results are applied to the simple inverse problem of identifying $\vec{\psi}$ from knowledge of $\mathbf{c}^{(L)}$. The minimization algorithm is briefly described. Its performance with most numerical examples is found to agree with the estimate based on the above minimum boundary defect.

SECTION 4 considers the ABP method and its application to shape identification. Given $L$, the approximate solution of the direct scattering problem is usually represented by the first card$[\Lambda(L)]$ entries of the vector $\mathbf{f}$. One method capable of estimating said entries under suitable hypotheses is the "**T** matrix scheme" [25], [18], which yields the vector $\mathbf{f}^{(W,L)}$. An affine mapping relates $\mathbf{f}^{(W,L)}$ to $\mathbf{c}^{(L)}$: it contains $\mathbf{W}^{(L)}$, the ABP operator. On its turn $\mathbf{W}^{(L)}$ is the product of some matrices, the entires of which are inner products of suitable basis functions in $L^2(\Gamma)$, hence depend on the shape of the obstacle. The properties of $\mathbf{W}^{(L)}$ known so far, the identification algorithm and some related open problems are presented. One result is relevant, which has no counterpart in other methods: the minimum value of $B^{(L)}$, when $\mathbf{W}^{(L)}\mathbf{f}^{(W,L)}$ replaces $\mathbf{c}^{(L)}$ in (1.3), is easily determined if $\vec{\psi}_{\text{ref}}$ is known, hence the performance of the identification algorithm with a given problem can be predicted. To the author's knowledge, this feature has neither been pointed out nor exploited before in connection with any existing shape identification method.

Finally, we discuss the computer code and the related results. The development and testing of a numerical algorithm usually consist of three stages: *i*) the computation of far field coefficients say $\mathbf{f}^{(L)}$, given $L, u$ and $\vec{\psi}_{\text{ref}}$; *ii*) the identification of $\vec{\psi}$ given $L, u$ and $\mathbf{f}^{(L)}$; *iii*) the evaluation of the results. In this case, several identification results are in agreement with theoretical prediction. In particular, the minimization rule is seen to become inconsistent, hence the algorithm is stopped at value of $B^{(L)}$ which is close (within one order of magnitude) to the minimum value determined in §4. The latter is independently computed beforehand from $\vec{\psi}_{\text{ref}}$, which generated the direct scattering data.

The ABP method, its connection with the forward propagator used by [7], the preliminary results obtained from scattering amplitude data and other features are briefly discussed in the final SECTION 5.

## 2. Direct and inverse scattering

**2.1. The scattered field.** The scattering problem considered herewith arises from acoustics, which provides the main motivation. The same equation governs the classical (non quantum) behaviour of the complex massless scalar field, which is obtained when the mass hyperboloid degenerates to the light cone. The scattered field, $v$, is the solution to an exterior Dirichlet boundary value problem for Helmholtz equation

$$(\nabla^2 + k^2)v = 0 \text{ in } \Omega^c := \mathbb{R}^3 \backslash (\Omega \cup \partial\Omega) \tag{2.1}$$

(2.2) $(v+u)|_\Gamma = 0$, on $\Gamma := \partial\Omega$; $r(\partial_r v - ikv) \to 0$ as $r := |\mathbf{x}| \to \infty$.

Here $\Omega$ is a bounded, open, simply connected domain, which is star - shaped w.r. to the origin. A unit amplitude plane wave is chosen as the incident field (superposition will be considered in §4)

$$u = \exp[i\mathbf{k}\cdot x]; \quad \mathbf{k} = \frac{2\pi}{\lambda}\begin{bmatrix} \sin\theta_{\text{in}}\,\cos\phi_{\text{in}} \\ \sin\theta_{\text{in}}\,\sin\phi_{\text{in}} \\ \cos\theta_{\text{in}} \end{bmatrix} \tag{2.3}$$

where $|\mathbf{k}| \equiv k$ and $\theta_{\text{in}}, \phi_{\text{in}}$ are the colatitude and azimuth incidence angles. Let the distance from the origin to a point $\mathbf{x} \in \Gamma$ be denoted by $r(\theta,\phi)$, $0 \le \theta \le \pi$, $0 \le \phi \le 2\pi$. The boundary $\partial\Omega \equiv \Gamma$ is assumed to be "smooth" i.e., $r(.,.) \in \mathcal{C}^{1,\alpha}([0,\pi]\times[0,2\pi]), \alpha > 0$. The (open) sphere of radius $R$ in $\mathbb{R}^3$, centered at the origin is denoted by $S_R$, the unit sphere in $n$ dimensions by $S^n$ and a general, simply connected, bounded domain with smooth boundary by $V$; the boundary of the latter is $\partial V \equiv \Sigma$.

Let $\{f_{l,m,p} | 0 \le m \le l \le \infty, p = 0,1\}$ stand for the set of far-field coefficients and $\{v_{l,m,p}\}$ for the set of outgoing spherical wavefunctions, where

$$v_{l,m,p} = h_l^{(1)}(k,r)P_l^m(\cos\theta)[(1-p)\cos m\phi + p\sin m\phi]g_{l,m}. \tag{2.4}$$

$h_l^{(1)}(.)$ is the $l$-th order spherical Hankel function of the 1st kind, $P_l^m(.)$ a Legendre associate function s.t., the corresponding normalization factor is

$$g_{l,m} := \sqrt{\epsilon_m \frac{2l+1}{4\pi}\frac{(l-m)!}{(l+m)!}} \tag{2.5}$$

with the Neumann factor $\epsilon_0 = 1, \epsilon_m = 2, m \neq 0$.

The following result is well - known.

PROPOSITION 2.1. (an application of THM. 14 of [16]). *Let $V \supset S_R$, bounded, $R = \max_{(\theta,\phi)} r(\theta,\phi)$ and $\partial V$ be sufficiently smooth. Then $v$ on $\partial V$ can be expanded by*

$$v = \sum_{l=0}^{\infty}\sum_{m=0}^{l}\sum_{p=0}^{1} f_{l,m,p} v_{l,m,p}. \tag{2.6}$$

The index triple $(l,m,p)$ is now replaced by $\lambda$ and the corresponding set denoted by $\Lambda$.

The asymptotic behaviour of $v$ is described by two basic Theorems, which are recalled without proof. Here $r \equiv |\mathbf{x}|$.

THEOREM 2.2. (existence of a *scattering amplitude* function [18, 26]). *There exists a function $F$ of the unit vector $\hat{\mathbf{x}}$ and of the incident wavevector $\mathbf{k}$, $F \in C^\infty(S^{n-1}\times\mathbb{R}^n\backslash\{0\}), n \ge 3 \cdot \ni \cdot$*

$$v = \frac{e^{ikr}}{r^{\frac{n-1}{2}}}F(\frac{\mathbf{x}}{\mathbf{r}},\mathbf{k}) + O(r^{-\frac{n+1}{2}}) \text{ as } r \to \infty \tag{2.7}$$

*uniformly for* $\hat{\mathbf{x}} \in S^{n-1}$ *and* $\mathbf{k} \in K \subset\subset \mathbb{R}^n \backslash \{0\}$.

THEOREM 2.3. (characterization of $F$ [15])

*i) The equivalence holds:* $F(\hat{\mathbf{x}}, \mathbf{k})$ *is a scattering amplitude* $\Leftrightarrow$ $\{\exists$ *a harmonic function* $H(\mathbf{x}, \mathbf{k})$, *analytic* $\forall x \in \mathbb{R}^n$; $H = F$ *on* $S^{n-1}$ *and there* $\exists$ *a non negative constant* $c \cdot \ni \cdot$

$$\int_{r=R} |H(\mathbf{x}, \mathbf{k})|^2 ds = o(\exp[2krc])\}. \tag{2.8}$$

*ii) If i) is met then* $\exists!\ v(\mathbf{x}, \mathbf{k})$, *which satisfies (2.1), (2.2) and (2.7) for* $r > c$.

The properties of the scattering amplitude lead to the uniqueness results for the inverse scattering problem, which are now summarized.

THEOREM 2.4. *The following conditions are sufficient for the uniqueness of the scatterer.*

*i)* (Schiffer, [14]) $F(\hat{\mathbf{x}}, \mathbf{k})$ *of* THM. *2.2 is completely known;*

*ii)* (Schiffer, [14]); also [4], [20]) $F(\hat{\mathbf{x}}, \mathbf{k})$ *is known* $\forall \hat{\mathbf{x}} \in S^{n-1}$ *and for* $a \leq k \leq b$, *with* $\hat{\mathbf{k}}$ *fixed;*

*iii)* [18] $F(\hat{\mathbf{x}}, \hat{\mathbf{k}}, k_0)$ *is known for one fixed wave number* $k_0$ *on the spherical patch:* $\hat{\mathbf{x}} \in \tilde{S} \subset S^{n-1}$ *for countably many distinct unit wave vectors* $\hat{\mathbf{k}} = \hat{\mathbf{k}}_1, \hat{\mathbf{k}}_i, \ldots$;

*iv)* (p 107 of [3]) *The obstacle is sound soft, is centered at the origin and is strictly contained in* $S_R$, *the scattering amplitude* $F(\hat{\mathbf{x}}, \hat{\mathbf{k}}, k_0)$ *is known for one fixed wave number* $k_0$, $\forall \hat{\mathbf{x}} \in S^2$ *and for* $N+1$ *distinct unit wave vectors* $\hat{\mathbf{k}} \in \{\hat{\mathbf{k}}_0, \ldots \hat{\mathbf{k}}_N\}$, *where*

$$N = \sum_{t_{nl} < kR} 2n + 1. \tag{2.9}$$

*Here* $t_{nl}$, *with* $l = 0, 1, \ldots$ *and* $n = 0, 1, \ldots$, *denotes the* $l$*-th order positive zero of the spherical Bessel function* $j_n(x)$.

As it can be seen, uniqueness conditions have evolved towards decreasing the required amount of information.

Of extreme interest for the applications considered below (§§3 and 4) is a COROLLARY to THM. 2.4 - *iv*.

COROLLARY 2.5. (e.g., p 107 of [3], p 100 of [20]) *Let* $\Gamma_1, \Gamma_2$ *be two obstacles s.t.,* $\Gamma_1, \Gamma_2 \subset S_R$, $kR < \pi$, *and s.t., their scattering amplitude functions coincide for one incident wave with wave number* $k$. *Then* $\Gamma_1 = \Gamma_2$.

In the general case, however, it is still an open problem (see e.g., p. 108 of [3]) whether $F(\hat{\mathbf{x}}, \hat{\mathbf{k}}, k_0)$ known $\forall \hat{\mathbf{x}} \in S^2$, $\hat{\mathbf{k}}$ and $k_0$ fixed, uniquely determines the scatterer or not.

**2.2. Complete families.** In order to establish the existence of a solution to the inverse problem and possibly construct one, several methods have been applied. The properties of the scattering amplitude suggested first to apply analytic continuation; more recently methods based on *approximation of the scattered field* have been put to work (an early review can be found in [21]; to the author's knowledge, the most up to date is the monograph [3]). The method to be described in §4 below belongs to this class. The next statements provide the necessary framework. Complex conjugation is denoted by *. Bold square brackets, **[ ]**, denote arrays. The bra-ket notation , $\langle .|.\rangle$, will be used for inner products in $L^2(\Gamma)$. The subscript $L^2(\Gamma)$ will be dropped from the norms whenever no confusion arises.

DEFINITION 2.6. (e.g., p 146 ff of [18]) Let $G$ be a Lyapunov surface (as defined e.g., in Ch. IV, §5.7 of [22]). Denote the a.e. null function by $\omega$. A family of linearly independent functions $\{v_\lambda|\lambda \in \Lambda\}$ is said to be *closed* in $L^2(G)$ provided the implication holds

$$\{\int_G \phi^* v_\lambda dG = 0, \forall\lambda \in \Lambda\} \Rightarrow \{\phi =_{\text{a.e.}} \omega\} \tag{2.10}$$

DEFINITION 2.7. (e.g., p 146 ff of [18]) Let $G$ and $\{v_\lambda|\lambda \in \Lambda\}$ be as above. The family is said to be *complete* in $L^2(G)$ provided $\forall\epsilon > 0, \forall\phi \in$ $\in L^2(G)$ there exist an approximation order $M(\epsilon)$ and a set of coefficients $\{c_\lambda^{(M(\epsilon))}(\phi),\ \lambda \in \Lambda(M(\epsilon))\}$, which depend both on $\phi$ and on $M(\epsilon)$ s.t.,

$$\|\phi - \sum_{\lambda\in\Lambda(M(\epsilon))} v_\lambda c_\lambda^{(M(\epsilon))}(\phi)\|^2_{L^2(G)} < \epsilon \tag{2.11}$$

As it can be expected, the two properties are related.

LEMMA 2.8. *Given linear independence, completeness is equivalent to closure.*

For the *proof* the Reader is referred to e.g., p 70 ff of [23]. It should be pointed out that the $\{v_\lambda|\lambda \in \Lambda\}$ need not be orthonormal for this property to hold. In fact, linear independence of the $v_\lambda$'s does not imply orthogonality, unless $G$ is the surface of a sphere.

The term "complete family" will be hereinafter used to denote a linearly independent, complete family. Although PROP. 2.1 has in general no counterpart for the boundary of domains, which are contained in $S_R$, some relevant results hold for any closed, sufficiently regular surface.

LEMMA 2.9. ([24], [18]; also [8], [11).

*Let $G$ be Lyapunov and denote outward normal differentiation on $G$ by $\partial_n(.)|_G$, then both $\{v_\lambda|\lambda \in \Lambda\}$ and $\{\partial_n v_\lambda|_G|\lambda \in \Lambda\}$ are complete families in $L^2(G)$.*

The next result applies to *real wave functions* defined by $u_\lambda := Re[v_\lambda]$.

LEMMA 2.10. [10,18]. *Denote the spectra of the interior Dirichlet and Neumann negative Laplacians by $\sigma(-\Delta_D)$, resp. $\sigma(-\Delta_N)$. The family of real waves $\{u_\lambda|\lambda \in \Lambda\}$ is a complete family in $L^2(G)$ if and only if $k^2 \notin \sigma(-\Delta_D)$. Similarly $\{\partial_n u_\lambda|_G|\lambda \in \Lambda\}$ is a complete family in $L^2(G)$ if and only if $k^2 \notin \sigma(-\Delta_N)$.*

Axially symmetric obstacles can be shown to yield scattered fields, where odd parity ($p = 1$ in (2.4)) far field coefficients all vanish. The parity index will therefore be dropped and the multiindex $\lambda$ will stand for the $\{l, m\}$ pair. In particular $\lambda = 0$ will be shorthand for $\lambda = \{0, 0\}$. Approximation of a given, finite order $L$ will mean that $l$ and $m$ range as follows: $0 \le m \le l \le L$ i.e., $\lambda \in \Lambda(L)$, a subset of $\Lambda$, as anticipated in §1, s.t., $\mathrm{card}[\Lambda(L)] = N + 1$, where $N := (L+2)\frac{L+1}{2}$.

Some finite dimensional matrices of inner products in $L^2(\Gamma)$ have to be introduced.

DEFINITION 2.11.

$$\mathbf{P}^{(L)} := [\langle v_\lambda | v_\mu \rangle]; \quad \mathcal{R}\mathbf{P}^{(L)} := [\langle v_\lambda | u_\mu \rangle]; \tag{2.12}$$

$$\mathbf{Q}^{(L)} := [\langle \partial_n u_\lambda | v_\mu \rangle]; \quad \mathcal{R}\mathbf{Q}^{(L)} := [\langle \partial_n u_\lambda | u_\mu \rangle]. \tag{2.13}$$

The next invertibility result is a straightforward consequence of the linear independence of $\{v_\lambda\}$:

THEOREM 2.12. [2,10]. $\forall L, \exists [\mathbf{P}^{(L)}]^{-1}$.

Some notation is needed for the subspaces spanned by the families introduced above:

$$X_1^{(L)} := \mathrm{span}\{v_\lambda | \lambda \in \Lambda(L)\}; \quad X_2^{(L)} := \mathrm{span}\{\partial_n v_\lambda | \lambda \in \Lambda(L)\}; \tag{2.14}$$

$$X_3^{(L)} := \mathrm{span}\{u_\lambda | \lambda \in \Lambda(L)\}; \quad X_4^{(L)} := \mathrm{span}\{\partial_n u_\lambda | \lambda \in \Lambda(L)\}; \tag{2.15}$$

Invertibility of the infinite dimensional counterparts of the above matrices was obtained by [10] under an additional hypothesis:

LEMMA 2.13. *If all involved bases are Schauder and resp.*

| | | |
|---|---|---|
| *$k$ is arbitrary* | *then* | $\exists[\mathbf{P}]^{-1}$ |
| $k^2 \notin \sigma(-\Delta_D)$ | *then* | $\exists[\mathcal{R}\mathbf{P}]^{-1}$ |
| $k^2 \notin \sigma(-\Delta_N)$ | *then* | $\exists[\mathbf{Q}]^{-1}$ |
| $k^2 \notin \{\sigma(-\Delta_D) \cup \sigma(-\Delta_N)\}$ | *then* | $\exists[\mathcal{R}\mathbf{Q}]^{-1}$. |

REMARK 2.1. The invertibility of $\mathcal{R}\mathbf{Q}^{(L)}$ cannot be established in general. In fact examples of obstacles can be produced [5] s.t., $\mathcal{R}\mathbf{Q}^{(L)}$ is *not* invertible.

## 3. The boundary coefficients

**3.1. The direct problem.** As a consequence of LEMMA 2.9, the problem of approximating $v|_\Gamma$ in the $L^2(\Gamma)$ norm can be stated.

PROBLEM 3.1. Given $u, \Gamma$ and the order $L(> 0)$, hence the set of indices $\Lambda(L)$, determine the boundary coefficients $\{c_\lambda^{(L)} | \lambda \in \Lambda(L)\}$ i.e. the vector $\mathbf{c}^{(L)}$ s.t., the approximate boundary scattered field $v^{(L)}|_\Gamma$ defined by EQ. 1.3 minimizes $B^{(L)}$ of (1.2).

The properties of the solution $\mathbf{c}^{(L)}$ i.e., existence, uniqueness and the asymptotic behaviour as $L \to \infty$ are summarized by the next two THEOREMS.

THEOREM 3.1. (existence and uniqueness)
*i) The vector* $\mathbf{c}^{(L)}$ *is the unique solution to the algebraic system*

$$\mathbf{P}^{(L)}\mathbf{c}^{(L)} = -\mathbf{g}^{(L)} \tag{3.1}$$

*where* $\mathbf{g}^{(L)} := [\langle v_\lambda | u\rangle]$.
*ii) The corresponding minimum defect is*

$$B^{(L)}(-[\mathbf{P}^{(L)}]^{-1}\mathbf{g}^{(L)}) = \|u_1^\perp\|^2_{L^2(\Gamma)}, \tag{3.2}$$

*where* $u_1^\perp$ *denotes the orthogonal complement of* $u$ *w.r. to* $X_1^{(L)}$ *of (2.14).*

*Proof.*
*i)* EQ. 3.1 is the normal equation obtained from applying the first order stationarity condition to $B^{(L)}$. The well-posedness of the algebraic system follows from THM. 2.12
*ii)* EQ. 3.2 is a straightforward consequence of the projection theorem related to the orthogonal decomposition $L^2(\Gamma) = X_1^{(L)} \oplus X_1^\perp$. □

REMARK 3.1. The argument in *i)* has been used before in [2] to establish the existence and uniqueness of the scattering matrix, $[s_{\lambda\mu}^{(L)}]$, which replaces $\mathbf{c}^{(L)}$ when the incident field is $v_\mu^*$, an incoming spherical wave.

THEOREM 3.2. (asymptotic behaviour)

$$\lim_{L\to\infty} \|u_1^\perp\| = 0; \tag{3.3}$$

$$\lim_{L\to\infty} |c_\lambda^{(L)} - f_\lambda| = 0 \tag{3.4}$$

*uniformly in* $0 \leq \lambda \leq \lambda_0$, *where* $\lambda_0$ *is arbitrarily large, fixed and* $f_\lambda$ *is the far field coefficient appearing in* EQ. 1.1.

*Proof.* The property of $u_1^\perp$ follows from the definition of completeness, DEF. 2.7. The other result is proven on p. 170 ff of [18]. □

REMARK 3.2. EQ. 3.3 states a pointwise convergence property for the boundary coefficients. It means that in practice the far-field coefficients, hence the entries of the scattering matrix, can be *approximately* computed by solving EQ. 3.1. The possibility (or impossibility) of *exactly* representing the surface scattered field by the expansion (1.1) is not affected by EQ. 3.4.

The shape function $r(\theta)$ of an axially symmetric obstacle is parameterized by fixing an order $I$ and choosing either trigonometric functions

$$\{\Psi_i(\theta), 0 \le i \le I\} = \{1, \cos\theta, \sin\theta, \cos 2\theta, \sin 2\theta, \dots\} \tag{3.5}$$

or Legendre polynomials

$$\Psi_i(\theta) = P_i^0(\cos\theta), 0 \le i \le I \tag{3.6}$$

s.t.,

$$\frac{1}{r^2}(\theta) = \sum_{i=0}^{I} \psi_i \Psi_i(\theta). \tag{3.7}$$

By introducing upper and lower bounds for $r(\theta)$, the set of *admissible shape parameters* is defined by

$$\Psi_{\text{ad}} := \{\vec{\psi} | \vec{\psi} \in \mathbb{R}^{I+1}; r \in \mathcal{C}^2([0,\pi]); 0 < r_L \le r \le r_H, \forall \vartheta \in [0,\pi]\}. \tag{3.8}$$

If trigonometric base functions are chosen, the smoothness of $r(.)$ specified by (3.8) implies some linear constraints between the parameters, which are not elaborated herewith.

Another consequence of axial symmetry is the block diagonal structure of the Gramian matrix $\mathbf{P}^{(L)}$ (FIG. 3.1): only those inner products $P^{(L)}_{\lambda\mu}$ do not vanish, where both basis functions have the same $m$. Moreover $\mathbf{P}^{(L)}$ is Hermitean conjugate, because the complex valued terms in $v_\lambda$ only depend on $l$.

The algorithm, which determines $\mathbf{c}^{(L)}$ given $u$ and $\vec{\psi}_{\text{ref}} \in \Psi_{\text{ad}}$, consists of two procedures: *i*) Gaussian quadrature over $n_p$ points in the interval $0 \le \theta \le \pi$, to compute all inner products and *ii*) an algebraic solver for EQ. 3.1 based on Kholetskii's method (e.g., [6]), where the properties of $\mathbf{P}^{(L)}$ substantially reduce storage. Once $\mathbf{g}^{(L)}$ and $\mathbf{c}^{(L)}$ are known, the r.h.s. of EQ. 3.2 is determined by

$$\|u_1^\perp\|^2 = \|u - u_1^{(L)}\|^2 = \|u\|^2 + \mathbf{g}^{(L)*} \cdot \mathbf{c}^{(L)}. \tag{3.9}$$

**3.2. The inverse problem.** If the roles of shape parameters and boundary coefficients are interchanged in the minimization of $B^{(L)}$, a very simple constructive procedure is obtained, which solves the following problem.

PROBLEM 3.2. Given $u, \mathbf{c}^{(L)}, \Psi_{\text{ad}}$ find $\vec{\psi}^\star \in \Psi_{\text{ad}}$ s.t.,

$$B^{(L)}(\vec{\psi}^\star; \mathbf{c}^{(L)}) = \min_{\Psi_{\text{ad}}} B^{(L)}(\vec{\psi}; \mathbf{c}^{(L)}). \tag{3.10}$$

The existence of at least one $\vec{\psi}^\star$ follows from the continuity of $B^{(L)}$ w.r. to $\vec{\psi}$ and the boundedness of $\Psi_{\text{ad}}$. A minimization algorithm has

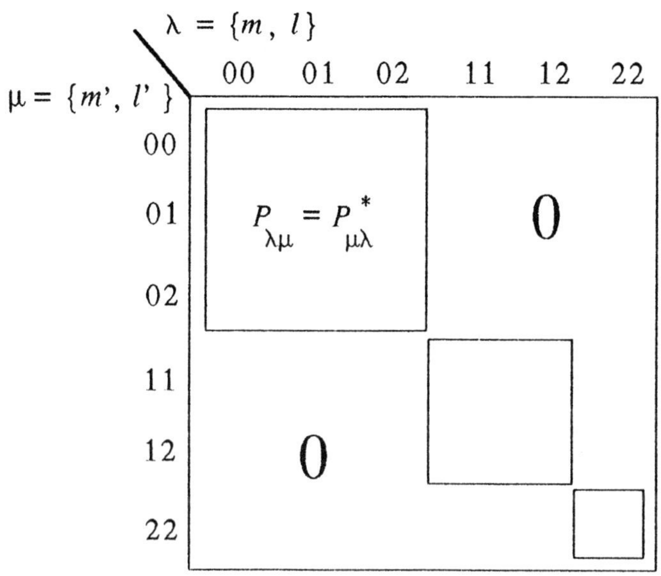


FIG. 3.1. *The block diagonal structure of the Gramian matrix* $\mathbf{P}^{(L)}, L = 2$.

been developed, which relies on the discrete gradient and Hessian of $B^{(L)}$ w.r. to $\vec{\psi}$ and implements Newton's rule to update the vector of shape parameters. Symbolic calculations have been carried out s.t., no numerical differentiation is required. The function gradient is e.g., the discrete counterpart of

$$(3.11)\nabla_{\vec{\psi}} B^{(L)} = 2Re[\langle v+u|\nabla_{\vec{\psi}}(v+u)\rangle] + \int_{S^2} |v+u|^2 \nabla_{\vec{\psi}} J(\vec{\psi}) d\theta d\phi,$$

where $J(\vec{\psi})$ is the Jacobian of $\Gamma$.

A line search procedure is also included in the algorithm, as in any optimization code, in order to determine the optimal updating step whenever Newton's rule is either inapplicable or has been unsuccessful. Repeated failure of the line search usually serves as a stopping test. This event is linked to the bound determined by EQ. 3.2: the latter predicts how far minimization will proceed with a given example. More precisely, when a direct problem is solved for a known $\vec{\psi}_{\text{ref}}$, the order $L$ determines how small $B^{(L)}$ can be or, in other words, how accurate the finite dimensional approximation to $v|_\Gamma$ is. When $\vec{\psi}_{\text{ref}}$ is unknown, approaching the value given by EQ. 3.2 means approaching the bound set by the numerical scheme. At that point approximation errors are expected to make the

TABLE 3.1
*Identification of Prolate Spheroids from Boundary Coefficients*

Common features:
$r(\theta=0)=1; I=7$, trigonometric parameterization of $r(\theta)$; $L=12; \mathbf{k}=\hat{\mathbf{z}}$.
$\epsilon\equiv$ eccentricity; $r(\pi/2)\equiv$ minor half axis;
no. of Gauss quadr. points: $n_p=64$
$N_{it}\equiv$ number of iterations until search failure
$B^{(L)}(\vec{\psi}^*)\equiv$ computed minimum defect after $N_{it}$ iterations

| $\epsilon^2$ | $r(\pi/2)$ | $\left\lvert\frac{P_{0LL}}{P_{000}}\right\rvert$ | $\lVert u_1^{\perp}\rVert^2_{L^2(\Gamma)}$ | $N_{it}$ | $B^{(L)}(\vec{\psi}^*)$ |
|---|---|---|---|---|---|
| $\frac{2}{3}$ | .577 | $(1.4/4\pi)\cdot 10^{29}$ | $.123\cdot 10^{-2}$ | 30 | $.103\cdot 10^{-2}$ |
| $\mathbf{\frac{4}{5}}$ | .477 | $(5.6/4\pi)\cdot 10^{31}$ | $.158\cdot 10^{-1}$ | 30 | $.773\cdot 10^{-2}$ |
| $\frac{5}{6}$ | .408 | $(4.9/4\pi)\cdot 10^{32}$ | $.269\cdot 10^{-1}$ | 30 | $.102\cdot 10^{-1}$ |
| $\frac{6}{7}$ | .378 | $(3.1/4\pi)\cdot 10^{33}$ | $.381\cdot 10^{-1}$ | 30 | $.113\cdot 10^{-1}$ |
| $\frac{7}{8}$ | .353 | $(1.5/4\pi)\cdot 10^{34}$ | $.486\cdot 10^{-1}$ | 30 | $.119\cdot 10^{-1}$ |
| $\mathbf{\frac{8}{9}}$ | .333 | $(6.1/4\pi)\cdot 10^{35}$ | $.581\cdot 10^{-1}$ | 30 | $.116\cdot 10^{-1}$ |
| $\mathbf{\frac{9}{10}}$ | .316 | $(2.1/4\pi)\cdot 10^{35}$ | $.663\cdot 10^{-1}$ | 30 | $.126\cdot 10^{-1}$ |
| $\mathbf{\frac{99}{100}}$ | .100 | $(2.0/4\pi)\cdot 10^{47}$ | $.755\cdot 10^{-1}$ | 14 | $.136\cdot 10^{-2}$ |

scheme inconsistent and stop the minimization algorithm: $\vec{\psi}$ is no longer updated. Most examples solved in this context confirm the prediction: the line search phase consistently fails and stops the iterations. Control based on the lower bound for $B^{(L)}$ has been deliberately disabled in the code, not to pass information from the direct to the inverse problem solver.

The identification of several shapes has been considered. TABLE 3.1, however, only reports about some prolate spheroids, illuminated from the South pole. More examples are described in §4 below. The initial guess is the unit sphere. In application of (3.5), spheroids are exactly parameterized by $I+1=4$. However, $I+1=8$ is chosen not to bias shape identification. Values of $\epsilon^2$ printed in boldface are documented by a figure (FIG. 3.2). As long as $\epsilon^2\leq\frac{2}{3}$, the identified and reference shapes almost coincide, hence they do not need to be displayed. The ratio $|P_{0LL}/P_{000}|$, which involves the last and 1st diagonal entries in the 1st block ($m=0$) of $\mathbf{P}^{(L)}$ is an estimate of the condition number of this matrix.

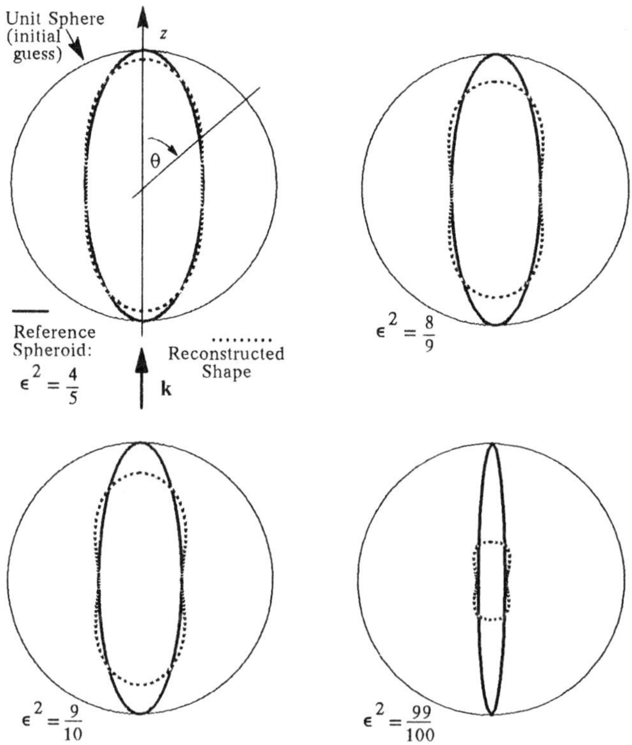


FIG. 3.2. *Synopsis of prolate spheroids identified from boundary coefficients.*

In FIG. 3.2 axial cross sections are shown of the unit sphere, the reference shape (solid line) and the reconstructed shape (broken line).

The infinite dimensional counterparts of these examples comply with the uniqueness of COR. 2.5: namely, *exact* knowledge of the boundary scattered field would imply knowledge of the scattering amplitude. As long as $R = 1$ and $k = 1$ all spheroids ought to be uniquely identified from just one incident wavevector.

When $L$ is finite and roundoff errors are introduced by the numerical code, no unique reconstruction shall be expected any longer. The identified shape may differ from the reference one and depend on the initial guess.

FIG. 3.2 indeed shows some of these facts. The quality of the reconstruction deteriorates progressively up to $\epsilon^2 = \frac{9}{10}$. Beyond this value, see last line of TAB. 3.1, the condition number makes the problem intractable and EQ. 3.2 ceases to be a reliable predictor of the computed minimum defect. Nonetheless, the minor half axis is estimated correctly, even at $\epsilon^2 = \frac{99}{100}$.

The purpose of the next two SECTIONS is to describe and discuss a shape identification method, which may be applicable to a realistic situation.

## 4. The approximate back propagation method

**4.1. Outline of the $T$ matrix scheme.** [18,25]. Eq. 1.1 relates the scattered field to the far field coefficients. The plane wave $u$ of Eq. 2.3 can be represented everywhere in $\Omega^c$ as the superposition $u = \sum_{\mu\in\Lambda} a_\mu u_\mu$, which, given $L$, is split as

$$u = \sum_{\mu\in\Lambda(L)} a_\mu u_\mu + \sum_{\mu\notin\Lambda(L)} a_\mu u_\mu := u_a^{(L)} + u_{\text{rem}}^{(L+1)}. \tag{4.1}$$

The purpose of the $\mathbf{T}$ matrix scheme is to approximately determine the far field coefficients from knowledge of $a_\mu$'s. The coefficients $f_\lambda, a_\mu$ can be shown to be given by the following inner products, where the kets are unknown, however.

$$f_\lambda = -\frac{ik}{4\pi}\langle u_\lambda|\partial_n(u+v)\rangle; \quad a_\mu = \frac{ik}{4\pi}\langle v_\mu|\partial_n(u+v)\rangle. \tag{4.2}$$

Assume $k^2 \notin \sigma(-\Delta_N)$ and replace the normal derivative of the total field at $\Gamma$ by its projection onto $X_4^{(L)}$

$$[\partial_n(u+v)|_\Gamma]_4^{(L)} := \sum_{\mu\in\Gamma(L)} \xi_\mu^{(L)}\partial_n u_\mu|_\Gamma; \tag{4.3}$$

then the approximate counterparts of Eqns. 4.2 are obtained

$$f_\lambda^{(T,L)} := -\frac{ik}{4\pi}\langle u_\lambda|\partial_n(u+v)_4^{(L)}\rangle; \quad a_\mu^{(T,L)} := \frac{ik}{4\pi}\langle v_\mu|\partial_n(u+v)_4^{(L)}\rangle. \tag{4.4}$$

By recalling (2.13) and using an obvious notation, the relationship between the vectors $\mathbf{a}^{(T,L)}$, $\mathbf{f}^{(T,L)}$ and $[\xi_\mu^{(L)}]$ becomes

$$\mathbf{a}^{(T,L)} = \frac{ik}{4\pi}\mathbf{Q}^{(L)\dagger}[\xi_\mu^{(L)}]; \quad \mathbf{f}^{(T,L)} = -\frac{ik}{4\pi}\mathcal{R}\mathbf{Q}^{(L)Trs}[\xi_\mu^{(L)}]. \tag{4.5}$$

If $\mathbf{Q}^{(L)}$ is invertible, then the $\mathbf{T}$ (or *transition*) *matrix* of order $L$ can be defined by

$$\mathbf{T}^{(L)} := -\mathcal{R}\mathbf{Q}^{(L)Trs} \cdot [\mathbf{Q}^{(L)\dagger}]^{-1}. \tag{4.6}$$

It relates $\mathbf{a}^{(T,L)}$ to $\mathbf{f}^{(T,L)}$. When $\mathbf{a}^{(T,L)}$, unknown, is replaced by $\mathbf{a} := = [a_\mu|\mu| \in \Lambda(L)]$, the Waterman approximate $\mathbf{f}^{(W,L)}$ to $[f_\lambda|_\lambda \in \Lambda(L)]$ is obtained

$$\mathbf{f}^{(W,L)} = \mathbf{T}^{(L)} \cdot \mathbf{a}. \tag{4.7}$$

Alternatives to this procedure exist and come e.g., from projecting $\partial_n v|_\Gamma$ alone either in $X_4^{(L)}$ or in $X_1^{(L)}$. Their purpose is however the same as that of Eq. 4.7.

Although it has been known for over twenty five years and proven very attractive from a computational viewpoint, the $\mathbf{T}$ matrix scheme leaves some unanswered questions about the invertibility of $\mathbf{Q}^{(L)}$ and convergence. Namely, as it happens with $\mathcal{R}\mathbf{Q}^{(L)}$ (see REM. 2.1), it is not known in general whether a given obstacle and a given $\mathbf{k}$ result is an invertible $\mathbf{Q}^{(L)}$. Moreover the relationship

$$\lim_{L\to\infty}(f_\lambda - f_\lambda^{(T,L)}) = \lim_{L\to\infty}(-\frac{ik}{4\pi}\langle u_\lambda|\partial_n(u+v)_4^{\perp}\rangle) = 0, \tag{4.8}$$

which immediately follows from (4.2), (4.4) and LEMMA 2.10, does **not** generally imply the convergence of $f_\lambda^{(W,L)}$ of EQ. 4.7 to $f_\lambda$ as $L\to\infty$, unless some of the involved bases are Schauder [13, 18].

**4.2. Shape identification from the approximate far field.** A comprehensive description of published methods is provided by [3]. Attention will be paid herewith only to a class of methods, which rely on defect minimization. For the sake of simplicity assume that the scattered far field is approximated to order $L$; as usual the obstacle is parameterized by $\vec{\psi}$, which is among the unknowns. Ramm [19] suggests to minimize $B^{(L)}(\vec{\psi};\mathbf{f}^{(L)})$ w.r. to $\vec{\psi}$ alone; Angell *et al.* [1] consider the minimization of the penalized function

$$B^{(L)}(\vec{\psi};\mathbf{b}^{(L)}) + \tau||\tilde{v}(\mathbf{b}^{(L)}) - \tilde{v}(\mathbf{f}^{(L)})||^2_{L^2(S^2)}, \tau > 0, \tag{4.9}$$

where $\tilde{v}(.)$ is the far scattered field and $\mathbf{b}^{(L)}$ is the additional unknown vector of boundary coefficients; the task of determining the latter by adjusting $\mathbf{f}^{(L)}$ is left to the second summand, the penalty term. The method described by Jones and Mao [7] can be interpreted as the minimization of

$$B^{(L)}(\vec{\psi};\mathbf{c}^{(L)}) + \tau||\tilde{v}(\mathbf{f}^{(J,L)}) - \tilde{v}(\mathbf{f}^{(L)})||^2_{L^2(S^2)}, \tau > 0 \tag{4.10}$$

where a relationship (forward propagation) is implied between $\mathbf{c}^{(L)}$, as defined in §3, and $\mathbf{f}^{(J,L)}$. More recently Onishi [17] solves the boundary integral equation of a direct problem and minimizes a regularized boundary defect by a homotopy method.

**4.3. The approximate back propagator** (ABP). Fix $L$, assume $\vec{\psi}_{ref}$ is known and $\mathbf{k}$ is given s.t., two conditions are fulfilled

$$i)\quad k^2 \notin \sigma(-\Delta_N); \quad ii)\quad \exists[\mathcal{R}\mathbf{Q}^{(L)}]^{-1}. \tag{4.11}$$

Then the following operator can be defined.

DEFINITION 4.1. (The approximate back propagator of order L)

$$\mathbf{W}^{(L)}(\vec{\psi}_{\text{ref}}) := [\mathbf{P}^{(L)}]^{-1}\cdot\mathcal{R}\mathbf{P}^{(L)}\cdot\mathbf{Q}^{(L)\dagger}\cdot[\mathcal{R}\mathbf{Q}^{(L)Trs}]^{-1}. \tag{4.12}$$

The properties of $\mathbf{W}^{(L)}$ are now stated.

THEOREM 4.2.

*i) The relationship between the far field coefficients* $\mathbf{f}^{(W,L)}$ *of* EQ. 4.7 *and the boundary coefficients* $\mathbf{c}^{(L)}$ *is*

$$\mathbf{c}^{(L)} = \mathbf{W}^{(L)} \cdot \mathbf{f}^{(W,L)} - [\mathbf{P}^{(L)}]^{-1} \cdot \mathbf{w}^{(L+1)}, \tag{4.13}$$

*where* $\mathbf{w}^{(L+1)} = [\langle v_\lambda | u_{rem}^{(L+1)} \rangle]$ *and* $u_{rem}^{(L+1)}$ *is defined by* EQ. 4.1.

*ii) The value of the boundary defect at the approximately back propagated coefficients of* EQ. 4.7 *is*

$$B^{(L)}(\vec{\psi}_{ref}\,; \mathbf{W}^{(L)} \cdot \mathbf{f}^{(W,L)}) = \|u_1^{\perp}\|^2_{L^2(\Gamma)} + \|(u_{rem}^{(L+1)})_1^{(L)}\|^2_{L^2(\Gamma)} \tag{4.14}$$

*where the rightmost term is obtained by projecting the remainder* $u_{rem}^{(L+1)}$ *onto* $X_1^{(L)}$.

*iii) If* $k^2 \notin \sigma(-\Delta_D)$ *then* $\lim_{L\to\infty} |\mathbf{w}^{(L+1)}| = 0$.

*Proof.*

*i*) Recall the decomposition of $u$ in EQ. 4.1 and rewrite $\mathbf{g}^{(L)}$ of EQ. 3.1 as

$$\mathbf{g}^{(L)} = [\langle v_\lambda | u_a^{(L)} \rangle + \langle v_\lambda | u_{\text{rem}}^{(L+1)} \rangle]. \tag{4.15}$$

EQNS. 4.1 and 2.12 yield

$$\mathbf{g}^{(L)} = \mathcal{R}\mathbf{P}^{(L)} \cdot \mathbf{a} + [\langle v_\lambda | u_{\text{rem}}^{(L+1)} \rangle] = \mathcal{R}\mathbf{P}^{(L)} \cdot \mathbf{a} + \mathbf{w}^{(L+1)}. \tag{4.16}$$

Substitute the r.h.s. of (4.16) into EQ. 3.1. Eliminate $\mathbf{a}$ by means of EQNS. 4.6 and 4.7 and obtain

$$\mathbf{P}^{(L)} \cdot \mathbf{c}^{(L)} = \mathcal{R}\mathbf{P}^{(L)} \cdot \mathbf{Q}^{(L)\dagger} \cdot [\mathcal{R}\mathbf{Q}^{(L)Trs}]^{-1} \cdot \mathbf{f}^{(W,L)} - \mathbf{w}^{(L+1)}. \tag{4.17}$$

Multiplication to the left by $[\mathbf{P}^{(L)}]^{-1}$ yields EQ. 4.13. This result is the main motivation behind DEF. 4.1.

*ii*) From EQNS. 4.12 and 4.7 it follows

$$\mathbf{P}^{(L)} \cdot \mathbf{W}^{(L)} \cdot \mathbf{f}^{(W,L)} = -\mathcal{R}\mathbf{P}^{(L)} \cdot \mathbf{a}. \tag{4.18}$$

Explicitly, given the definition of $\mathcal{R}\mathbf{P}^{(L)}$, this is the linear algebraic system

$$\langle v_\lambda | \sum_{\mu\in\Lambda(L)} [\mathbf{W}^{(L)} \cdot \mathbf{f}^{(W,L)}]_\mu v_\mu \rangle = -\langle v_\lambda | \sum_{\mu\in\Lambda(L)} a_\mu u_\mu \rangle, \forall \lambda \in \Lambda(L). \tag{4.19}$$

It can be read as the normal equation originated from the least squares problem

$$\min_{\mathbf{b}^{(L)}} \| \sum_{\lambda\in\Lambda(L)} b_\lambda^{(L)} v_\lambda + u_a^{(L)} \|^2_{L^2(\Gamma)}, \tag{4.20}$$

the solution to which is known beforehand: $\mathbf{b}^{(L)} = \mathbf{W}^{(L)} \cdot \mathbf{f}^{(W,L)}$. These are the coefficients, which yield the projection of $u_a^{(L)}$ onto $X_1^{(L)}$ i.e. ,

$$\sum_{\lambda \in \Lambda(L)} [\mathbf{W}^{(L)} \cdot \mathbf{f}^{(W,L)}]_\lambda v_\lambda = -(u_a^{(L)})_1^{(L)}. \tag{4.21}$$

The corresponding value of the boundary defect is therefore $\|u - (u_a^{(L)})_1^{(L)}\|^2_{L^2\Gamma}$.

By decomposing $u$ it follows

$$\begin{aligned} u - (u_a^{(L)})_1^{(L)} &= u_1^{(L)} + u_1^{\perp} - (u_a^{(L)})_1^{(L)} = \\ &= (u_a^{(L)})_1^{(L)} + (u_{\text{rem}}^{(L+1)})_1^{(L)} + u_1^{\perp} - (u_a^{(L)})_1^{(L)}. \end{aligned} \tag{4.22}$$

Since the two kets left are mutually orthogonal, EQ. 4.14 is obtained.

*iii*) This is immediate consequence of LEMMA 2.10. □

**4.4. The ABP algorithm and the numerical results.** A relationship has been established between some far field and boundary coefficients; if this piece of information is included into the definition of a boundary defect, the heuristic penalty term in e.g., the objective functions (4.9) and (4.10) can be dropped. Moreover, boundary coefficients shall no longer be counted as unknowns. Given a (tentative) vector $\vec{\psi}$, back propagation of the generic data $\mathbf{f}^{(L)}$, which need not be $\mathbf{f}^{(W,L)}$, by $\mathbf{W}^{(L)}$ will try to insure approximate compliance with the representation of the scattered field on $\Gamma$. Hence the counterpart of PBM. 3.2 can be stated.

PROBLEM 4.1. Given $u, \mathbf{f}^{(L)}$, $\Psi_{\text{ad}}$ find $\vec{\psi}^* \in \Psi_{\text{ad}}$ s.t.,

$$B^{(L)}(\vec{\psi}^*; \mathbf{W}^{(L)}(\vec{\psi}^*) \cdot \mathbf{f}^{(L)}) = \min_{\Psi_{\text{ad}}} B^{(L)}(\vec{\psi}; \mathbf{W}^{(L)}(\vec{\psi}) \cdot \mathbf{f}^{(L)}). \tag{4.23}$$

The existence of at least one $\vec{\psi}^*$ follows as with PBM 3.2.

The minimization algorithm is based on the same principles and rules as described in §3, although its complexity is increased by the back propagation procedure. Namely the incident field, the basis functions $\{v_\lambda\}$, $\mathbf{W}^{(L)}$ and the Jacobian all depend on the unknown shape parameters and shall be differentiated accordingly, for the steepest descent or Newton rule to apply. The criteria to predict the algorithm performance and to design numerical examples described in §3.2 continue to hold. In particular, if the data are $\mathbf{f}^{(W,L)}$, EQ. 4.14 replaces EQ. 3.2.

In all numerical tests the second summand on the r.h.s. of EQ. 4.14 has negligible value w.r. to the first one.

Identification results by ABP applied to prolate spheroids under the same conditions ($I, \mathbf{k}, L$, initial guess) as in TAB. 3.1 and FIG. 3.2 are shown in FIG. 4.1, together with the corresponding values of $N_{it}$ and $B^{(L)}(\vec{\psi}^*)$.

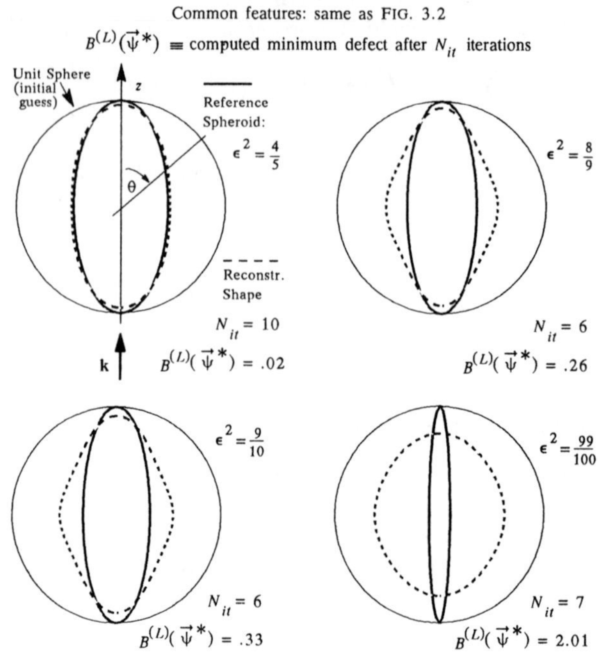


FIG. 4.1. *Identification of prolate spheroids from* $\mathbf{f}^{(W,L)}$ *far field coefficients.*

All computations reported herewith have been carried out by IBM RT PC 6150/125 machines. When $\epsilon^2 \geq \frac{4}{5}$, the computed minimum defect is consistently larger ($\times 3$) than the estimated value; at the same time reconstruction becomes poorer: minor half axes are e.g., overestimated. The result at $\epsilon^2 = \frac{99}{100}$ is meaningless. These negative aspects may be ascribed to accumulating numerical errors from the inversion and multiplication of poorly conditioned matrices.

Again, by COR. 2.5, uniqueness would hold only in the infinite dimensional case. The loss of uniqueness due to approximation ($L = 12$) shows up e.g., by the dependence of the optimal shape on the initial guess: see FIG. 4.2.

Results with oblate spheroids are similar and are not reported for the sake of conciseness.

Other shapes have been considered, among which: the bullet, the peanut and the cell. The corresponding $\mathbf{f}^{(W,L)}$ data have been obtained either from one or two incident plane waves $(\mathbf{k}_1, \mathbf{k}_2)$, with the same wavenumber. In the latter case the ABP algorithm is easily adapted: both $u$ and $v$ become superpositions of two fields. No relevant modifications affect the code. TAB. 4.1, TAB. 4.2 and FIG. 4.3 respectively provide the trigonometric shape parameter vectors, some relevant data and the reconstruction results.

TABLE 4.1
*Trigonometric Shape Parameters*

| Shape | $I$ | $r(0)$ | $r(\pi/2)$ | $\vec{\psi}_{\text{ref}}$ |
|---|---|---|---|---|
| bullet | 5 | .577 | $\sim .75$ | $\{2,0,0,0,0,1\}$ |
| peanut | 7 | 1.0 | .49 | $\{2,0,0,-1.6,0,0,0,.6\}$ |
| cell | 11 | .77 | 1.4 | $\{1.4,0,-.04,+.85,0,0,0,-.3,0,0,0,-.3\}$ |

TABLE 4.2
*Shape Identification from Two Incident Wavevectors and* $\mathbf{f}^{(W,L)}$ *Data*

$N_{it} \equiv$ number of iterations until search failure
$B^{(L)}(\vec{\psi}^*) \equiv$ computed minimum defect after $N_{it}$ iterations

| Shape | $I$ | $(\mathbf{k}_1, \mathbf{k}_2)$ | $L$ | $\left\lvert\frac{P_{0LL}}{P_{000}}\right\rvert$ | $B^{(L)}$ (EQ. 4.14) | $N_{it}$ | $B^{(L)}(\vec{\psi}^*)$ |
|---|---|---|---|---|---|---|---|
| bullet | 7 | $(\hat{\mathbf{z}}, -\hat{\mathbf{z}})$ | 12 | $(1.4/4\pi)\cdot 10^{30}$ | .0343 | 19 | .0247 |
| peanut | 7 | $(\hat{\mathbf{z}}, \hat{\mathbf{x}})$ | 12 | $(4.5/4\pi)\cdot 10^{31}$ | .0435 | 9 | .0336 |
| cell | 11 | $(\hat{\mathbf{z}}, -\hat{\mathbf{z}})$ | 9 | $(4.8/4\pi)\cdot 10^{19}$ | .362 | 6 | 13.2 |

The maximum to minimum radius ratios are approximately 2 for all three shapes. The quality of the identified shapes can be predicted from the estimates of conditio n numbers (column 5, TAB. 4.2): said values are comparable to those in lines 1 to 3, column 3 of TAB. 3.1.

Same notation as in FIG. 4.1.

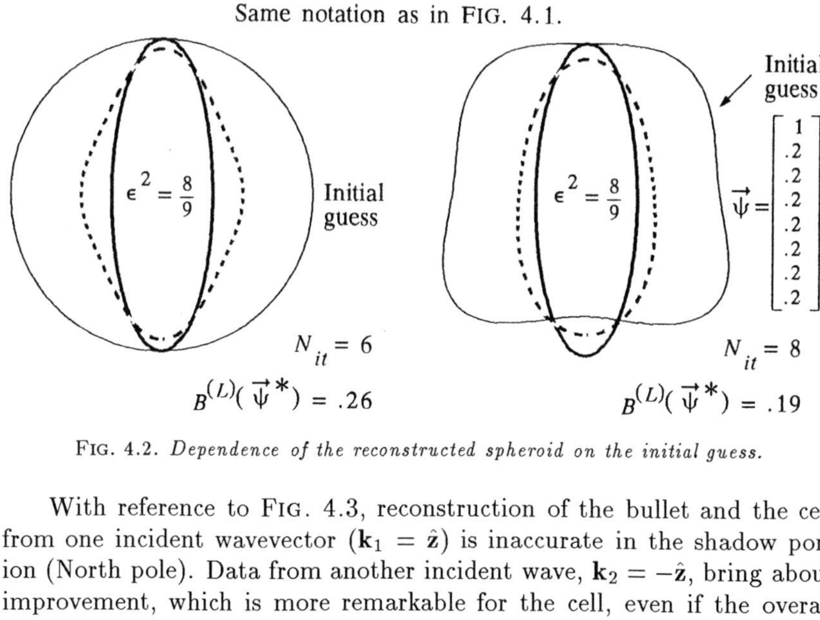


FIG. 4.2. *Dependence of the reconstructed spheroid on the initial guess.*

With reference to FIG. 4.3, reconstruction of the bullet and the cell from one incident wavevector ($\mathbf{k}_1 = \hat{\mathbf{z}}$) is inaccurate in the shadow port ion (North pole). Data from another incident wave, $\mathbf{k}_2 = -\hat{\mathbf{z}}$, bring about improvement, which is more remarkable for the cell, even if the overall quality remains poor. In other words, supplementary information makes up, in part, for $L$-depe ndent lack of uniqueness. No improvement is noticed when reconstructing the peanut.

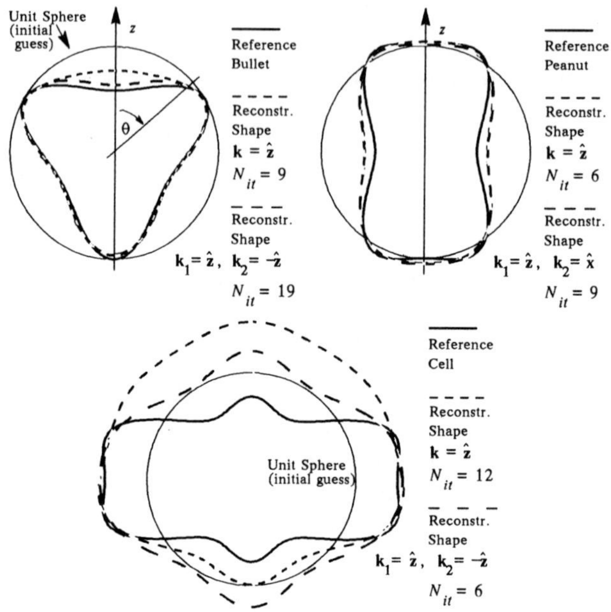


FIG. 4.3. *Shape identification from* $\mathbf{f}^{(W,L)}$ *data.*

**5. Discussion and conclusion.** The ABP method relies on a relation between far field and boundary coefficients, which replaces the penalty term in the objective function to be minimized. Emphasis has been placed herewith on the ABP counterpart of EQNS. 4.9 and 4.7. Other similar relations can be met in connection with different objective functions. It can be shown that e.g., (4.10) leads to the following affine map between boundary and far field coefficients of order $L$

$$\mathbf{f}^{(J,L)} = \mathcal{R}\mathbf{L}^{(L)}\mathbf{c}^{(L)} - \frac{ik}{4\pi}[\langle u_\lambda | \partial_n u \rangle], \tag{5.1}$$

where

$$\mathcal{R}\mathbf{L}^{(L)} = [\langle \partial_n v_\lambda | u_\mu \rangle]. \tag{5.2}$$

EQ. 5.1 provides an interesting alternative to EQ. 4.7 for the approximate computation of far field coefficients, $\mathbf{f}^{(J,L)}$. This may find application to validate the method of §4 or even to replace the ABP defined there.

The ABP operator exists, provided the admissible values of $k^2$ are restricted according to (4.11.$i$) and (4.11.$ii$) holds. Since the shape parameters are updated all the time during minimization, either condition may fail to be met at some stage and stop the iteration. The problem has not yet been thoroughly investigated.

In the previous SECTIONS the $\mathbf{f}^{(W,L)}$ far field coefficients have always been the starting point. This has certainly simplified the development and testing of identification algorithms and codes. In order to make problems more realistic, the data must be the values of the scattering amplitude at some grid points on $S^2$, from which the far field coefficients shall be estimated by means of inner products. Denote the corresponding result by $\tilde{\mathbf{f}}^{(L)}$: the values of its entries generally differ from those of $\mathbf{f}^{(W,L)}$ and affect reconstruction. If an identification method is numerically robust, results shall remain satisfactory. One parameter, which is expected to control robustness is $L$: namely higher order coefficients are more sensitive to estimation errors. FIG. 5.1 provides some preliminary, positive evidence: $L$ has been set to 6 and $n_p$ to 32 (instead of 64 used in §4).

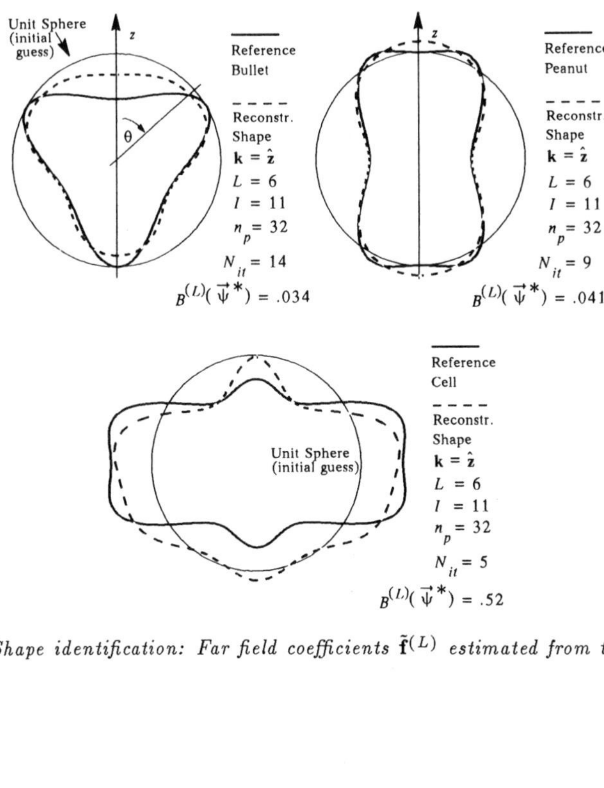


FIG. 5.1. *Shape identification: Far field coefficients* $\tilde{\mathbf{f}}^{(L)}$ *estimated from the scattering amplitude.*

In order to complete the picture, it would be interesting to estimate $\tilde{\mathbf{f}}^{(L)}$ from the scattering amplitude yielded by methods other than $\mathbf{T}$ matrix: e.g., Nystroem's [3,p. 68], Kleinman's et al. [9]. This has not yet been investigated.

Another problem, which has not been addressed herewith is limited aperture reconstruction [3,p. 134].

## REFERENCES

[1] T.S. Angell, R.E. Kleinman, G.F. Roach, *An Inverse Transmission Problem for the Helmholtz Equation*, Inverse Problems, **3** (1987), pp. 149–80.

[2] R.G. Barantsev, V.V. Kozachek, *Issledovanie Matrizy Rasseyaniya na Ob'ektach Slozhnoi Formy (Investigation of the Scattering Matrix for Bodies of Complicated Form)*, Vestnik Leningr. Univers., # 7 (1968), pp. 71–77.

[3] D. Colton, R. Kress, *Inverse Acoustic and Electromagnetic Scattering Theory*, Springer-Verlag: Berlin, 1992.

[4] D. Colton, B. Sleeman, *Uniqueness Theorems for the Inverse Problem of Acoustic Scattering*, IMA Journal of Applied Mathematics, **31** (1983), pp. 253–59.

[5] A.G. Dallas, *The Waterman Algorithm for Approximation of the Transition Matrix in Scattering by Obstacles*, talk delivered at the XXII General Assembly of the International Union of Radio Science, Tel Aviv, Aug. 24-Sep. 2, 1987.

[6] B. Demidovitch, I. Marom, *Elements de Calcul Numerique*, MIR: Moscow, 1979.

[7] D.S. Jones, X.Q. Mao, *A Method for Solving the Inverse Problem in Soft Acoustic Scattering*, Inverse Problems, **5** (1989), pp. 731–48.

[8] R.E. Kleinman, G.F. Roach, Proc. Royal Soc. of London, **A 383** (1982), pp. 313–32.

[9] R.E. Kleinman, G.F. Roach, L.S. Schuetz, J. Shirron, *An Iterative Solution to Acoustic Scattering by Rigid Objects*, Journal of the Acoust. Soc. Amer., **84** (1988), pp. 385-91.

[10] R.E. Kleinman, G.F. Roach, S.E.G. Stroem, *The Null Field Method and Modified Green Functions*, Proc. Royal Soc. of London, **A 394** (1984), pp. 121–36.

[11] V.V. Kravtsov, *Integral'nye Uravneniya v Zadachakh Diffraktsii. Sborn.: Vychislitelnye Metody i Programmirovanie (Integral Equations of Diffraction Problems. Coll.: Computational Methods and Programming)*, Vestn. Tsentr. Moskow. Gosud. Univers. # V (1966), pp. 260–93.

[12] R. Kress, *Numerical Methods in Inverse Acoustic Obstacle Scattering*, in D. Colton, R. E. Ewing, W. Rundell Eds., *Inverse Problems in Partial Differential Equations*, pp. 59–70, SIAM: Philadelphia, 1990.

[13] G. Kristensson, A.G. Ramm, S.E.G. Stroem, *Convergence of the* **T** *Matrix Approach in Scattering Theory. II*, J. Mathem. Phys., **24** (1983), pp. 2619–31.

[14] P.D. Lax, R.S. Phillips, *Scattering Theory*, Academic Press: New York, 1967.

[15] C. Mueller, *Radiation Patterns and Radiation Fields*, Arch. Rat. Mech. and Analysis, **4** (1955), pp. 235–46.

[16] C. Mueller, *Foundations of the Mathematical Theory of Electromagnetic Waves*, Springer-Verlag: Berlin, 1969.

[17] K. Onishi, *Numerical Identification in Inverse Scattering Problems*, in M. Yamaguti et al., Eds., *Inverse Problems in Engineering Sciences*, pp. 79–85, Springer-Verlag: Tokyo, 1991.

[18] A.G. Ramm, *Scattering by Obstacles*, Reidel: Dordrecht, 1986.

[19] A.G. Ramm, *A Method for Solving Inverse Diffraction Problems*, Inverse Problems, **3** (1987), pp. L23–25.

[20] A.G. Ramm, *Multidimensional Inverse Scattering Problems*, Longman: Harlow, 1992.

[21] B. Sleeman, *The Inverse Problem of Acoustic Scattering*, IMA J. of Applied Mathem., **29** (1982), pp. 113–42.

[22] A.N. Tichonov, A.A. Samarskii, *Equazioni della Fisica Matematica*, MIR: Moscow, 1981 (Italian translation from: A.N. Tichonov, A.A. Samarskii, *Uravneniya Matematiceskoi Fiziki*, Nauka: Moscow, 1977).

[23] V.A. Trenoguine, *Analyse Fonctionnelle*, MIR: Moscow, 1985.

[24] I.N. Vekua, *O Polnote Sistemy Metagarmoniceskikh Funkstii (On the Completeness of Systems of Meta-harmonic functions)*, Doklady Ak. Nauk SSSR, **90**, # 5 (1953), pp. 715–18.

[25] P.C. Waterman, *New Formulation of Acoustic Scattering*, Journal of the Acoust. Soc. Amer., **45** (1969), pp. 1417–29.

[26] C.H. Wilcox, *Scattering Theory for the d' Alembert Equation in Exterior Domains*, Springer-Verlag: Berlin, 1975.

# ON A VARIATIONAL EQUATION FOR THIN SHELLS*

M.C. DELFOUR† AND J.P. ZOLÉSIO‡

**Abstract.** We present a mathematical construction of a boundary equation for thin shells from classical elasticity by using the tangential differential calculus and the oriented distance function. We specify the appropriate function spaces and give an existence and uniqueness theorem. The assumptions on the displacement field around a mean surface are analogous to the ones found in the work of W.T. Koiter and P.M. Naghdi.

**1. Introduction.** The object of this paper is to construct a *boundary differential equation* for thin shells using the *tangential differential calculus* (cf. for instance the recent book by J. Sokolowski and J.-P. Zolésio [14]) and the *oriented distance function* which provides a direct access to geometric properties of a surface such as the normal, the curvature tensor and its derivatives (cf. M.C. Delfour and J.-P. Zolésio [4], [5]). A theorem on the existence and uniqueness of solutions in an appropriate function space framework is given for models which are related to the work of W.T. Koiter [9] and P.M. Naghdi [12]. The question of existence of solutions for the latter models is an important issue and extremely interesting results have been recently obtained by M. Bernadou, P.G. Ciarlet and B. Miara [1]. However our variational boundary differential equation is significantly different from the traditional models of thin shells and this is the basic contribution of this paper. We never use local bases for the tangent spaces. We rather make use of global operators defined on the mean surface (boundary) such as the *tangential gradient*, the *tangential divergent*, the *Laplace-Beltrami operator*, the *"bi-Laplace-Beltrami" operator*, etc. Following practices for plate equations models (cf. for instance J. Lagnese and J.-L. Lions [11]), we have kept terms up to the third order in the thickness. In our view this model is mathematically more tractable than existing models. It offers certain technical advantages in dealing with theoretical issues. However for the numerical approximation of the solution it is not clear that such an approach will yield better or newer algorithms since local bases will have to be introduced in the approximation process. One direction to explore would be the techniques developed in the *boundary element method* or related techniques, but this question is beyond the scope of this paper (cf.

* The two authors acknowledge the support of the Institute for Mathematics and its Applications while attending the IMA Workshop on Control and Optimal Design of Distributed Parameter Systems (November 9–13, 1992).

† Centre de recherches mathématiques et Département de mathématiques et de statistique, Université de Montréal, C.P. 6128 Succ. Centre-ville, Montréal, Québec, Canada H3C 3J7. The research of the first author has been supported in part by National Sciences and Engineering Research Council of Canada operating grant A-8730 and by a FCAR grant from the Ministère de l'Éducation du Québec.

‡ Institut Non Linéaire de Nice, 1361 route des lucioles, 06904 Sophia-Antipolis Cédex, France.

for instance G. Chen and J. Zhou [2]).

Consider an oriented surface $\Gamma$ in $\mathbb{R}^N$ which is sufficiently smooth. The shell will be made up of level curves around $\Gamma$ up to a thickness $h$, that is the thickness of the shell will be $2h$. It will be assumed that the parameter $h$ is small compared to the curvature tensor of $\Gamma$ and that the material has a linear elastic behaviour. Moreover since the elastic energy to be minimized with respect to the displacement field is strictly convex, we obtain the existence and uniqueness of solutions in an appropriate function space defined on $\Gamma$.

For simplicity we assume that the *mean surface* $\Gamma$ of the shell is the boundary of a smooth domain $\Omega$ and that $\overline{\Omega}$ is compact. The boundary $\Gamma$ is oriented by choosing the direction of the normal $n$ to the boundary $\Gamma$ pointing outward from $\Omega$. By construction the oriented surface $\Gamma$ has no boundary. We shall also make use of the oriented distance function

$$b_\Omega(x) = \begin{cases} d_\Gamma(x), & x \in \overline{\complement\Omega} \\ -d_\Gamma(x), & x \in \overline{\Omega}, \end{cases} \tag{1.1}$$

where $\complement\Omega = \{x \in \mathbb{R}^N : x \notin \Omega\}$ and $d_\Gamma$ is the usual distance function to $\Gamma$. In the sequel the function $b_\Omega$ will be denoted by $b$.

All the geometrical information on $\Gamma$ is captured by the function $b$. In particular we shall make use of the matrix $D^2 b$ of second order partial derivatives of $b$ whose eigenvalues are the *principal curvatures* of the surface $\Gamma$ and whose trace is the *mean curvature*

$$H = \text{tr}\,(D^2 b) = \Delta b. \tag{1.2}$$

The reader is referred to M.C. Delfour and J.-P. Zolésio [4][5] for a detailed study of the properties of the function $d_\Omega$ and to Gilbarg and Trudinger [8] for the study of curvature via distance functions. We shall also make use of the rate of change on the level surfaces

$$\Gamma_z = \{x \in \mathbb{R}^N : b(x) = z\} \tag{1.3}$$

"parallel to $\Gamma$".

In the above reference the reader will find that for $k \geq 2$ a domain has a $C^k$ boundary $\Gamma$ if and only if the oriented distance function $b_\Omega$ is $C^k$ in a neighbourhood of $\Gamma$. If we assume that $\Gamma$ is of class $C^2$, then there exists $h > 0$ sufficiently small such that

$$\Omega_h = \{x \in \mathbb{R}^N : |b(x)| < h\} \tag{1.4}$$

is a neighbourhood of $\Gamma$ where $b$ is $C^2$. This will be the *shell* made up of level surfaces $\Gamma_z, -h < z < h$.

It is useful to recall that the trace of $\nabla b$ on a level curve $\Gamma_z$ coincides with the unit outward normal field while $\Delta b$ is the mean curvature along the level curve $\Gamma_z$. The projection mapping $p$ onto $\Gamma$ can also be obtained directly from the oriented distance function $b$. When $\Gamma$ is $C^2$, the function $b$ is $C^2$ in $\overline{\Omega}_h$ and the map $p : \Omega_h \to \Gamma$ is given by

$$p(x) = x - b(x)\, \nabla b(x) \tag{1.5}$$

and its Jacobian matrix by

$$Dp(x) = I_d - b(x)\, D^2b(x) - \nabla b(x)\, {}^*\nabla b(x) = P(x) - b(x)\, D^2b(x)\,, \tag{1.6}$$

where ${}^*\nabla b(x)$ is the transpose of the vector $\nabla b(x)$, $I_d$ is the identity matrix and $P(x)$ is the *linear projector* onto the *tangential plane* $T_x\Gamma$ at the point $x$ of $\Gamma$. The *tangential gradient*

$$\nabla_\Gamma u = \nabla U\big|_\Gamma - \frac{\partial U}{\partial n}\, n \tag{1.7}$$

defined for a function $u$ on $\Gamma$ via an extension $U$.

In particular

$$\nabla(u \circ p) = Dp\,(\nabla_\Gamma u) \circ p = [I - b\, D^2 b](\nabla_\Gamma u) \circ p\,, \tag{1.8}$$

where "$\circ$" denotes the composition of the corresponding maps. In addition we shall also use the tangential Jacobian matrix of a vector $v : \Gamma \to \mathbb{R}^N$. It is convenient to define its transpose

$$D_\Gamma^* v = (\nabla_\Gamma v_1, \dots, \nabla_\Gamma v_N) \tag{1.9}$$

in terms of the column tangential gradients. In particular

$$D_\Gamma^*(v \circ p) = Dp(\nabla_\Gamma v_1, \dots, \nabla_\Gamma v_N) \circ p = Dp(D_\Gamma^* v) \circ p \tag{1.10}$$

and

$$D(v \circ p) = (D_\Gamma v) \circ p\, Dp\,. \tag{1.11}$$

The above identities will be used in Section 4.

**2. Displacement field and flow of grad $b$.** The shell has been characterized as a domain $\Omega_h$ made up of level curves which are parallel to the mean surface $\Gamma$ and the surface $\Gamma$ is the boundary of a domain $\Omega$ of class $C^2$. Associate with each point $x$ of the shell a displacement vector of the following form

$$V(x) = u \circ p(x) + b(x)\, \ell \circ p(x) + w \circ p(x)\, \nabla b(x), \quad x \in \Gamma, \tag{2.1}$$

where $p$ is the projection operator from $\Omega_h$ onto $\Gamma$. The mappings $u$ and $\ell$ in $H^1(\Gamma; \mathbb{R}^N)$ are vector valued functions such that

$$u(x) \cdot \nabla b(x) = \ell(x) \cdot \nabla b(x) = 0, \quad x \in \Gamma, \tag{2.2}$$

that is, at each point $x$ on $\Gamma$, $u(x)$ and $\ell(x)$ belong to the tangent space $T_x\Gamma$ at $x$. Finally $w$ is a scalar function in $H^1(\Gamma)$ along the vector $\nabla b(x) = n(p(x))$. For $x$ on $\Gamma, b(x) = 0$ and

$$V(x) = u(x) + w(x)\, n(x) \tag{2.3}$$

and $V(x)$ can be identified with a pair $(\vec{u}\ (x), w(x))$ in $\mathbb{R}^{N-1} \times \mathbb{R}$ in the orthonormal basis at $x$ determined by the unit outward normal $n(x)$ and an appropriate basis for the tangent space $T_x\Gamma$ at $x$. For $x \in \Gamma_z, z \neq 0$, there is an additional tangential displacement $b(x)\,\ell\circ p(x)$ to the level curve $\Gamma_z$ which is proportional to $z = b(x)$.

The vector $\nabla b$ is $C^1$ in $\Omega_h$ and we can consider the *flow mapping* $T_z(\nabla b)$, defined by

$$T_z(\nabla b)(X) = x(z), \tag{2.4}$$

where

$$\begin{cases} \frac{dx}{dz}(z) &= \nabla b(x(z)),\ |z| < h\,, \\ x(0) &= X\,. \end{cases} \tag{2.5}$$

It is an homeomorphism from $\Gamma$ onto $\Gamma_z$ and from $\Gamma_s$ onto $\Gamma_{s+z}$ for $|s+z| < h$ and $|s| < h$. As $\nabla b$ is an autonomous vector field we have the group property:

$$T_{s+z}(\nabla b) = T_s(\nabla b) \circ T_z(\nabla b) = T_z(\nabla b) \circ T_s(\nabla b). \tag{2.6}$$

Therefore the Jacobian matrices satisfy the following chain rule

$$DT_{s+z}(\nabla b) = DT_s(\nabla b) \circ T_z(\nabla b)\, DT_z(\nabla b) \tag{2.7}$$

so that

$$\begin{aligned} \frac{\partial}{\partial z} DT_z(\nabla b) &= \frac{\partial}{\partial s} DT_{s+z}(\nabla b)\bigg|_{s=0} \\ &= D^2 b \circ T_z(\nabla b)\, DT_z(\nabla b). \end{aligned} \tag{2.8}$$

The above technique will be used to compute several derivatives which will be used elsewhere in the paper. For details the reader is referred to the book of J. Sokolowski and J.-P. Zolésio [14].

Using the notation $T_z = T_z(\nabla b)$, it is readily seen that

$$b(T_z(X)) = b(X) + z \tag{2.9}$$

$$\nabla(b \circ Tx) = \nabla b\,. \tag{2.10}$$

and necessarily

$$ {}^*DT_z \nabla b = \nabla b \,. \tag{2.11} $$

Moreover

$$ D^2 b \circ T_z DT_z = D^2 b \tag{2.12} $$

and from (2.8)

$$ DT_z = I + z\, D^2 b \,. \tag{2.13} $$

Hence $\det DT_z$ is a polynomial of degree $N-1$

$$ \det DT_z = 1 + zH + \sum_{\ell=2}^{N-1} z^\ell K_\ell \,, \tag{2.14} $$

where

$$ H = \ \mathrm{tr}\ D^2 b = \Delta b \tag{2.15} $$

is the mean curvature (up to a multiplicative constant) and $K_{N-1}$ is the Gauss curvature on $\Gamma$. The term in $z^N$ is zero since

$$ |\nabla b|^2 = 1 \quad \Rightarrow \quad D^2 b \, \nabla b = 0 \quad \Rightarrow \quad \det D^2 b = 0 \,. $$

Finally from (2.11)

$$ {}^*DT_z^{-1}\, n = {}^*\ DT_z^{-1}\, \nabla b = \nabla b = n, \quad \text{on } \Gamma \,. \tag{2.16} $$

**3. Integration over $\Omega_h$.** The next ingredient is the use of Federer's decomposition of the measure on $\Omega_h$: for any sufficiently smooth function $f$, say in $H^1(\Omega_h)$,

$$ \int_{\Omega_h} f(x)\, dx = \int_{-h}^{h} dz \left\{ \int_{\Gamma_z} f \, d\Gamma_z \right\}, \tag{3.1} $$

where $d\Gamma_z$ is the surface measure of $\Gamma_z$. The decomposition of a measure along the level curves has been used by R. Temam [1] in the context of Plasma Physics problems. After a change of variable using the transformation $T_z$ we get

$$ \int_{\Gamma_z} f \, d\Gamma_z = \int_{\Gamma} f \circ T_z(\nabla b)\, j(z)\, d\Gamma \,, \tag{3.2} $$

where in view of (2.14) and (2.16)

$$ j(z) = \det(DT_z)\ \|{}^* DT_z^{-1}\, n\| = \det\,(DT_z), \quad \text{on } \Gamma \,, \tag{3.3} $$

and

$$ j(z) = 1 + zH + \frac{z^2}{2}\ K_2 + 0(z^3) \,. \tag{3.4} $$

**4. The deformation tensor.** From the decomposition (2.1) of the displacement vector field $V$ in $\Omega_h$, a direct computation yields the following expression on the level curves $\Gamma_z$

$$\begin{aligned}\varepsilon(V) &= \varepsilon_\Gamma(u)\circ p + b\,\varepsilon_\Gamma(\ell)\circ p \\ &\quad +\tfrac{1}{2}\left[\nabla b\,\ell^*\circ p + \ell\circ p\,{}^*\nabla b\right] \\ &\quad +\tfrac{1}{2}\left[\nabla b(\nabla_\Gamma^* w)\circ p + (\nabla_\Gamma w)\circ p\,{}^*\nabla b\right] \\ &\quad + w\circ p\, D^2 b \qquad \text{in } \Omega_h,\end{aligned} \tag{4.1}$$

where $\nabla_\Gamma w$ is the *tangential gradient* of $w$ on $\Gamma$ and $\varepsilon_\Gamma(u)$ is the *tangential deformation tensor* defined by

$$2\,\varepsilon_\Gamma(u) = {}^*D_\Gamma u + D_\Gamma u\,. \tag{4.2}$$

The *tangential Jacobian matrix* is defined through its transposed matrix as

$${}^*D_\Gamma u = (\nabla_\Gamma u_1, \dots, \nabla_\Gamma u_N)\,, \tag{4.3}$$

where $\nabla_\Gamma$ is the tangential gradient defined in (1.8). Expression (4.1) contains five terms and the term tr $(\varepsilon(V)\cdot\varepsilon(V))$ will contain fifteen terms. In order to simplify the computations, and only to simplify the computations, we assume that the tangent field $\ell$ is related to $w$ through the following classical assumption

$$\ell = -\nabla_\Gamma w \quad \text{on } \Gamma\,. \tag{4.4}$$

It can be checked that (4.4) implies that in the deformation process

$$X \longmapsto X + V(X)$$

the transverse line

$$\{x + z\,\nabla b(x) : |z| < h\}$$

remains, at the first order in $u$ and $w$, perpendicular to the transformed mean surface

$$\{X + V(X) : X \in \Gamma\}\,.$$

Under (4.4) the deformation tensor reduces to the following expression

$$\varepsilon(V) = \varepsilon_\Gamma(u)\circ p - b\,\varepsilon_\Gamma(\nabla_\Gamma w)\circ p + w\circ p\, D^2 b. \tag{4.5}$$

Notice that assumption (4.4) is essentially the assumption used in the model of W.T. Koiter [9,10] and related works while the general model uses the assumption found in the model of P.M. Naghdi [12,13] and related works.

**5. Computation of the quadratic energy $W$.** Define

$$W = \int_{\Omega_h} \sigma \cdot\cdot\, \varepsilon(V)\, dx - \int_{\Omega_h} f \cdot v\, dx\,, \tag{5.1}$$

where $\sigma$ is the *stress tensor* and $f$ the loading of the shell. Assume that $f$ belongs to $H^1(\Omega_h)$ so that the trace is well defined on the surface $\Gamma$. As we assume linear elasticity we choose to simplify the computation by using the following Hooke's law

$$\sigma = 2\mu\, \varepsilon(V)\,. \tag{5.2}$$

Now recall the measure decomposition (3.1), (3.2) and the expression (4.5) of the deformation tensor. Recall also the construction of the flow $T_\mu = T_\mu(\nabla b)$ in Section 2 and notice that

$$p \circ T_\mu = I_d \quad \text{and} \quad b \circ T_\mu = \mu \ \text{ on } \Gamma\,. \tag{5.3}$$

In addition note that from

$$(\nabla b) \circ p = \nabla b \ \text{ on } \Omega_h \tag{5.4}$$

we easily get

$$D^2 b = \varepsilon_\Gamma(\nabla b) \circ p \tag{5.5}$$

so that, in view of (5.3), the term $D^2 b \circ T_z$ simplifies:

$$D^2 b \circ T_z(\nabla b) = \varepsilon_\Gamma(\nabla b)\,. \tag{5.6}$$

With the above observations

$$V \circ T_z = u + z\ell + w\, \nabla b$$

and

$$\varepsilon(V) \circ T_z = \varepsilon_\Gamma(u) + z\, \varepsilon_\Gamma(\ell) + w\, \varepsilon_\Gamma(\nabla b).$$

In the special case where $\ell = -\nabla_\Gamma w$

$$W = \int_{-h}^{h} dz \int_\Gamma d\Gamma \{2\mu \varepsilon_\Gamma(V) \circ T_z \cdot\cdot\, \varepsilon_\Gamma(V) \circ T_z - f \circ T_z \cdot V \circ T_z\} j(z)$$

where

$$\begin{aligned} &\varepsilon_\Gamma(V) \circ T_z \cdot\cdot\, \varepsilon_\Gamma(V) \circ T_z j(z) \\ &= [A_0 + zA_1 + z^2 A_2] \times \left[1 + zH + \tfrac{z^2}{2} K_2 + o(z^3)\right] \end{aligned}$$

and

$$\begin{aligned} A_0 &= \varepsilon_\Gamma(u) \cdot\cdot\, \varepsilon_\Gamma(u) + w^2 \varepsilon_\Gamma(\nabla b) \cdot\cdot\, \varepsilon_\Gamma(\nabla b) + 2w\varepsilon_\Gamma(\nabla b) \cdot\cdot\, \varepsilon_\Gamma(u) \\ A_1 &= -2\varepsilon_\Gamma(\nabla_\Gamma w) \cdot\cdot\, [\varepsilon_\Gamma(u) + w\varepsilon_\Gamma(\nabla b)] \\ A_2 &= \varepsilon_\Gamma(\nabla_\Gamma w) \cdot\cdot\, \varepsilon_\Gamma(\nabla_\Gamma w). \end{aligned}$$

Finally up to the second order in $z$

$$\begin{aligned} &\varepsilon_\Gamma(V) \cdot T_z \cdot\cdot\, \varepsilon_\Gamma(V) \cdot T_z \\ &= A_0 + z(A_1 + HA_0) + z^2 \left(A_2 + HA_1 + \tfrac{K_2}{2} A_0\right) + o(z^3)\,. \end{aligned}$$

Similarly if we assume that the force $f$ is of the form

$$f = F \circ p \ \text{ on } \ \Omega_h, \quad F \in H^1(\Gamma; \mathbb{R}^N)$$

then we get

$$\begin{aligned} &(f \circ T_z) \cdot (V \circ T_z) j(z) \\ &= F \cdot (u - z\nabla_\Gamma w + w\nabla b) \left[1 + zH + \tfrac{z^2}{2} K_2 + o(z^3)\right] \\ &= F \cdot \{u + w\nabla b + z(-\nabla_\Gamma w + H(u + w\nabla b)) \\ &\qquad + z^2 \left(\tfrac{K_2}{2}(u + w\nabla b) - H\nabla_\Gamma w\right) + o(z^3)\}\,. \end{aligned}$$

Finally after integration with respect to $z$ and keeping terms up to $h^3$ we obtain

$$\begin{aligned} W_0 &= \int_{-h}^{h} dz \int_\Gamma d\Gamma\, \varepsilon_\Gamma(V) \circ T_z \cdot\cdot\, \varepsilon_\Gamma(V) \circ T_z\, j(z) \\ &= 2\mu \int_\Gamma \left\{2hA_0 + 2\tfrac{h^3}{3}(A_2 + HA_1 + \tfrac{K_2}{2} A_0)\right\} d\Gamma \\ W_0 &= 2\mu \int_\Gamma \Big\{\left(2h + \tfrac{h^3}{3} K_2\right) [\varepsilon_\Gamma(u) + w\varepsilon_\Gamma(\nabla b)] \cdot\cdot\, [\varepsilon_\Gamma(u) + w\varepsilon_\Gamma(\nabla b)] \\ &\qquad + 2\tfrac{h^3}{3} \varepsilon_\Gamma(\nabla_\Gamma w) \cdot\cdot\, \varepsilon_\Gamma(\nabla_\Gamma w) \\ &\qquad - 4\tfrac{h^3}{3} H\, \varepsilon_\Gamma(\nabla_\Gamma w) \cdot\cdot\, [\varepsilon_\Gamma(u) + w\varepsilon_\Gamma(\nabla b)]\Big\} d\Gamma \end{aligned}$$

and

$$\begin{aligned} W_1 &= \int_{-h}^{h} dz \int_\Gamma d\Gamma (f \circ T_z) \cdot (V \circ T_z) j(z) \\ &= \int_\Gamma \left\{\left(2h + \tfrac{h^3}{3} K_2\right) F \cdot (u + w\nabla b) - 2\tfrac{h^3}{3} HF \cdot \nabla_\Gamma w\right\} d\Gamma\,. \end{aligned}$$

To complete the model it remains to specify the space in which $(u, w)$ live. Define the vectorial space

$$\begin{aligned} \mathcal{N} = \{(u, w) \in H^1(\Gamma; \mathbb{R}^N) \times H^2(\Gamma)\ & u \cdot n = 0 \ \text{ on } \ \Gamma \\ & \varepsilon_\Gamma(u) + w\varepsilon_\Gamma(n) = 0 \ \text{ on } \ \Gamma \\ & \varepsilon_\Gamma(\nabla_\Gamma w) = 0 \ \text{ and } \ \nabla w \cdot n = 0 \ \text{ on } \ \Gamma\}\,. \end{aligned}$$

Then we shall minimize the energy functional $W = W_0 + W_1$ over the following Hilbert space

$$\mathcal{H} = \left\{ (u, w) \in H^1(\Gamma; \mathbb{R}^N) \times H^2(\Gamma) \quad \begin{array}{l} u \cdot n = 0 \quad \text{on} \quad \Gamma \\ \nabla w \cdot n = 0 \end{array} \right\} / \mathcal{N}$$

with the norm

$$|||(u, w)||| = \left\{ \int_\Gamma ||\varepsilon_\Gamma(\nabla_\Gamma w)||^2 + ||\varepsilon_\Gamma(u) + w\varepsilon_\Gamma(n)||^2 d\Gamma \right\}^{1/2}$$

which is equivalent to the quotient norm on the space

$$\{(u, w) \in H^1(\Gamma; \mathbb{R}^N) \times H^2(\Gamma)\; u \cdot n = \nabla w \cdot n = 0 \quad \text{on} \quad \Gamma\}$$

by $\mathcal{N}$.

Expressions for $W_0$ and $W_1$ suggest to introduce the new variable

$$e = u + w\nabla b$$

whose tangential and normal components are precisely $u$ and $w$

$$u = e_\Gamma = e - (e \cdot n)n, \quad w = e \cdot n$$

so that

$$u = e - w\, n$$

and

$$\nabla_\Gamma w = \nabla_\Gamma(e \cdot n) = \nabla(e \cdot \nabla b).$$

The terms $W_0$ and $W_1$ can now be expressed in term of $e$:

$$\begin{aligned} W_0 \;=\; & 2\mu \textstyle\int_\Gamma \Big\{ \left(2h + \frac{h^3}{3} K_2\right) \Big[ ||\varepsilon_\Gamma(e)||^2 + \frac{1}{2} ||\nabla_\Gamma(e \cdot n)||^2 \\ & \qquad\qquad - \varepsilon_\Gamma(e) \cdot\cdot \left({}^*\nabla_\Gamma(e \cdot n)n + {}^*n\nabla_\Gamma(e \cdot n)\right) \Big] \\ & + 2\frac{h^3}{3} ||\varepsilon_\Gamma(\nabla_\Gamma(e \cdot n)||^2 \\ & - 4\frac{h^3}{3} H \Big[ \varepsilon_\Gamma(\nabla_\Gamma(e \cdot n)) \cdot\cdot\, \varepsilon_\Gamma(e) \\ & \qquad - \frac{1}{2}\varepsilon_\Gamma(\nabla_\Gamma(e \cdot n)) \,\cdot\cdot \\ & \qquad ({}^*\nabla_\Gamma(e \cdot n)n + {}^*n\nabla_\Gamma(e \cdot n)) \Big] \Big\} d\Gamma \\ W_1 \;=\; & \textstyle\int_\Gamma \left\{ \left(2h + \frac{h^3}{3} K_2\right) F \cdot e - 2\frac{h^3}{3} H F \cdot \nabla_\Gamma(e \cdot n) \right\} d\Gamma, \end{aligned}$$

where $H$ and $K_2$ are the coefficients of the polynomial (2.14).

The underlying spaces are now

$$H = \{e \in H^1(\Gamma;\mathbb{R}^N)\ e\cdot n \in H^2(\Gamma), \nabla_\Gamma(e\cdot n)\cdot n = 0\}$$

and

$$N = \{e \in H\ \varepsilon_\Gamma(e) = 0, \varepsilon_\Gamma(\nabla_\Gamma(e\cdot n)) = 0 \text{ on } \Gamma).$$

The decomposition

$$e \longmapsto (u,w) = (e-(e\cdot n)n, e\cdot n)$$

is an isomorphism between $H$ and $\mathcal{H}$ and $N$ and $\mathcal{N}$. The quotient norm in $H/N$ is equivalent to

$$|||e||| = \left\{\int_\Gamma \|\varepsilon_\Gamma(\nabla_\Gamma(e\cdot n))\|^2 + \|\varepsilon_\Gamma(e)\|^2 d\Gamma\right\}^{1/2}.$$

THEOREM 5.1. *Assume that there exists $\alpha > 0$ such that $2h+\frac{h^3}{3}K_2 \geq \alpha$ on $\Gamma$. The displacement vector $e$ is the unique minimizing element in $H/N$ of the functional $W(e)$ over all $e$ in $H/N$, where*

$$\begin{aligned} W(e) \;=\; 2\mu \int_\Gamma \Bigg\{ & \left(2h + \frac{h^3}{3}K_2\right)\Big[\|\varepsilon_\Gamma(e)\|^2 + \frac{1}{2}\|\nabla_\Gamma(e\cdot n)\|^2 \\ & \qquad - \varepsilon_\Gamma(e)\cdot\cdot({}^*\nabla_\Gamma(e\cdot n)n + {}^*n\nabla_\Gamma(e\cdot n))\Big] \\ & +2\frac{h^3}{3}\|\varepsilon_\Gamma(\nabla_\Gamma(e\cdot n)\|^2 \\ & -4\frac{h^3}{3}H\Big[\varepsilon_\Gamma(\nabla_\Gamma(e\cdot n))\cdot\cdot\,\varepsilon_\Gamma(e) \\ & \qquad -\frac{1}{2}\varepsilon_\Gamma(\nabla_\Gamma(e\cdot n))\cdot\cdot({}^*\nabla_\Gamma(e\cdot n)n + {}^*n\nabla_\Gamma(e\cdot n))\Big]\Bigg\}d\Gamma \\ & + \int_\Gamma \left\{\left(2h+\frac{h^3}{3}K_2\right)F\cdot e - 2\frac{h^3}{3}HF\cdot\nabla_\Gamma(e\cdot n)\right\}d\Gamma. \end{aligned}$$

□

It is interesting to note that $W(e)$ only depends on the two explicit variables $e$ and $\ell = \nabla_\Gamma(e\cdot n)$.

**6. The shell variational equation.** In view of Theorem 5.1, the shell variational equation is given by the first order necessary condition

$$e \in H/N, \quad dW(e;\varphi) = 0, \quad \forall \varphi \in H/N,$$

where $dW(e;\varphi)$ is the directional derivative of $W(e)$ at $e$ in the direction $\varphi$ given by the expression

$$
\begin{aligned}
2\mu \int_\Gamma \Bigg\{ & \left(2h + \frac{h^3}{3} K_2\right) \Big[2\varepsilon_\Gamma(e) \cdot\cdot\, \varepsilon_\Gamma(\varphi) + \nabla_\Gamma(e\cdot n) \cdot\cdot\, \nabla_\Gamma(\varphi \cdot n) \\
& \qquad -\varepsilon_\Gamma(e) \cdot\cdot\, ({}^*\nabla_\Gamma(\varphi\cdot n) n + {}^*n \nabla_\Gamma(\varphi\cdot n)) \\
& \qquad -\varepsilon_\Gamma(\varphi) \cdot\cdot\, ({}^*\nabla_\Gamma(e\cdot n) n + {}^*n \nabla_\Gamma(e\cdot n))\Big] \\
& +4\frac{h^3}{3}\, \varepsilon_\Gamma(e\cdot n)) \cdot\cdot\, \varepsilon_\Gamma(\nabla_\Gamma(\varphi\cdot n)) \\
& -4\frac{h^3}{3}\, H \Big[\varepsilon_\Gamma(\nabla_\Gamma(e\cdot n)) \cdot\cdot\, \varepsilon_\Gamma(\varphi) + \varepsilon_\Gamma(\nabla_\Gamma hh\varphi \cdot n)) \cdot\cdot\, \varepsilon_\Gamma(e) \\
& \qquad -\frac{1}{2}\, \varepsilon_\Gamma(\nabla_\Gamma(e\cdot n)) \cdot\cdot\, ({}^*\nabla_\Gamma(\varphi\cdot n) n + {}^*n\, \nabla_\Gamma(\varphi\cdot n)) \\
& \qquad -\frac{1}{2}\, \varepsilon_\Gamma(\nabla_\Gamma(\varphi\cdot n)) \cdot\cdot\, ({}^*\nabla_\Gamma(e\cdot n) n + {}^*n\, \nabla_\Gamma(e\cdot n))\Big]\Bigg\}\, d\Gamma\ . \\
+\int_\Gamma \Bigg\{ & \left(2h + \frac{h^3}{3} K_2\right) F\cdot\varphi - 2\frac{\partial f}{\partial n} + Hf\Big]\, V\cdot n\, d\Gamma
\end{aligned}
$$

for a scalar function $f$ and a vector function $V$. In particular

$$\int_\Gamma (\nabla_\Gamma f \cdot V + f \mathrm{div}_\Gamma V)\, d\Gamma = \int_\Gamma H f\, V\cdot n\, d\Gamma$$

since $\nabla_\Gamma f \cdot n = 0$ by definition of the tangential gradient.

Using the above constructions we can introduce the "vectorial tangential divergent" $\overrightarrow{\mathrm{div}_\Gamma}\, \sigma$ for any symmetrical matrix $\sigma$ as a vector field whose $i$-th component is given by

$$(\overrightarrow{\mathrm{div}_\Gamma}\, \sigma)_i = \mathrm{div}_\Gamma(\sigma_{i,\cdot})$$

where $\sigma_{i,\cdot}$ is considered as an $\mathbb{R}^n$–vector. With this definition we obtain the following integration by parts formula for a vector field $e$ and a matrix $\sigma$:

$$\int_\Gamma [\varepsilon_\Gamma(e) \cdot\cdot\, \sigma + e\cdot \overrightarrow{\mathrm{div}_\Gamma}\, \sigma]\, d\Gamma = \int_\Gamma H(\sigma\cdot n)\cdot e\, d\Gamma.$$

With the above formula it is possible to obtain an equation for $e$.

**7. Discussion.** It is our conjecture that the model presented in §4 under assumption (4.4) is directly related to the model of W.T. Koiter [9,10] probably with a few extra terms which are usually neglected. It can be regarded as a generalization of plate equations models (cf. J. Lagnese and J.-L. Lions [11]) since it involves operators of the types *"Laplace-Beltrami"* and *"bi-Laplace-Beltrami"*. As for plates we only keep terms up to the third order in the thickness of the shell.

Similarly all the computations could be carried out without assumption (4.4) and we again conjecture that our model would be related to the model of P.M. Naghdi [12,13] again with a few extra terms which are usually neglected. Direct comparison of the two models is unfortunately beyond the scope of this paper.

*Note added in the proofs.* This paper is a first attempt to illustrate how our techniques could be used to obtain an equation for thin shells. The convenient assumption (4.4) is precisely the Kirchhoff-Love assumption for plates as given in P. Germain [7]. More elaborated and pertinent results are now available in Delfour and Zolésio [3], [6]. M. Bernadou (INRIA, France) pointed out that our approach belongs to the family of intrinsic methods along the lines of P. Germain [7] and R. Valid [16].

## REFERENCES


[1] M. Bernadou, P.G. Ciarlet, and B. Miara, *Exitence theorems for two-dimensional linear shell theories*, J. of Elasticity (to appear). (Report R 92018, Laboratoire d'Analyse Numérique, Université Pierre et Marie Curie, Paris, France, October 1992).

[2] G. Chen and J. Zhou, *Boundary Element Methods*, Academic Press, London, San Diego, New York, Boston, Sydney, Tokyo, Toronto 1992.

[3] M.C. Delfour and J.P. Zolésio, *A boundary differential equation for thin shells*, J. Differential Equations (to appear).

[4] M.C. Delfour and J.P. Zolésio, *Functional analytic methods in shape analysis*, in *Boundary Control and Variation*, Chapter 6 (ed., J.P. Zolésio) Marcel Dekker, New York 1994, pp. 105–189.

[5] M.C. Delfour and J.P. Zolésio, *Shape analysis via oriented distance functions*, J. Functional Analysis (1994), (in press).

[6] M.C. Delfour and J.P. Zoleésio, *Tangential differential equations for dynamical thin/shallow shells*, report CRM-2177, Université de Montréal, May 1994.

[7] P. Germain, *Mécanique, tomes I and II*, (Ellipses, éditeur) Paris 1986.

[8] D. Gilbarg and N.S. Trudinger, *Elliptic Partial Differential Equations of Second Order*, Springer-Verlag, Berlin, Heidelberg, New York, Tokyo 1983.

[9] W.T. Koiter, *On the nonlinear theory of thin elastic shells*, Proc. Kon. Ned. Akad. Wetensch **B69** (1966), pp. 1–54.

[10] W.T. Koiter, *On the foundations of the linear theory of thin elastic shells*, Proc. Kon. Ned. Akad. Wetensch **B73** (1970), pp. 169–195.

[11] J.E. Lagnese and J.-L. Lions, *Modelling, analysis and control of thin plates*, RMA, Vol. 6, Masson, Paris, Milan, Barcelona, Mexico 1988.

[12] P.M. Naghdi, *Foundations of elastic theory*, in *Progress in Solid Mechanics*, vol. 4, North-Holland, Amsterdam, 1963, pp. 1–90.

[13] P.M. Naghdi, *The theory of shells and plates*, in *Handbuch der Physik*, vol. VI a-2, Berlin, 1972, pp. 425–640.

[14] J. Sokolowski and J.-P. Zolésio, *Introduction to Shape Optimization: Shape Sensitivity Analysis*, Springer-Verlag, New York, Berlin, Heidelberg 1992.

[15] R. Temam *Monotone rearangement of a function and the Grad-Mercier equation of plasmas physics*, in *Recent methods in non linear analysis and application*, (eds., E. De Giorgi, E. Magenes and U. Mosco), Pitagora Editrice, Bologna 1979.

[16] R. Valid, *Mechanics of Continuous Media and Analysis of Structures*, North-Holland, Amsterdam, New York 1981.

# ORIENTED DISTANCE FUNCTIONS IN SHAPE ANALYSIS AND OPTIMIZATION*

MICHEL C. DELFOUR† AND JEAN-PAUL ZOLÉSIO‡

**Abstract.** The object of this paper is twofold. We first present constructions which induce topologies on subsets of a fixed domain or *hold-all* $D$ in $\mathbb{R}^N$ by using set parametrized functions in an appropriate function space. Secondly we study the role of the family of *oriented distance functions* (also known as *algebraic* or *signed distance functions*) in the analysis of shape optimization problems. They play an important role in the introduction of topologies which retain the classical geometric properties associated with sets: convexity, exterior normals, mean curvature, $C^k$ boundaries, etc.

**1. Introduction.** In Shape Optimization problems the independent variable is a geometric domain in the Euclidean space $\mathbb{R}^N$. This class of problems is significantly different from problems where the independent variable is a vector of parameters or functions in a vector space set–up.

To obtain existence theorems the choice of a topology is very critical and strongly dependent on the problem at hand. In an optimization problem the topology must simultaneouly provide some semicontinuity of the shape functional to be optimized and some compactness of the family of domains over which it is to be optimized.

The fundamental difficulty is that the classical geometry with Lipschitzian, $C^k$ or $C^\infty$ domains is not very well suited to induce a topology for this class of problems. Moreover for sets of domains the convenience of the vector space structure is lost. When dealing with a family of compact subsets of a compact hold-all $D$, one of the topologies which naturally comes to mind is the Hausdorff topology. With that topology the set of all nonempty compact subsets of the compact hold-all $D$ is compact. However this topology is so weak that very few interesting shape functionals will be semicontinuous.

A very fruitful approach has been provided by *Geometric Measure Theory* where the topology of the domains is obtained from the the $L^p(D)$-topology on the corresponding characteristic functions. This cumulated in the theory of *Finite Perimeter or Caccioppoli Sets* which provided a

---

* The two authors also acknowledge the support of the Institute for Mathematics and its Applications while attending the IMA Workshop on Control and Optimal Design of Distributed Parameter Systems (November 9–13, 1992).

† Centre de recherches mathématiques et Département de mathématiques et de statistique, Université de Montréal, C.P. 6128 Succ. Centre-ville, Montréal, Québec, Canada, H3C 3J7. The research of the first author has been supported in part by a Killam fellowship from Canada Council, National Sciences and Engineering Research Council of Canada operating grant A-8730 and by a FCAR grant from the Ministère de l'Éducation du Québec.

‡ Institut Non Linéaire de Nice, Faculté des Sciences, 1361 route des lucioles, 06904 Sophia-Antipolis Cédex, France.

complete solution to the celebrated Plateau's[1] problem. The reader is referred to the book of E. Giusti [25] for a modern treatment of this topic.

The underlying technique behind this approach is to embed equivalence classes of measurable subsets $\Omega$ of the measurable hold-all $D$ in the space $L^p(D)$ for some $p$, $1 \le p < \infty$, via the characteristic function $\chi_\Omega$. This identification makes it possible to define a metric on the set of equivalence classes of measurable domains. In that sense we could say that Henri Lebesgue has introduced one of the fundamental idea to put a topology not on sets but rather on equivalence classes of sets.

The Hausdorff metric is the result of a similar construction. The equivalence classes are made up of subsets of $D$ with the same closure and the distance function embeds this set of equivalence classes in the space $C(D)$ of continuous functions on $D$. It is also known (cf. C. Dellacherie [16]) that the set of such distance functions is closed in $C(D)$. The distance function plays a very important role in *Set-Valued Analysis* and *Nonsmooth Analysis* where it arises in the definition of the various topologies and cones associated with nonsmooth domains (cf. J.P. Aubin and H. Frankowska [3], F.H. Clarke [8]).

This general idea of embedding equivalence classes of domains into a function space via a domain dependent function is very appealing. Another such function is the *support function* of *Convex Analysis* (cf. R.T. Rockafellar [33]) which provides a topology for equivalence classes of subsets of $D$ with the same closed convex hull.

In this paper we shall consider *oriented distance functions* which play an important role in the description of the geometric properties of domains. They are also known in the literature as *algebraic* or *signed distance functions*. We shall see that the smoothness of the oriented boundary functions in a neighbourhood of the boundary of a given set is equivalent to the smoothness of the boundary of that set. Similarily the convexity of the function is equivalent to the convexity of the corresponding set. In addition their gradient and their Laplacian are directly related to the normal and the mean curvature on the boundary of the set.

In §§ 2, 3, and 4 we shall present the basic constructions and the general approach we shall adopt, and review and sharpen currently available results. When the elements of the Hessian matrix of second order partial derivatives of the distance function are measures, we show that the corresponding set is a Caccioppoli set and new compactness theorems are obtained. In §5 we shall then investigate the *oriented distance functions* which retain the nice properties of the distance functions but also generate the standard geometric properties associated with sets: convexity, exterior normals, mean curvature, $C^k$ boundaries, ... For instance a set is convex if and only if its oriented distance function is convex; it has a boundary

[1] In honor of the Belgian physicist and professor J. Plateau [31] (1801-1883) who did experimental observations on the geometry of soap films.

of class $C^k$, $k \geq 2$, if and only if its oriented distance function is of class $C^k$ in a neighbourhood of its boundary. Such functions are closed in $C(D)$ but also in $W^{1,p}(D)$, $1 \leq p < \infty$ with the weak or the strong topology. For a convergent sequence in the strong topology the corresponding sequences of characteristic functions for the closure of the sets, for the closure of their complement and for their boundary all converge in $L^p(D)$. Another side result is that the family of equivalence classes of subsets of $D$ whose boundary has a zero Lebesgue measure in $\mathbb{R}^N$ is closed in $W^{1,p}(D)$ when $D$ is compact and in $W^{1,p}_{loc}(D)$ when $D$ is only closed. This provides a link with Geometric Measure Theory. When the domain is sufficiently smooth then the trace of the Laplacian of the oriented distance function on the boundary coincides with the *mean curvature*. In fact the Hessian matrix of second order partial derivatives contains the principal *curvatures*.

Finally the following interesting fact comes out of our analysis: the set of equivalence classes of convex subsets of a compact hold-all $D$ is relatively compact in all the topologies we shall consider in this paper (Theorem 2.2 (ii), Corollary 2.5, Theorem 3.10, Theorem 4.2 (ii), Theorem 5.13 (ii)).

Some of the results in this paper have been announced in M.C. Delfour and J.P. Zolésio [12] and all the proofs and details can be found in M.C. Delfour and J.P. Zolésio [14].

**2. Characteristic functions in $L^p$ spaces and measurable domains.** There are many ways to formulate or relax a shape optimization problem. In this section we look at the underlying topologies which arise when the problem can be recast as an optimization problem over a family of characteristic functions of measurable sets. We assume that the variable domains $\Omega$ are (Lebesgue) measurable subsets of a fixed *hold-all* $D$ which is a measurable domain in $\mathbb{R}^N$. The *characteristic function* of a measurable subset $\Omega$ of $D$ is the function $\chi$ from $D$ to $\mathbb{R}$ which is equal to 1 in $\Omega$ and 0 outside. A measurable function $\chi$ is a characteristic function if and only if the following identity is verified

$$\chi(x)[1 - \chi(x)] = 0 \quad \text{in } D. \tag{2.1}$$

The measurable set $\Omega$ associated with such a $\chi$ is given by $\{x \in D : \chi(x) = 1\}$. These considerations also extend to equivalence classes od measurable subsets of $D$ and condition (2.1) must be verified almost everywhere in $D$. Many *shape optimization* problems such as the *transmission problem* can be relaxed to the set of equivalence classes of characteristic functions

$$X(D) = \{\chi : \chi \text{ is measurable and } \chi(x)[1 - \chi(x)] = 0, \quad \text{a.e. in } D\}. \tag{2.2}$$

However this set is not convex and weak $L^2$–limits converge in the larger set

$$\text{co } X(D) = \{\chi : \chi \text{ is measurable and } \chi(x) \in [0,1], \quad \text{a.e. in } D\}. \tag{2.3}$$

This type of solution is usually related to the appearance of a *microstructure* and arises from the fact that $X(D)$ is not closed in the weak topology of $L^2(D)$.

The first and basic remark is that the usual weak convergences of characteristic functions are equivalent. Let $D$ be a measurable subset of $\mathbb{R}^N$ with finite measure and let $p$, $1 \le p < \infty$, be a real number. Given any sequence of characteristic functions $\{\chi_n\}$, the following statements are equivalent

a) $\chi_n$ converges in $L^\infty(D)$-weak $\star$,
b) $\chi_n$ converges in $L^p(D)$-weak,
c) $\chi_n$ converges in $\mathcal{D}(D)'$,

where $\mathcal{D}(D)'$ is the space of scalar distributions on $D$. For c) it is assumed that $D$ is open. Moreover if $\chi_n$ converges in $L^p(D)$-strong for some $p$, $1 \le p < \infty$, then it converges in $L^p(D)$-strong for all $p$, $1 \le p < \infty$ and its limit is itself a characteristic function, that is an element of $X(D)$.

THEOREM 2.1. *Let $D$ be a measurable (resp. bounded measurable) subset of $\mathbb{R}^N$. Then $X(D)$ is closed in $L^p_{loc}(D)$-strong (resp. $L^p(D)$-strong) for all $p$, $1 \le p < \infty$.*

Obviously co $X(D)$ is also closed for all topologies between the $L^\infty(D)$-weak$\star$ topology and the $\mathcal{D}(D)'$ topology. In view of this equivalence we shall adopt the following terminology for the set $X(D)$ of characteristic functions of all measurable subsets of a measurable subset $D$ of $\mathbb{R}^N$ with finite measure.

We shall say that a sequence $\{\chi_n\}$ in $X(D)$ is *weakly convergent* if it converges for some topology between the $L^\infty(D)$-weak$\star$ and the $\mathcal{D}(D)'$. By analogy we shall also say that a sequence $\{\chi_n\}$ in $X(D)$ is *strongly convergent* if it converges in $L^p(D)$-strong for some $p$, $1 \le p < \infty$. If the measurable subset $D$ of $\mathbb{R}^N$ does not have finite measure, we shall say that a sequence $\{\chi_n\}$ in $X(D)$ is *locally strongly convergent* if it converges in $L^p_{loc}(D)$-strong for some $p$, $1 \le p < \infty$.

Theorem 2.1 says that if we are interested in distinguishing between equivalent classes of Lebesgue measurable domains the strong topology in $L^p(D)$ is the natural topology where $X(D)$ is closed. For $D$ with finite measure, this induces the metric

$$\rho(\Omega_1, \Omega_2) = \|\chi_{\Omega_1} - \chi_{\Omega_2}\|_{L^1(D)}$$

which is the measure of the symmetric difference

$$\Omega_1 \Delta \Omega_2 = (\Omega_1 \setminus \Omega_2) \cup (\Omega_2 \setminus \Omega_1),$$

where $A \backslash B = \{x \in A : x \notin B\}$. So it is conceivable that depending on the problem other appropriate measures could be constructed to "see" specific features of equivalence classes of measurable sets.

One consequence of the above considerations is that the weak convergence of characteristic functions is not quite appropriate if we are looking for the existence of maximizers or minimizers which are characteristic functions. So it would seem reasonable to work with the strong topology for which $X(D)$ is closed. The following theorem says that an arbitrary element of $X(D)$ can be approximated in the strong topology by a sequence of characteristic functions of $C^\infty$-domains, and that the subset of characteristic functions of convex sets is closed.

THEOREM 2.2. *Let $\Omega$ be an arbitrary Borel set in $\mathbb{R}^N$.*
*(i) There exists a sequence $\{\Omega_n\}$ of open $C^\infty$ domains in $\mathbb{R}^N$ such that*

$$\chi_{\Omega_n} \to \chi_\Omega \quad \text{in } L^1_{\text{loc}}(\mathbb{R}^N).$$

*(ii) The set $\mathcal{C}(D)$ of equivalence classes $[\Omega]$ of convex measurable domains $\Omega$ in $D$ is closed in $X(D)$ for the strong $L^p$-topology, $1 \le p < \infty$.*

The lack of a priori smoothness on $\Omega$ may introduce technical difficulties in the formulation of some boundary value problems. However it is possible to relax such boundary value problems from a smooth simply connected domain $\Omega$ to a measurable domain (cf. J.P. Zolésio [45]). For instance consider the homogeneous Dirichlet problem

$$\text{(2.4)} \qquad \begin{cases} -\Delta y &= f \text{ in } \Omega \\ y &= 0 \text{ on } \Gamma \end{cases}$$

over a simply connected domain $\Omega$ with a boundary $\Gamma$ of class $C^1$ and associate with its solution $y = y(\Omega)$ the cost functional

$$\text{(2.5)} \qquad J(\Omega) = \frac{1}{2} \int_\Omega |y - g|^2 \, dx$$

with the additional constraints on the volume and perimeter of $\Omega$:

$$\text{(2.6)} \qquad \int_\Omega dx = \pi, \quad \text{and} \quad \int_\Gamma d\gamma \le 4\,\pi.$$

It was convenient to assume that $\Omega$ was a smooth simply connected domain. However there is a priori no reason to assume that an eventually optimal (minimizing) domain $\Omega^*$ be of class $C^1$. From the analysis point of view it is usually more convenient to relax constraints on the smoothness of $\Omega$ and work with the largest possible class of domains. Of course this relaxation must be done in a way which preserves the meaning of the underlying function spaces and the well-posedness of the original problem.

In problem (2.4)-(2.5)-(2.6) we first have to make sense of the Sobolev space for measurable subsets $\Omega$ of $D$. Assume that $D$ is a fixed bounded smooth open domain in $\mathbb{R}^N$. So the Sobolev space $H^1_0(D)$ is well defined in the usual way and we can introduce for each measurable $\Omega$ in $D$ the linear subspace

$$\text{(2.7)} \qquad H^1_0(\Omega) = \left\{\varphi \in H^1_0(D) : (1 - \chi_\Omega)\nabla\varphi = 0 \text{ a.e. in } D\right\}.$$

THEOREM 2.3. *$H_0^1(\Omega)$ as defined in (2.7) is a closed subspace of $H_0^1(D)$ and hence a Hilbert space.*

Given a measurable subset $\Omega$ of $D$ we can now consider the variational problem

$$\text{(2.8)}\qquad \begin{cases} \text{to find } y(\Omega) \in H_0^1(\Omega) \text{ such that} \\ \forall \varphi \in H_0^1(\Omega), \quad \int_D \nabla y \cdot \nabla \varphi \, dx = \int_D \chi_\Omega \, f \, \varphi \, dx \end{cases}$$

and the associated cost function

$$\text{(2.9)}\qquad J(\Omega) = h(\chi_\Omega, y(\Omega)),$$

where

$$\text{(2.10)}\qquad h(\chi_\Omega, \varphi) = \frac{1}{2} \int_D \chi_\Omega |y(\Omega) - g|^2 \, dx.$$

This problem is now well-defined and its restriction to smooth simply connected domains coincides with the original problem (2.4)-(2.5)-(2.6). Indeed if $\Omega$ is a simply connected domain with a boundary $\Gamma$ of class $C^1$ then $\Gamma$ has a zero measure in $\mathbb{R}^N$ and the definition (2.7) of $H_0^1(\Omega)$ coincides with the usual one:

$$\text{(2.11)}\qquad H_0^1(\Omega) = \left\{\varphi \in H_0^1(D) : (1 - \chi_\Omega)\varphi(x) = 0 \text{ a.e. in } D\right\},$$

where $H_0^1(\Omega)$ is identified with the subspace of functions in $H_0^1(D)$ with support in $\Omega$. From G. Stampacchia [38] we know that

$$\nabla\varphi(x) = 0 \text{ a.e. in } D\backslash\overline{\Omega} \text{ and } \varphi \in H^1(D\backslash\overline{\Omega})$$

is equivalent to

$$\varphi|_{D\backslash\overline{\Omega}} = \text{constant}.$$

Then $\Omega$ being simply connected, $\Omega \subset D$ and $\varphi|_{\partial D} = 0$ imply that the constant is equal to zero. Therefore the two definitions are equivalent for $C^1$ simply connected domains and problem (2.6)-(2.8)-(2.9)-(2.10) is a well-defined extension of problem (2.4)-(2.5)-(2.6). In general when $\Omega$ contains holes, then this constant can be different from hole to hole.

We complete this section by quoting some results for *finite perimeter sets* which have been introduced by R. Caccioppoli [7], E. De Giorgi [10] in the context of *Plateau's problem*. Consider measurable subsets $\Omega$ of a fixed bounded open domain $D$ in $\mathbb{R}^N$ and consider their characteristic functions $\chi_\Omega$ which belong to $X(D)$. Then $\chi_\Omega$ is an $L^1(D)$-function

$$\text{(2.12)}\qquad \|\chi_\Omega\|_{L^1(D)} = \int_D \chi_\Omega \, dx = \text{meas } \Omega \leq \text{meas } D < \infty$$

with a distributional gradient

$$(2.13) \qquad <\nabla\chi_\Omega, \vec{\varphi}>_{\mathcal{D}} = -\int_D \chi_\Omega \operatorname{div} \vec{\varphi}\, dx, \quad \forall\, \vec{\varphi} \in (\mathcal{D}(D))^N.$$

We consider measurable subsets $\Omega$ of $D$ whose characteristic function belongs to the space $BV(D)$ of $L^1$-functions $f$ on $D$ with a distributional gradient $\nabla f$ in the space of (vectorial) bounded measures, that is

$$(2.14) \qquad \vec{\varphi} \mapsto <\nabla f, \vec{\varphi}>_{\mathcal{D}} = -\int_D f \operatorname{div} \vec{\varphi}\, dx : \mathcal{D}^0(D;\mathbb{R}^N) \to \mathbb{R}$$

is continuous with respect to the space of continuous functions in $D$ with compact support in $D$. Introduce the seminorm (cf. J. Dieudonné [17])

$$(2.15) \qquad \|\nabla f\|_{M_1(D)^N} = \sup_{\substack{\vec{\varphi} \in \mathcal{D}^0(D;\mathbb{R}^N) \\ \|\vec{\varphi}\|_C \le 1}} <\nabla f, \vec{\varphi}>_{\mathcal{D}'}$$

where the space $\mathcal{D}^0(D;\mathbb{R}^N)$ is endowed with the norm

$$\|\vec{\varphi}\|_C = \sup_{x \in D} |\vec{\varphi}(x)|_{\mathbb{R}^N},$$

$M_1(D) = \mathcal{D}^0(D)'$, the topological dual of $\mathcal{D}^0(D)$, and

$$\nabla f \in \mathcal{L}(\mathcal{D}^0(D;\mathbb{R}^N), \mathbb{R}) \cong \mathcal{L}(\mathcal{D}^0(D), \mathbb{R})^N = M_1(D)^N.$$

We denote by $BV(D)$ the space

$$(2.16) \qquad BV(D) = \left\{ f \in L^1(D) : \nabla f \in M_1(D)^N \right\}$$

endowed with the norm

$$(2.17) \qquad \|f\|_{BV(D)} = \|f\|_{L^1(D)} + \|\nabla f\|_{M_1(D)^N}.$$

For more details and properties see also F. Morgan [29, p. 117], and H. Federer [23, §4.5.9].

When the characteristic function of a measurable subset of $D$ belongs to $BV(D)$, the *perimeter of $\Omega$ (relative to $D$)* is defined as

$$(2.18) \qquad P_D(\Omega) = \|\nabla\chi_\Omega\|_{M_1(D)^N}.$$

and

$$(2.19) \qquad m(\Gamma) = P_D(\Omega) + m(\Gamma \cap \partial D).$$

The interest behind this construction is twofold. Firstly it extends the notion of perimeter of a set to measurable sets; secondly this framework provides a compactness result which is useful in obtaining existence of optimal domains.

THEOREM 2.4. *(Cf. E. Giusti [25, Theorem 1.19, p. 17], V.G. Maz'ja [28, Theorem 6.1.4, p. 300, Lemma 1.4.6 p. 62], C.B. Morrey, Jr. [30, Theorem 3.4.4, p. 75]) Assume that $D$ is a bounded open domain in $\mathbb{R}^N$ with a Lipschitzian boundary $\partial D$. Let $\{\Omega_n\}$ be a sequence of measurable domains in $D$ with finite perimeter with respect to $D$. If there exists a constant $c > 0$ such that*

$$\forall\ n,\ P_D(\Omega_n) \leq c, \tag{2.20}$$

*then there exists a subsequence $\{\Omega_{n_k}\}$ and a measurable set $\Omega$ in $D$ such that*

$$P_D(\Omega) \leq \liminf_{k\to\infty}\ P_D(\Omega_{n_k}) \leq c. \tag{2.21}$$

*Moreover*

$$\chi_{\Omega_{n_k}} \to \chi_\Omega \text{ in } L^1(D) \text{ as } k \to \infty, \tag{2.22}$$

*and $\nabla\chi_{\Omega_{n_k}}$ "converges in measure" to $\nabla\chi_\Omega$ in $M_1(D)^N$, that is, for all $\vec{\varphi}$ in $\mathcal{D}^0(D, \mathbb{R}^N)$,*

$$\lim_{k\to\infty} < \nabla\chi_{\Omega_{n_k}}, \vec{\varphi} >_{M_1(D)^N} \to < \nabla\chi_\Omega, \vec{\varphi} >_{M_1(D)^N}. \tag{2.23}$$

COROLLARY 2.5. *Let $D$ verify the assumptions of Theorem 2.4. If $\{\Omega_n\}$ is a sequence of convex measurable domains in $D$, then there exists a constant $c > 0$ such that*

$$\forall n,\ P_D(\Omega_n) \leq c,\ \ \mathrm{m}(\Omega_n) \leq c$$

*and we can find a convex measurable set $\Omega$ in $D$ for which the conclusions of Theorem 2.4 are verified.*

This corollary says that the set of characteristic functions of convex measurable domains in a bounded open subset $D$ of $\mathbb{R}^N$ with Lipschitzian boundary is compact in $L^p(D)$, $1 \leq p < \infty$, since $X(D)$ is already closed.

It is important to recall that even if a set $\Omega$ in $D$ has a finite perimeter $P_D(\Omega)$, its *relative boundary* could have a non zero $N$-dimensional Lebesgue measure $m(\Omega)$. To illustrate this point the reader is referred to Example 2.1 in [14] which is an adaptation of Example 1.10 in E. Giusti [25, p. 7].

**3. Distance functions and associated topologies.** The first metric that usually comes to the mind is the Hausdorff metric. It turns out that the distance function plays a role similar to the characteristic function. The distance function embeds equivalence classes of subsets $A$ of a closed hold-all $D$ with the same closure $\overline{A}$ into the space $C(D)$ of continuous functions. In addition the metric induced by this construction is equal to the Hausdorff metric when $D$ is compact. The Hausdorff topology has many much desired properties. In particular for $D$ compact the set of equivalence classes of nonempty subsets $A$ of $D$ is compact. Yet the volume functional will not be continuous with respect to that topology. However distance functions also belong to $W^{1,p}(D)$ which provides a more interesting topology. In fact we shall see that when the elements of the Hessian matrix of second order derivatives are measures the closure of the corresponding domain is a set of locally finite perimeter.

Associate with each nonempty subset $A$ of $\mathbb{R}^N$ the distance function

$$d_A(x) = \inf_{y \in A} |y - x|, \quad x \in \mathbb{R}^N.$$

If $d_A$ is finite in $\mathbb{R}^N$, then $\overline{A} \neq \varnothing$ and a fortiori $A \neq \varnothing$. We recall the following properties of distance functions (cf. J. Dugunji [18, p. 185, § 4.2 and Theorem 4.3]):

(i) $d_A(x) = 0 \iff x \in \overline{A}$

(ii) $\overline{A} = \{x \in \mathbb{R}^N : d_A(x) = 0\}$

(iii) the map $x \mapsto d_A(x)$ is uniformly Lipschitz continuous and

$$\forall\, x, y, \quad |d_A(y) - d_A(x)| \leq |y - x|$$

(iv) $d_A$ is differentiable almost everywhere and

$$|\nabla d_A(x)| \leq 1, \quad \text{a. e. in } \mathbb{R}^N$$

(cf. H. Federer [23, Rademaker's theorem §3.16]).

(v) $\varnothing \neq \overline{A} \subset \overline{B} \iff d_A \geq d_B$.

(vi) $d_{A \cup B} = \min\{d_A, d_B\}$.

From the above properties the distance function induces a map $A \mapsto d_A : \mathcal{P}_+(\mathbb{R}^N) \to C(\mathbb{R}^N)$ from the set $\mathcal{P}_+(\mathbb{R}^N)$ of nonempty subsets of $\mathbb{R}^N$ to the space $C(\mathbb{R}^N)$ of continuous functions on $\mathbb{R}^N$ endowed with the Fréchet topology of uniform convergence on compact subsets of $\mathbb{R}^N$. Moreover

$$\forall\, B \text{ such that } A \subset B \subset \overline{A}, \text{ then } \forall\, x \in \mathbb{R}^N,\ d_A(x) = d_B(x) = d_{\overline{A}}(x)$$

(cf. J. Dugunji [18, Problem 4.3c]) and in particular

$$\forall\, A, \forall\, B \text{ such that } \overline{A} = \overline{B}, \quad d_A = d_{\overline{A}} = d_{\overline{B}} = d_B.$$

**3.1. The set of distance functions in $C(D)$.** Let $D$ be a closed nonempty subset of $\mathbb{R}^N$. We can associate with each nonempty subset $A$ of $D$ the equivalence class

$$[A] = \{B : \forall\, B,\ B \subset D \text{ and } \overline{B} = \overline{A}\}$$

and consider the set

$$\mathcal{F}(D) = \{[A] : \forall A, \ \varnothing \neq A \subset D\}.$$

The map

$$[A] \mapsto d_A : \mathcal{F}(D) \to C(D)$$

is now injective since

$$d_A = d_B \quad \Leftrightarrow \quad \overline{A} = \overline{B} \quad \Leftrightarrow \quad [A] = [B].$$

Here $C(D)$ is endowed with the Fréchet topology of uniform convergence on compact subsets of $D$. So $\mathcal{F}(D)$ can be identified with the subset of distance functions

$$C_d(D) = \{d_A : \forall A, \ \varnothing \neq A \subset D\} \subset C(D).$$

To complete the construction it remains to prove that $C_d(D)$ is closed in $C(D)$. For compact $D$'s this result can be found in C. Dellacherie [16, p. 42, Theorem 2 and p. 43, Remark 1]. Moreover in that case we can introduce the metric

$$\rho(A, B) = \sup_{x \in D} |d_A(x) - d_B(x)|$$

on $\mathcal{F}(D)$ which is equal to the usual Hausdorff metric

$$\rho_H(A, B) = \max\left\{\sup_{x \in B} d_A(x), \sup_{y \in A} d_A(y)\right\}.$$

When $D$ is closed but not necessarily compact, the space $C(D)$ of continuous functions on $D$ is endowed with the Fréchet topology of uniform convergence on compact subsets $K$ of $D$. This topology is defined by the family of seminorms

$$q_K(f) = \max_{x \in K} |f(x)|, \quad \forall K \text{ compact} \subset D.$$

It is metrizable since the topology induced by the family of seminorms $\{q_K\}$ is equivalent to the one generated by the subfamily $\{q_{K_k}\}_{k \geq 1}$, where the compact sets $\{K_k\}_{k \geq 1}$ are chosen as follows:

$$K_k = \{x \in D : d_{\complement D}(x) \geq \frac{1}{k} \text{ and } |x| \leq k\}, \quad k \geq 1$$

(cf. J. Horváth [26, Example 3, p. 116]). Moreover the topology on $C(D)$ is equivalent to the topology defined by the metric

$$\delta(f, g) = \sum_{k=1}^{\infty} \frac{1}{2^k} \frac{q_{K_k}(f - g)}{1 + q_{K_k}(f - g)}.$$

We shall see below that $C_d(D)$ is a closed subset of $C(D)$. This justifies the introduction of the following metric on $\mathcal{F}(D)$

$$\rho_\delta(A,B) = \delta(d_A, d_B) = \sum_{k=1}^{\infty} \frac{1}{2^k} \frac{q_{K_k}(d_A - d_B)}{1 + q_{K_k}(d_A - d_B)}$$

which can be seen as a generalization of the Hausdorff metric for subsets of a closed unbounded domain $D$.

THEOREM 3.1. *Let $D$ be a nonempty closed (resp. compact) subset of $\mathbb{R}^N$.*
*(i) The set $C_d(D)$ is closed in $C(D)$ (resp. $C^0(D)$),*
*(ii) and $\rho_\delta$ (resp. $\rho$) defines a metric on $\mathcal{F}(D)$. Moreover when $D$ is compact the metrics $\rho$ and $\rho_H$ are equal.*

*Proof.* Cf. C. Dellacherie [16, p. 42, Theorem 2 and p. 43, Remark 1]. □

When $D$ is compact the set $\mathcal{F}(D)$ enjoys many more interesting properties.

THEOREM 3.2. *Let $D$ be a nonempty compact subset of $\mathbb{R}^N$ and $\mathcal{F}(D)$ the corresponding set of equivalence classes of nonempty subsets of $D$. Then*
*(i) For each closed nonempty subset $F$ of $D$ the set*

$$\{[A] \in \mathcal{F}(D) : \forall A,\ F \subset \overline{A}\}$$

*is closed.*
*(ii) If $S$ is open (resp. closed), then the sets*

$$\begin{aligned} I(S) &= \{[A] \in \mathcal{F}(D) : \forall A,\ \varnothing \neq \overline{A} \subset S\} \\ J(S) &= \{[A] \in \mathcal{F}(D) : \forall A,\ \overline{A} \cap S \neq \varnothing\} \end{aligned}$$

*are open (resp. closed).*
*(iii) $C_d(D)$ (and hence $\mathcal{F}(D)$) is compact in $C^0(D)$.*

*Proof.* Cf. J. Dugunji [18, Chapter XI, Problem 4.6, p. 253] or J. Dieudonné [17, p. 61, Problem 3]. □

This theorem has many interesting corollaries. For instance it can be used to say something about the function which gives the number of connected components of $\overline{A}$ (cf. T.J. Richardson [32] for an application to image segmentation).

COROLLARY 3.3. *Let $D$ be a nonempty compact subset of $\mathbb{R}^N$ and $\mathcal{F}(D)$ the corresponding set of equivalence classes of nonempty subsets of $D$. Associate with an equivalence class $[A]$ the number*

$$\#([A]) = \text{number of connected components of } \overline{A}.$$

*Then the map*

$$[A] \mapsto \#([A]) : \mathcal{F}(D) \to \mathbb{R}$$

*is lower semicontinuous. In particular for a fixed number $c \geq 0$ the subset*

$$\{d_A \in C_d(D) : \#([A]) \leq c\}$$

*is also compact in $C^0(D)$.*

*Remark* 3.4. In the partial differential equation literature one usually deals with open sets. To accomodate this point of view, consider the set of relatively open subsets $O$ of a closed hold-all $D$ and associate with each $O$ the distance function of its complement $\complement_D O$ with respect to $D$. Then we can repeat the above and the following considerations with obvious changes. This approach was used by J.-P. Zolésio [43, §1.3, p. 405] in the context of Free Boundary Problems. We shall see in § 5 that the oriented distance function will implicitlly contain both points of view. Also one of the participant in the workshop brought to our attention recent results by V. Šverák [39] who used the set of distance functions of the complement $\complement_D O$ with respect to $D$ of open sets $O$ in $D$ such that $\#([\complement_D O]) \leq c$ for some fixed $c > 0$. His main result is that in dimension 2 the convergence of a sequence $\{O_n\}$ to $O$ of such sets implies the convergence of the corresponding projection operators $\{P_{O_n} : H_0^1(D) \to H_0^1(O_n)\}$ to $P_O : H_0^1(D) \to H_0^1(O)$, where the projection operators are directly related to an homogeneous linear boundary problem on the corresponding domains $\{O_n\}$ and $O$. In dimension one the constraint on the number of components can be dropped. This result is useful in situations where the shape optimization problem can be relaxed to the distance function $d_{\complement_D O}$. □

It was pointed out to us that the distance function is extensively used in the study and approximation of front propagation (cf. M. Falcone [20], M. Falcone and R. Ferretti [21], M. Falcone, T. Giorgi and P. Loreti [22], M. Bardi and M. Falcone [4], G. Barles [5]).

### 3.2. Projection operators and differentiability of $d_A$

DEFINITION 3.5. Given a nonempty subset $A$ of $\mathbb{R}^N$ and a point $x$ in $\mathbb{R}^N$, denote by $\Pi_A(x)$ the set of projections of $x$ onto $\overline{A}$

$$\Pi_A(x) \stackrel{\text{def}}{=} \{z \in \overline{A} : |z - x| = d_A(x)\}.$$

The elements of $\Pi_A(x)$ will be called projections and denoted by $P_A(x)$. The subscript $A$ will be dropped whenever no confusion arises.

The directional derivative of the distance function can be computed by using a theorem on the differentiability of the min with respect to a parameter. It is directly related to the support function of the set $\Pi_A(x)$.

LEMMA 3.6. *Let $A$ be a nonempty subset of $\mathbb{R}^N$ and $x$ a point in $\mathbb{R}^N$. Define*

$$f(x) = \frac{1}{2}[x^2 - d_A^2(x)].$$

(i) *The set $\Pi_A(x)$ is nonempty, compact and for all $x$ in $\overline{\complement A}$, $\Pi_A(x) \subset \partial A$.*

(ii) *For all $v$ in $\mathbb{R}^N$*

$$\begin{aligned} dd_A^2(x;v) &= \lim_{t\searrow 0} \frac{d_A^2(x+tv)-d_A^2(x)}{t} &&= \inf_{z\in\Pi_A(x)} 2(x-z)\cdot v \\ df(x;v) &= \lim_{t\searrow 0} \frac{f(x+tv)-f(x)}{t} &&= \sup_{z\in\Pi_A(x)} z\cdot v \\ &= \sigma_{\Pi_A(x)}(v) &&= \sigma_{co\,\Pi_A(x)}(v), \end{aligned}$$

*where $\sigma_{\Pi_A(x)}$ is the support function of the set $\Pi_A(x)$ and $co\,\Pi_A(x)$ is the convex hull of $\Pi_A(x)$.*

The distance function is differentiable almost everywhere with respect to the Lebesque measure on $\mathbb{R}^N$. So for almost all $x$ in $\mathbb{R}^N$, the set $\Pi_A(x)$ contains a unique element $P_A(x) \in \overline{A}$. In general the smoothness of the boundary $\partial A$ does not imply that $\Pi_A(x)$ be a singleton everywhere in $\mathbb{R}^N$. For instance for the unit sphere $A = \{x \in \mathbb{R}^N : |x| = 1\}$ the boundary $\partial A$ is $C^\infty$ but $\Pi_A(0) = A$. However for convex sets $A$, $\Pi_A(x)$ is always a singleton (cf. F.A. Valentine [41]).

THEOREM 3.7. *Let $A$ be a nonempty subset of $\mathbb{R}^N$. (i) If $\nabla d_A(x)$ exists at a point $x$ in $\mathbb{R}^N$, then there exists a unique $Px \in \overline{A}$ such that*

$$d_A(x) = |Px - x| \quad \text{and} \quad \nabla d_A(x) = \begin{cases} 0 \quad , & \text{if } x \in \overline{A} \\ \frac{x-Px}{|x-Px|}, & \text{if } x \notin \overline{A} \end{cases}$$

(cf. for instance F.H. Clarke [8, Proposition 2.5.4, p. 66]).
*(ii) The function $d_A^2(x)$ is Gâteaux differentiable at $x$ if and only if $\Pi_A(x)$ is a singleton. In that case*

$$\nabla d_A^2(x) = 2(x - Px).$$

*If, in addition, $Px \neq x$, then $d_A(x)$ is Gâteaux differentiable at $x$.*

*Remark* 3.8. If $\nabla d_A(x)$ exists, then

$$\nabla d_A^2(x) = \left\{ \begin{array}{ll} 0, & x \in \overline{A} \\ 2(x - P_A x), & x \notin \overline{A} \end{array} \right\}$$

and, in view of Theorem 3.7, $|\nabla d_A(x)| = 0$ in $\overline{A}$ and 1 outside of $\overline{A}$. So $\nabla d_A(x)$ is *always discontinuous* across the boundary $\partial A$. However $\nabla d_A^2(x)$

can be continuous if the projection $P(x)$ is continuous. We shall see that we can get around this difficulty by considering the oriented distance function in § 5. □

**3.3. Characterization of convex sets.** In the convex case the distance function is differentiable everywhere and this property can be used to characterize the convexity of a set.

THEOREM 3.9. *Let $A$ be a nonempty convex subset of $\mathbb{R}^N$. Then*

(i) *for each $x \in \mathbb{R}^N$ there exists a unique $P_A(x)$ in $\overline{A}$ such that $d_A(x) = |P_A(x) - x|$, $\Pi_A(x) = \{P_A(x)\}$ is a singleton and*

$$\forall x,\ \forall y \in \mathbb{R}^N, \quad |P_A(y) - P_A(x)| \leq |y - x|$$

(cf. for instance E.H. Zarantonello [42, pp. 237–246] or J.P. Aubin [2, p. 24, Example 4.3]),

(ii) *$d_A^2$ belongs to $C^1_{loc}(\mathbb{R}^N)$ and even to $W^{2,\infty}_{loc}(\mathbb{R}^N)$.*

THEOREM 3.10.

(i) *Let $A$ be a nonempty subset of $\mathbb{R}^N$. Then*

$$\overline{A} \text{ is convex} \quad \Leftrightarrow \quad d_A \text{ is convex.}$$

(ii) *The subset of all distance functions of nonempty convex subsets of the closed hold-all $D$ is closed in $C(D)$ and compact in $C^0(D)$ if, in addition, $D$ is bounded.*

**3.4. $W^{1,p}$-topology for distance functions and relation with characteristic functions.** An important question which arises when defining a new topology for equivalence classes of subsets of a hold-all $D$ is its relationship with the topology induced by the $L^p$-norm on characteristic functions. For the Hausdorff convergence it is already known that we do not necessarily get this $L^p$-convergence as can easily be seen from Example 3.1 in [14]. In fact the volume functional is only upper semicontinuous with respect to the Hausdorff topology.

To get the full continuity we need the $L^p$-convergence of the gradients of the distance functions which are related to the characteristic functions of the closure of the associated sets.

*Notation* 3.11. In the sequel the closed hold-all $D$ will be a subset of $\mathbb{R}^N$ with a nonempty interior int $D$ such that

$$\overline{\text{int}\ D} = D \quad \text{and} \quad \chi_{\text{int}\ D} = \chi_D.$$

Some of the function spaces involved will be defined on either $D$ or *int* $D$. However for simplicity we shall use $D$ in both cases unless it is not clear from the context. Recall that a distance function defined on an open set continuously extends to its closure since it is Lipschitzian and hence uniformly continuous. □

THEOREM 3.12. *Let $D$ be a closed domain in $\mathbb{R}^N$ with a nonempty interior and a locally Lipschitz boundary $\partial D$. Fix $p$, $1 \le p < \infty$.*
*(i) For each nonempty set $A$ in $D$*

$$\chi_{\overline{A}} = 1 - |\nabla d_A|, \quad \text{a.e. in } D$$

*and the map*

$$d_A \mapsto \chi_{\overline{A}} : W^{1,p}_{loc}(D) \to L^p_{loc}(D)$$

*is "Lipschitz continuous": for all compact subsets $K$ of $D$ and nonempty subsets $A_1$ and $A_2$ of $D$*

$$\|\chi_{\overline{A_2}} - \chi_{\overline{A_1}}\|_{L^p(K)} \le \|\nabla d_{A_2} - \nabla d_{A_1}\|_{L^p(K)} \le \|d_{A_2} - d_{A_1}\|_{W^{1,p}(K)}.$$

*(ii) The set $C_d(D)$ is closed in $W^{1,p}_{loc}(D)$.*
*(iii) The set $C_d(D)$ is closed in $W^{1,p}_{loc}(D)$-weak. If $D$ is bounded, then $C_d(D)$ is compact in $W^{1,p}(D)$-weak.*
*(iv) Given a sequence $\{d_{A_n}\}$ and $d_A$ in $C_d(D)$, then*

$$d_{A_n} \to d_A \quad \text{in } W^{1,p}_{loc}(D)\text{-strong} \quad (\text{resp. } W^{1,p}(D)\text{-strong})$$

*if and only if*

$$d_{A_n} \rightharpoonup d_A \quad \text{in } W^{1,p}_{loc}(D)\text{-weak} \qquad (\text{resp. } W^{1,p}(D)\text{-weak})$$
$$\chi_{\overline{A_n}} \to \chi_{\overline{A}} \quad \text{in } L^p_{loc}(D)\text{-strong} \qquad (\text{resp. } L^p(D)\text{-strong})$$

*(v) Assume that, in addition, $D$ is compact. Let $\{d_{A_n}\}$ be a sequence in $C_d(D)$ such that $\nabla d_{A_n} \in BV(D)^N$ and*

$$\exists\, c > 0, \forall n \ge 1, \quad \|\partial_{ij} d_{A_n}\|_{M_1(D)} \le c, \quad 1 \le i,j \le N.$$

*Then there exists a subsequence, still denoted $\{d_{A_n}\}$, and $d_A$ in $C_d(D)$ such that*

$$d_{A_n} \to d_A \quad \text{in } W^{1,p}(D)\text{-strong}$$

*and for all $\varphi \in \mathcal{D}^0(D)$*

$$\lim_{n\to\infty} < \partial_{ij} d_{A_n}, \varphi > = < \partial_{ij} d_A, \varphi >, \quad 1 \le i,j \le N.$$

*In particular*

$$\|\partial_{ij} d_A\|_{M_1(D)} \le c, \quad 1 \le i,j \le N.$$

*Remark* 3.13. As we shall see in the proof of Theorem 3.12, if a sequence of elements $\{d_{A_n}\}$ converges in $W^{1,p}$-strong for some $p$, $1 \le p < \infty$, it converges in $W^{1,p}$-strong for all $p$, $1 \le p < \infty$. So we shall say that $\{d_{A_n}\}$ *converges strongly*. Similarly if a sequence of elements $\{d_{A_n}\}$ converges in $W^{1,p}$-weak for some $p$, $1 \le p < \infty$, it converges in $W^{1,p}$-weak for all $p$, $1 \le p < \infty$. This is also equivalent to the convergence in the $W^{1,\infty}$-weak$\star$ (when $W^{1,\infty}$ is considered as a subspace of $L^\infty \times L^\infty$) and the $(\mathcal{D}(D)^N)'$ topologies. This situation is analogous to what we have seen for characteristic functions and we shall say that $\{d_{A_n}\}$ *converges weakly*. □

*Remark* 3.14. In view of part (iv) of Theorem 3.12 an optimization problem with respect to the characteristic functions $\chi_A$ of sets $A$ in $D$ for which we have the continuity with respect to $\chi_A$ can be transformed into an optimization problem with respect to $d_{\complement_D A}$ in $W^{1,1}(D)$ since

$$\chi_{D\cap \operatorname{int} A} = |\nabla d_{\complement_D A}|.$$

This is the class of relatively open sets $A$ in $D$. For instance this would apply to the transmission problem (2.8) to (2.10) in §2. □

Part (v) of Theorem 3.12 provides a sufficient condition to obtain the strong $W^{1,p}(D)$ convergence. Is the existence of the Hessian matrix of second order partial derivatives of $d_A$ a realistic property and what are those second derivatives? In order to answer this question, it is useful to consider a few simple examples. This will be followed by a theorem which will summarize some of our observations.

*Example* 3.15. *Half plane in* $\mathbb{R}^2$
Consider the domain

$$A = \{(x_1, x_2) \, : \, x_1 \le 0\}, \quad \partial A = \{(x_1, x_2) \, : \, x_1 = 0\}.$$

It is readily seen that

$$d_A(x_1, x_2) = \max\{x_1, 0\}$$

$$\nabla d_A(x_1, x_2) = \begin{cases} (0,0), & x_1 < 0 \\ (1,0), & x_1 > 0 \end{cases}$$

$$< \partial_{11} d_A, \varphi >= \int_{\partial A} \varphi \, dx$$

$$\partial_{12} d_A = \partial_{21} d_A = \partial_{22} d_A = 0$$

$$< \Delta d_A, \varphi >= \int_{\partial A} \varphi \, dx.$$

Thus $\Delta d_A$ is the surface measure on the boundary $\partial A$. □

*Example* 3.16. *Ball of radius* $R > 0$ *in* $\mathbb{R}^2$

Consider the domain

$$A = \{x \in \mathbb{R}^2 \,:\, |x| \le R\}, \quad \partial A = \{x \in \mathbb{R}^2 \,:\, |x| = R\}.$$

Clearly

$$d_A(x) = \min\{0, |x| - R\}, \quad \nabla d_A(x) = \begin{cases} \frac{x}{|x|}, & |x| > R \\ (0,0), & |x| < R \end{cases}$$

$$< \partial_{11} d_A, \varphi >= \int_0^{2\pi} R\cos^2(\theta)\, \varphi\, d\theta + \int_{\complement A} \frac{x_2^2}{(x_1^2 + x_2^2)^{3/2}}\, dx$$

$$< \partial_{22} d_A, \varphi > \int_0^{2\pi} R\sin^2(\theta)\, \varphi\, d\theta + \int_{\complement A} \frac{x_1^2}{(x_1^2 + x_2^2)^{3/2}}\, dx$$

$$< \partial_{12} d_A, \varphi >= \partial_{21} d_A, \varphi >= \int_0^{2\pi} R\, \cos(\theta)\, \sin(\theta)\, \varphi\, d\theta + \int_{\complement A} \frac{x_2 x_1}{(x_1^2 + x_2^2)^{3/2}}\, dx$$

$$< \Delta d_A, \varphi >= \int_{\partial A} \varphi\, dx + \int_{\complement A} \frac{1}{(x_1^2 + x_2^2)^{1/2}}\, dx.$$

We see that $\Delta d_A$ contains the boundary measure on $\partial A$ plus a term which looks like the volume integral of the mean curvature. □

*Example* 3.17. *Unit square in* $\mathbb{R}^2$

Consider the domain

$$A = \{x = (x_1, x_2) \,:\, |x_1| \le 1,\, |x_2| \le 1\}.$$

Since $A$ is symmetrical with respect to both axes, it is sufficient to specify $d_A$ in the first quadrant. We use the notation $Q_1$, $Q_2$, $Q_3$ and $Q_4$ for the four quadrants in the counterclockwise order and $c_1$, $c_2$, $c_3$, and $c_4$ for the four corners of the square in the same order. We also divide the plane in three regions

$$\begin{aligned} D_1 &= \{(x_1, x_2) \,:\, |x_2| \le \min\{1, |x_1|\}\} \\ D_2 &= \{(x_1, x_2) \,:\, |x_1| \le \min\{1, |x_2|\}\} \\ D_3 &= \{(x_1, x_2) \,:\, |x_1| \ge 1 \text{ and } |x_2| \ge 1\}. \end{aligned}$$

Hence

$$d_A(x) = \begin{cases} \min\{x_2 - 1, 0\}, & x \in D_2 \cap Q_1 \\ |x - c_1|, & x \in D_3 \cap Q_1 \\ \min\{x_1 - 1, 0\}, & x \in D_1 \cap Q_1 \end{cases}$$

$$\nabla d_A(x) = \begin{cases} (0,1) & x \in D_2 \cap Q_1 \text{ and } x_2 > 1 \\ \frac{x-c_1}{|x-c_1|}, & x \in D_3 \cap Q_1 \\ (1,0), & x \in D_1 \cap Q_1 \text{ and } x_1 > 1 \\ (0,0), & x \in Q_1,\ x_1 < 1 \text{ and } x_2 < 1. \end{cases}$$

$$< \partial_{11} d_A, \varphi >= \sum_{i=1}^{4} \int_{D_3 \cap Q_i} \frac{(x_2 - c_{i,2})^2}{|x - c_i|^2} \varphi\, dx + \int_{\partial A \cap Q_i \cap D_1} \varphi\, dx$$

$$< \partial_{22} d_A, \varphi > \sum_{i=1}^{4} \int_{D_3 \cap Q_i} \frac{(x_1 - c_{i,1})^2}{|x - c_i|^2} \varphi\, dx + \int_{\partial A \cap Q_i \cap D_2} \varphi\, dx$$

$$< \partial_{12} d_A, \varphi >= \partial_{21} d_A, \varphi >= \sum_{i=1}^{4} \int_{D_3 \cap Q_i} \frac{(x_2 - c_{i,2})(x_1 - c_{i,1})}{|x - c_i|^2} \varphi\, dx$$

$$< \Delta d_A, \varphi >= \sum_{i=1}^{4} \int_{D_3 \cap Q_i} \frac{1}{|x - c_i|} \varphi\, dx + \int_{\partial A} \varphi\, dx.$$

Notice that the structure of the Laplacian is similar to the ones observed in the previous examples. □

For locally piecewise $C^2$-domains, we can conjecture that the surface measure on the boundary $\partial A$ is contained in $\Delta d_A$. Recall that this surface measure was obtained from the gradient of the characteristic function of the corresponding set in the context of finite perimeter sets. Here we know that $\chi_{\overline{A}} = 1 - |\nabla d_A|$ and we have the following result which provides the first connection between distance functions and finite perimeter sets.

THEOREM 3.18. *Assume that $D$ is a compact domain with nonempty interior such that $\overline{\text{int}\, D} = D$ and a locally Lipschitzian boundary. For each domain $A \subset D$ such that $\{d_A\}$ belongs to the space*

$$BV^2(D) = \{f \in W^{1,1}(D) \,:\, \nabla f \in BV(D)^N\},$$

*$\overline{A}$ is a finite perimeter set, that is*

$$\nabla \chi_{\overline{A}} \in M_1(D)^N.$$

**4. Support function.** Another interesting embedding is provided by the support function of Convex Analysis. Here the set of equivalence classes of domains will consist in nonempty bounded domains with the same closed convex hull.

Let $A$ be a nonempty subset of $\mathbb{R}^N$. The *indicator function* of $A$ is

$$\psi_A(x) = \left\{ \begin{array}{ll} 0 \ , & if\ x \in A \\ +\infty, & if\ x \notin A \end{array} \right\}.$$

The *support function* $\sigma_A$ of $A$ is the conjugate $\psi_A^*$ of $\psi_A$

$$\sigma_A(x^*) = \psi_A^*(x^*) = \sup_{x \in \mathbb{R}^N} [x \cdot x^* - \psi_A(x)], \ x^* \in \mathbb{R}^N$$

or simply

$$\sigma_A(x^*) = \sup_{x \in A} \ x \cdot x^*.$$

The support functions of subsets in $\mathbb{R}^N$ form a very special class of functions from $\mathbb{R}^N$ to $\mathbb{R} \cup \{+\infty\}$.

LEMMA 4.1. (cf. J.P. Aubin [2, p. 40, Proposition 3.8])

(i) *Given any nonempty subset $A$ of $\mathbb{R}^N$, the support function $\sigma_A$ is convex, lower semicontinuous and positively homogeneous on $\mathbb{R}^N$.*

(ii) *Conversely given any convex, lower semicontinuous and positively homogeneous function $\sigma$ on $\mathbb{R}^N$, $\sigma$ is the support function of the closed convex set*

$$A_\sigma = \left\{ x \in \mathbb{R}^N : \forall\ x^* \in \mathbb{R}^N, \ x \cdot x^* \leq \sigma(x^*) \right\}.$$

(iii) *$\sigma_A$ is uniformly Lipschitz continuous if and only if $A$ is bounded in $\mathbb{R}^N$.*

(iv) *For any $\alpha \geq 0$, $\beta \geq 0$ and nonempty sets $A$ and $B$*

$$\alpha\sigma_A + \beta\sigma_B = \sigma_{\alpha A + \beta B}.$$

Denote by $C_+(\mathbb{R}^N)$ the set of all positively homogeneous and continuous functions on $\mathbb{R}^N$ and consider the embedding

$$f \mapsto f|_B : C_+(\mathbb{R}^N) \to C^0(B),$$

where $B$ is the closed unit ball. Let $C_+(B)$ be the image of $C_+(\mathbb{R}^N)$. By construction this map is bijective from $C_+(\mathbb{R}^N)$ onto $C_+(B)$ and there is a one to one correspondance between $C_+(\mathbb{R}^N)$ and the subset $C_+(B)$ of $C^0(B)$ endowed with the norm

$$||f|| = \sup_{|x^*| \leq 1} |f(x^*)|.$$

In particular

$$\|\sigma_A\| = \sup_A |x|.$$

So for $A$ bounded this defines a map

$$A \mapsto \sigma_A \mapsto \sigma_A|_B : \mathcal{B}(\mathbb{R}^N) \to C_+(\mathbb{R}^N) \to C^0(B),$$

where $\mathcal{B}(\mathbb{R}^N)$ denotes the set of all bounded nonempty subsets of $\mathbb{R}^N$. For a convex lower semicontinuous function $f : \mathbb{R}^N \to \mathbb{R} \cup \{+\infty\}$, the biconjugate function $f^{**}$ is equal to $f$ (cf. J.P. Aubin [2, Theorem 3.1, p. 30]). By convexity of $\sigma_A$

$$\sigma_A^* = \psi_A^{**} = \psi_{\overline{co}\, A} \Rightarrow \sigma_A = \sigma_{\overline{co}\, A}$$

(cf. Ekeland-Temam [19, Examples 4.3, p.18]). This induces on $\mathcal{B}(\mathbb{R}^N)$ equivalences classes $[A]$ of domains with the same support function $\sigma_A$

$$[A] = \{B \subset \mathbb{R}^N : \overline{co}\, B = \overline{co}\, A,\ B \text{ bounded and nonempty}\}.$$

If we denote by $\mathcal{K}(\mathbb{R}^N)$ the set of all such equivalence classes we obtain the following embedding

$$[A] \to \sigma_A \mapsto \sigma_A|_B : \mathcal{K}(\mathbb{R}^N) \to C_+(\mathbb{R}^N) \to C^0(B).$$

Denote by $\Sigma(\mathbb{R}^N)$ the subset of all convex functions in $C_+(\mathbb{R}^N)$. In view of Lemma 4.1 (ii), it coincides with all the support functions of nonempty bounded subsets of $\mathbb{R}^N$. If we denote by $\Sigma(B)$ the image of $\Sigma(\mathbb{R}^N)$ in $C^0(B)$, then $\Sigma(B)$ is the subset of all bounded convex, positively homogeneous and continuous functions on $B$.

THEOREM 4.2.

*(i)* *$\Sigma(B)$ is a closed convex cone with vertex 0 in $C^0(B)$.*

*(ii)* *Any bounded subset of $\Sigma(B)$ is relatively compact in $C^0(B)$. In particular for a sequence of nonempty subsets of a fixed bounded domain $D$, there exists a nonempty set $A \subset \overline{co}\, D$ and a subsequence of these subsets such that the corresponding subsequence of support functions converges to $\sigma_A$ in $C^0(B)$. If the sets of the sequence are convex, then $A$ can be chosen convex.*

**5. Oriented distance functions and geometry of domains.** The distance function $d_A$ provides a good description of a domain $A$ from the "outside", but does not say much about its "inside" where it is identically zero. To get a complete description of a nonempty set $A$ in a fixed closed domain $D$ of $\mathbb{R}^N$, it is natural to take into account its complement

$$\complement A = \{y \in \mathbb{R}^N : y \notin A\}.$$

When both $A$ and $\complement A$ are nonempty, we introduce the *oriented distance function*,

$$b_A(x) = d_A(x) - d_{\complement A}(x), \quad \forall\, x \in \mathbb{R}^N. \tag{5.1}$$

This can also be seen as an "algebraic distance" to the boundary. Noting that $b_{\complement A} = -b_A$, it means that we have chosen the negative sign for the interior of $A$ and the positive sign for the interior of its complement.

By construction

$$b_A(x) = \begin{cases} d_A(x) > 0, & x \in \mathbb{R}^N \backslash \overline{A} = \text{ int } \complement A \\ 0, & x \in \overline{A} \cap \overline{\complement A} = \partial A \\ -d_{\complement A}(x) < 0, & x \in \mathbb{R}^N \backslash \overline{\complement A} = \text{ int } A. \end{cases} \tag{5.2}$$

LEMMA 5.1. *Let $A$ be a nonempty subset of $\mathbb{R}^N$. Then*

*(i)* $A \neq \varnothing$ and $\complement A \neq \varnothing \iff \partial A \neq \varnothing$.

*(ii)* $|b_A| = \max\{d_A, d_{\complement A}\} = d_{\partial A}$,

*where $d_{\partial A}$ is the function which is equal to the minimum distance from $x$ to $\partial A$.*

So the nonemptyness hypothesis is equivalent to assuming that the boundary $\partial A$ is nonempty. This excludes $\varnothing$ and $\mathbb{R}^N$. The zero function $b_A(x) = 0$, $\forall x \in \mathbb{R}^N$, corresponds to the equivalence class of sets $A$ such that

$$b_A = 0 \quad \Leftrightarrow \quad d_A = d_{\complement A} \text{ in } \mathbb{R}^N \quad \Leftrightarrow \quad \overline{A} = \overline{\complement A}.$$

But

$$\begin{aligned} &\operatorname{int} \complement A = \complement \overline{A} = \complement \overline{\complement A} = \operatorname{int} A \quad \text{and} \quad A \cap \complement A = \varnothing \\ &\Rightarrow \operatorname{int} \complement A = \operatorname{int} A = \varnothing \quad \Rightarrow \quad \partial A = \overline{\complement A} = \overline{A} = \mathbb{R}^N. \end{aligned}$$

This class of sets is not empty. For instance choose the subset of points of $\mathbb{R}^N$ with rational coordinates.

**5.1. Basic properties.** This oriented distance function enjoys properties similar to those of $d_A$.

THEOREM 5.2. *Let $A$ be a subset of $\mathbb{R}^N$ with nonempty boundary $\partial A$.*

*(i) The function $b_A$ is Lipschitz continuous and*

$$\forall\, x, y \in \mathbb{R}^N, \quad |b_A(y) - b_A(x)| \leq |y - x|. \tag{5.3}$$

*(ii) For all $x \in \partial A$, $\nabla b_A^2(x)$ exists and is equal to 0. For all $x \notin \partial A$, $\nabla b_A^2(x)$ exists if and only if $\nabla b_A(x)$ exists.*

*(iii) If $\nabla b_A(x)$ exists, then there exists a unique $P_{\partial A}(x) \in \partial A$ such that*

$$
b_A(x) = \left\{ \begin{array}{rl} |P_{\partial A}(x) - x|, & \textit{if } x \in \operatorname{int} \complement A \\ 0 \qquad\quad & , \textit{if } x \in \partial A \\ - \quad |P_{\partial A}(x) - x|, & \textit{if } x \in \operatorname{int} A \end{array} \right\}, \tag{5.4}
$$

$$
\nabla b_A(x) = \frac{x - P_{\partial A}(x)}{b_A(x)}, \quad \textit{if } x \notin \partial A, \tag{5.5}
$$

and

$$
\frac{1}{2}\, \nabla b_A^2(x) = x - P_{\partial A}(x). \tag{5.6}
$$

*Moreover for almost all $x \in \partial A$, $\nabla b_A(x) = \nabla d_{\partial A}(x) = 0$.*

(iv) *Given an integer $k \geq 1$ and a neighbourhood $U(x)$ of a point $x \in \partial A$, then $b_A^2 \in C^k(U(x))$ if and only if for all $y \in U(x)$ the projection $P_{\partial A}(y)$ is unique and $P_{\partial A} \in C^{k-1}(U(x))$.*

(v) *For all subsets $A$ and $B$ of $\mathbb{R}^N$ with nonempty boundaries $\partial A$ and $\partial B$*

$$
\overline{A} \subset \overline{B} \text{ and } \operatorname{int} A \subset \operatorname{int} B \quad \Longleftrightarrow \quad b_B \leq b_A.
$$

*Remark* 5.3. In general $\nabla d_A(x)$ and $\nabla d_{\complement A}(x)$ do not exist for $x \in \partial A$. This is readily seen by constructing the directional derivatives for the half space

$$
A = \{(x_1, x_2) \in \mathbb{R}^2 : x_1 \leq 0\} \tag{5.7}
$$

at the point $(0,0)$. Nevertheless $\nabla b_A(0,0)$ exists and is equal to $(1,0)$ which is the outward unit normal at $(0,0) \in \partial A$ to $A$. Note also that for all $x \in \partial A$, $|\nabla b_A(x)| = 1$. This is possible since $m(\partial A) = 0$. If $\nabla b_A(x)$ exists, then it is easy to verify that $|\nabla b_A(x)| \leq 1$. □

*Remark* 5.4. The uniqueness of the projection $P_{\partial A}(x)$ at $x$ is directly related to the existence of $\nabla b_A^2(x)$ since

$$
P_{\partial A}(x) = x - \frac{1}{2}\, \nabla b_A^2(x). \tag{5.8}
$$

In other words $P_{\partial A}(x)$ is the gradient of the functional

$$
f(x) = \frac{1}{2}\, [|x|^2 - b_A^2(x)] \tag{5.9}
$$

and all the properties of $P_{\partial A}$ can be obtained from those of $\nabla b_A^2$. □

DEFINITION 5.5. Given a subset $A$ of $\mathbb{R}^N$ with a nonempty boundary $\partial A$ and a point $x$ in $\mathbb{R}^N$, the set of projections of $x$ onto $\partial A$ associated with $b_A(x)$ is defined as

$$(5.10) \qquad \Pi_{\partial A}(x) = \begin{cases} \Pi_A(x), & \text{if } x \in \operatorname{int} \complement A \\ \{x\}, & \text{if } x \in \partial A \\ \Pi_{\complement A}(x), & \text{if } x \in \operatorname{int} A. \end{cases}$$

The choice of notation is compatible with the fact that $\Pi_{\partial A}(x)$ coincides with the set of projections of $x$ onto $\partial A$ for the distance function $d_{\partial A}$ (cf. Lemma 5.1). By definition

$$(5.11) \qquad \forall\, P_{\partial A}(x) \in \Pi_{\partial A}(x), \quad |x - P_{\partial A}(x)| = |b_A(x)| = d_{\partial A}(x).$$

In fact the oriented distance function coincides with the *signed distance* (cf. N. Shimakura [36, p. 268]) up to a change of sign since

$$b_A(x) = \begin{cases} d_{\partial A(x)}, & \text{if } x \in \overline{\complement A} \\ -d_{\partial A(x)}, & \text{if } x \in \overline{A} \end{cases}$$

We have the analogue of Lemma 3.6 in § 3.

LEMMA 5.6. *Let $A$ be a subset of $\mathbb{R}^N$ with a nonempty boundary $\partial A$ and $x \in \mathbb{R}^N$. Then for all $v$ in $\mathbb{R}^N$*

$$(5.12) \qquad db_A^2(x; v) = \inf_{z \in \Pi_{\partial A}(x)} 2(x - z) \cdot v = 2\left[x \cdot v - \sigma_{\Pi_{\partial A}(x)}(v)\right],$$

*where*

$$(5.13) \qquad \sigma_{\Pi_{\partial A}(x)}(v) = \sup_{z \in \Pi_{\partial A}(x)} z \cdot v \quad (= \sigma_{co\ \Pi_{\partial A}(x)}(v) = df(x; v))$$

*is the support function to the set $\Pi_{\partial A}(x)$ and co $\Pi_{\partial A}(x)$ its convex hull.*

**5.2. Oriented boundary distance functions in $C(D)$ and $W^{1,p}(D)$.** Let $D$ be a fixed nonempty closed domain in $\mathbb{R}^N$. Given two domains $A$ and $B$ in $D$ with nonempty boundaries $\partial A$ and $\partial B$, then

$$b_A = b_B \quad \Leftrightarrow \quad \overline{A} = \overline{B} \text{ and } \overline{\complement A} = \overline{\complement B} \quad \Leftrightarrow \quad \overline{A} = \overline{B} \text{ and } \partial A = \partial B.$$

Associate with each subset $A$ of $D$ which has a nonempty boundary $\partial A$ the equivalence class

$$[A]_b = \{B : \forall B,\ B \subset D,\ \overline{B} = \overline{A} \text{ and } \partial A = \partial B\}$$

and define the family of equivalence classes

$$\mathcal{F}_b(D) = \{[A]_b : \forall A,\ A \subset D \text{ and } \partial A \neq \varnothing\}.$$

Notice that the equivalence classes induced by $b_A$ are finer than those induced by $d_A$ since both the closures and the boundaries of the respective sets must coincide. As in the case of $d_A$ we identify $\mathcal{F}_b(D)$ with the set

$$C_b(D) = \{b_A : \forall A,\ A \subset D \text{ and } \partial A \neq \varnothing\}$$

of oriented boundary distances in $C(D)$ through the embedding

$$[A]_b \mapsto b_A : \mathcal{F}_b(D) \to C_b(D).$$

We shall see below that $C_b(D)$ is closed in $C(D)$. Again this justifies the introduction of the following metrics on $\mathcal{F}_b(D)$

$$\begin{aligned} \rho(A,B) &= \sup_{x\in D} |b_A(x) - b_B(x)|, && \text{for } D \text{ compact} \\ \rho_\delta(A,B) &= \delta(b_A, b_B) = \textstyle\sum_{k=1}^{\infty} \frac{1}{2^k} \frac{q_{K_k}(b_A - b_B)}{1+q_{K_k}(b_A-b_B)}, && \text{for } D \text{ unbounded.} \end{aligned}$$

We now have the equivalent of Theorem 3.1.

THEOREM 5.7. *Let $D$ be a nonempty closed (resp. compact) subset of $\mathbb{R}^N$.*

*(i) The set $C_b(D)$ is closed in $C(D)$ and $W^{1,p}_{loc}(D)$ (resp. $C^0(D)$ and $W^{1,p}(D)$), for $p$, $1 \le p < \infty$,*
*(ii) The function $\rho_\delta$ (resp. $\rho$) defines a metric on $\mathcal{F}_b(D)$.*
*(iii) If $D$ is compact, then $C_b(D)$ (and hence $\mathcal{F}_b(D)$) is compact in $C^0(D)$.*
*(iv) The set $C_b(D)$ is closed (resp. compact) in $W^{1,p}_{loc}(D)$-weak (resp. $W^{1,p}(D)$-weak).*

We now have the analogue of Theorem 3.2 and its corollary for $A$, $\complement A$ and $\partial A$. In the last case it takes the following form (to be compared with T.J. Richardson [32, Lemma 3.2, p. 44] and S.R. Kulbarni, S. Mitter and T.J. Richardson [27] for an application to image segmentation).

COROLLARY 5.8. *Let $D$ be a nonempty compact subset of $\mathbb{R}^N$ and $\mathcal{F}_b(D)$ the corresponding set of equivalence classes of subsets of $D$ with a nonempty boundary. Then*
*(i) For each subset $F$ of $D$ the set*

$$\{[A]_b \in \mathcal{F}_b(D) : \forall A,\ F \subset \partial A\}$$

*is closed.*
*(ii) If $S$ is open (resp. closed), then the sets*

$$\begin{aligned} I(S) &= \{[A]_b \in \mathcal{F}_b(D) : \forall A,\ \varnothing \neq \partial A \subset S\} \\ J(S) &= \{[A]_b \in \mathcal{F}_b(D) : \forall A,\ \partial A \cap S \neq \varnothing\} \end{aligned}$$

*are open (resp. closed).*

*(iii) Associate with an equivalent class $[A]_b$ the number*

$$\#_b([A]_b) = \text{number of connected components of } \partial A.$$

*Then the map*

$$[A]_b \mapsto \#_b([A]_b) \, : \, \mathcal{F}_b(D) \to \mathbb{R}$$

*is lower semicontinuous.*

*Remark* 5.9. Notice that if $b_{A_n} \to b_A$ in $C(D)$, then

$$d_{A_n} = \max\{0, b_{A_n}\}, \quad d_{\complement A_n} = \max\{0, -b_{A_n}\}, \quad d_{\partial A_n} = |b_{A_n}|$$

converge to $d_A$, $d_{\complement A}$ and $d_{\partial A}$ in $C(D)$. □

We shall now introduce an important subfamily of equivalence classes of domains whose boundary has a zero Lebesgue measure in $\mathbb{R}^N$. Define the following subsets of $\mathcal{F}_b(D)$ and $C_b(D)$:

$$\begin{aligned} \mathcal{F}_b^0(D) &= \{[A]_b \, : \, \forall A, \, A \subset D, \, \partial A \neq \varnothing, \, m(\partial A) = 0\} \\ C_b^0(D) &= \{b_A \, : \, \forall A, \, A \subset D, \, \partial A \neq \varnothing, \, m(\partial A) = 0\}. \end{aligned}$$

The next theorem says that for a sequence of nonempty sets $\{A_n\}$ for which $\{b_{A_n}\}$ converges in the $W^{1,p}$-topology there exists a set $A$ with nonempty boundary for which the characteristic functions of $\overline{A_n}$, $\overline{\complement A_n}$ and $\partial A_n$ converge to the characteristic functions of $\overline{A}$, $\overline{\complement A}$ and $\partial A$. In particular sequences of sets whose boundary has a zero measure converge to sets whose boundary has a zero measure.

With the convention on $D$ as described in Notation 3.11, we have the following theorem.

THEOREM 5.10. *Let $D$ be a closed (resp. compact) domain in $\mathbb{R}^N$ with nonempty interior such that $\overline{\text{int}\, D} = D$ and a locally Lipschitzian boundary.*

*(i) For each set $A$ in $D$ with nonempty boundary*

$$\chi_{\partial A} = 1 - |\nabla b_A|, \quad \text{a.e. in } D$$

*and the set*

$$\partial^* A = \{x \in \partial A \, : \, \nabla b_A(x) \text{ exists and } |\nabla b_A(x)| = 1\}$$

*has zero $N$-dimensional Lebesgue measure. Moreover for all $p$, $1 \le p \le \infty$, the maps*

$$\begin{aligned} b_A &\mapsto \chi_{\partial A} \, : \, W^{1,p}_{loc}(D) \to L^p_{loc}(D) \text{ (resp. } W^{1,p}(D) \to L^p(D)) \\ b_A &\mapsto \chi_{\overline{A}} \, : \, W^{1,p}_{loc}(D) \to L^p_{loc}(D) \text{ (resp. } W^{1,p}(D) \to L^p(D)) \\ b_A &\mapsto \chi_{\overline{\complement A}} \, : \, W^{1,p}_{loc}(D) \to L^p_{loc}(D) \text{ (resp. } W^{1,p}(D) \to L^p(D)) \end{aligned}$$

*are "Lipschitz continuous": for all compact subsets $K$ of $D$ and nonempty subsets $A_1$ and $A_2$ of $D$*

$$\|\chi_{\partial A_2} - \chi_{\partial A_1}\|_{L^p(K)} \le \|\nabla b_{A_2} - \nabla b_{A_1}\|_{L^p(K)} \le \|b_{A_2} - b_{A_1}\|_{W^{1,p}(K)}$$
$$\|\chi_{\overline{A_2}} - \chi_{\overline{A_1}}\|_{L^p(K)} \le \|\nabla b_{A_2} - \nabla b_{A_1}\|_{L^p(K)} \le \|b_{A_2} - b_{A_1}\|_{W^{1,p}(K)}$$
$$\|\chi_{\overline{\complement A_2}} - \chi_{\overline{\complement A_1}}\|_{L^p(K)} \le \|\nabla b_{A_2} - \nabla b_{A_1}\|_{L^p(K)} \le \|b_{A_2} - b_{A_1}\|_{W^{1,p}(K)}.$$

*(ii) The set $C_b^0(D)$ is closed in $W^{1,p}_{loc}(D)$ (resp. $W^{1,p}(D)$) for all $p$, $1 \le p < \infty$.*
*(iii) For any $b_A$ and any sequence $\{b_{A_n}\}$ in $C_b(D)$*

$$b_{A_n} \to b_A \quad \text{in } W^{1,p}_{loc}(D)\text{-strong} \quad (\text{resp. } W^{1,p}(D)\text{-strong})$$

*if and only if*

$$b_{A_n} \rightharpoonup b_A \quad \text{in } W^{1,p}_{loc}(D)\text{-weak} \quad (\text{resp. } W^{1,p}(D)\text{-weak})$$
$$\chi_{\overline{A_n}} \to \chi_{\overline{A}} \quad \text{in } L^p_{loc}(D)\text{-strong} \quad (\text{resp. } L^p(D)\text{-strong})$$
$$\chi_{\overline{\complement A_n}} \to \chi_{\overline{\complement A}} \quad \text{in } L^p_{loc}(D)\text{-strong} \quad (\text{resp. } L^p(D)\text{-strong}).$$

*(iv) Assume that $D$ is compact. Let $\{b_{A_n}\}$ be a sequence in $C_b(D)$ such that $\nabla b_{A_n} \in BV(D)^N$ and*

$$\exists\, c > 0, \forall n \ge 1, \quad \|\partial_{ij} b_{A_n}\|_{M_1(D)} \le c, \quad 1 \le i, j \le N,$$

*where $\partial_{ij} b_{A_n}$ is the second order partial derivative with respect to $x_i$ and $x_j$. Then there exists a subsequence, still denoted $\{b_{A_n}\}$, and $b_A$ in $C_b(D)$ such that*

$$b_{A_n} \to b_A \quad \text{in } W^{1,p}(D)\text{-strong}$$

*and for all $\varphi \in \mathcal{D}^0(D)$*

$$\lim_{n\to\infty} < \partial_{ij} b_{A_n}, \varphi > = < \partial_{ij} b_A, \varphi >, \quad 1 \le i, j \le N.$$

*In particular*

$$\|\partial_{ij} b_A\|_{M_1(D)} \le c, \quad 1 \le i, j \le N.$$

*Remark* 5.11. Part (iii) of Theorem 5.10 is to be interpreted carefully. If we only know that $\{b_{A_n}\}$ converges in $W^{1,p}(D)$-weak and that the characteristic functions $\{\chi_{\overline{A_n}}\}$ and $\{\chi_{\overline{\complement A_n}}\}$ converge in $L^p$-strong, it does not mean that there exists a $b_A$ such that $\{b_{A_n}\}$ converges to $b_A$ in $W^{1,p}(D)$-strong. This is readily seen from Example 2.1 in [14] where the sequence $\{\chi_{\overline{A_n}}\}$ converges strongly in $L^p$. In addition $m(\partial A_n) = 0$ and $\{\chi_{\overline{\complement A_n}}\}$ also converges strongly. Now $D$ is bounded and we can always find a subsequence of $\{b_{A_n}\}$ which weakly converges in $W^{1,p}(D)$. If that subsequence was strongly convergent in $W^{1,p}(D)$, then we would have, according to Theorem 5.10 (i), the strong convergence of $\{\chi_{\partial A_n}\}$ to $\{\chi_{\partial A}\}$. But we show in [14, Example 2.1] that for all $n \geq 1$,

$$m(\partial A_n) = 0 < m(\partial A).$$

Therefore the $L^p$-convergence of the characteristic functions and the convergence in measure of their gradients do not imply the strong convergence in $W^{1,p}(D)$. □

*Remark* 5.12. It is readily seen that for sufficiently smooth domains, the subset $\partial^* A$ of the boundary $\partial A$ coincides with the reduced boundary of finite perimeter sets. □

**5.3. Characterization of convex sets.** The next interesting property which was completely characterized by the distance function $d_A$ was the convexity of $\overline{A}$. This characterization remains true with $b_A$ in place of $d_A$.

THEOREM 5.13.
*(i) Let $A$ be a subset of $\mathbb{R}^N$ with a nonempty boundary $\partial A$. Then*

$$\overline{A} \text{ convex} \iff b_A \text{ convex in } \mathbb{R}^N. \tag{5.14}$$

*(ii) Let $D$ be a closed (resp. compact) domain in $\mathbb{R}^N$. The subset of all oriented boundary distance functions for nonempty convex subsets of $D$ is closed in $C(D)$ and $W^{1,p}_{loc}(D)$ (resp. compact in $C^0(D)$ and $W^{1,p}(D)$), $1 \leq p < \infty$.*

**5.4. Characterization of the smoothness of the boundary $\partial A$.** When the oriented boundary distance function exists in a neighbourhood of a point $x$ of the boundary $\partial A$ of a smooth domain $A$, then for $t$ small enough

$$\nabla b_A(x + tn(x)) = n(x) = \nabla b_A(x - tn(x)) \tag{5.15}$$

where $n(x)$ is the unit outward normal to $A$ in $x$. This suggests that $\nabla b_A(x) = n(x)$ and that $\nabla b_A$ is an extension field of the outward unit normal from $\partial A$ to a neighbourhood of $\partial A$. Another consequence of this observation is that

$$\Delta b_A(x) = \text{div}\ (\nabla b_A(x)) \tag{5.16}$$

becomes a candidate for the extension of the "mean curvature" from $\partial A$ to that same neighbourhood of $\partial A$ (cf. J.P. Zolésio [43,44], D. Gilbarg and N.S. Trudinger [24] and §5.5). It turns out that the smoothness of the boundary $\partial A$ is directly related to the smoothness of the function $b_A$ in a neighbourhood of $\partial A$. In the next two theorems we establish the equivalence for domains with a $C^k$ boundary, $k \geq 2$, and a partial result for $k = 1$ when the normal is Lipschitzian. This link between $b_A$ and classical geometric concepts clearly shows its advantages over the one-sided distance function $d_A$. In particular a domain $A$ can now be described by the level curves of $b_A$.

THEOREM 5.14. *Let $k \geq 1$ be an integer and $A$ be a domain in $\mathbb{R}^N$ with a nonempty boundary $\partial A$. If for each $x$ in $\partial A$ there exists a neighbourhood $V(x)$ of $x$ where $b_A$ is of class $C^k$, then the boundary $\partial A$ is of class $C^k$ and for each $x$ in $\partial A$ there exists a neighbourhood $U(x)$ of $x$ where*

$$\Pi_{\partial A}(y) = \{P_{\partial A}(y)\} \tag{5.17}$$

*is a singleton and*

$$y \mapsto P_{\partial A}(y) : V(x) \to \mathbb{R}^N \tag{5.18}$$

*is a $C^{k-1}$ mapping.*

The following result and its proof provide a partial converse to the previous theorem. The result of Theorem 5.15 can essentially also be found in D. Gilbarg and S. Trudinger [24, Lemma 14.16, p. 355].

THEOREM 5.15. *Let $A$ be a domain in $\mathbb{R}^N$ with a nonempty boundary $\partial A$ of class $C^k$ for $k \geq 2$. Then for each $x$ on $\partial A$, there exists a neighbourhood of $x$ where $b_A$ is $C^k$ and the boundary projection $P_{\partial A}$ is $C^{k-1}$.*

*Remark* 5.16. For $k \geq 2$ a domain $A$ in $\mathbb{R}^N$ has a nonempty boundary $\partial A$ of class $C^k$ if and only if the function $b_A$ is $C^k$ in a neighbourhood of $\partial A$. For $k = 1$ we have half of the result but it is not clear that $b_A$ is $C^1$ in a neighbourhood of $\partial A$ when the boundary of the domain is only $C^1$. □

Nevertheless we have the following additional result.

THEOREM 5.17. *Let $A$ be a domain in $\mathbb{R}^N$ with a nonempty boundary $\partial A$ of class $C^1$.*
*(i) For all $x \in \partial A$, there exists a neighbourhood $V(x)$ of $x$ such that*

$$\forall y \in V(x),\ \forall z \in \Pi_{\partial A}(y), \quad y - z = b_A(y)n(z). \tag{5.19}$$

*(ii) If, in addition, the normal $n$ is locally Lipschitz continuous, then for all $x \in \partial A$, there exists a neighbourhood $W(x)$ of $x$ such that*

$$\forall y \in W(x), \quad \Pi_{\partial A}(y) = \{P_{\partial A}(y)\} \text{ is a singleton} \tag{5.20}$$

*and the map*

$$y \mapsto P_{\partial A}(y) \ :\ W(x) \to \mathbb{R}^N \tag{5.21}$$

*is Lipschitz continuous. Moreover*

$$\forall y \in W(x), \quad \nabla b_A(y) = n(P_{\partial A}(y)), \tag{5.22}$$

*$\nabla b_A$ is Lipschitz continuous in $W(x)$ and $b_A \in W^{2,\infty}(W(x))$.*

*Remark* 5.18. So $b_A$ is $C^1$ in a neighbourhood of $\partial A$ if $\partial A$ is $C^1$ and the projection $P_{\partial A}$ is $C^0$ in a neighbourhood of $\partial A$. □

**5.5. Mean curvature of the boundary.** In this section we make the connection between the classical notion of *mean curvature* $H$ of the boundary $\partial A$ and the Laplacian $\Delta b_A$ of $b_A$. This question was studied independently and in different contexts by J.P. Zolésio [43], [44, p. 1103–1104] and by D. Gilbarg and S. Trudinger [24, Appendix 14.6, pp. 354–357].

We first recall some results by J.P. Zolésio [44, pp. 1103–1104]. Let $\Omega$ be a domain in $\mathbb{R}^N$ with a boundary $\Gamma = \partial\Omega$ of class $C^2$. The *tangential divergent* of $V$ in $C^1(\Gamma; \mathbb{R}^N)$ is defined as

$$div_\Gamma V = div\widetilde{V} - D\widetilde{V}n \cdot n\,|_\Gamma$$

where $\widetilde{V}$ is a $C^1$ extension of $V$ in an open neighbourhood $U$ of $\Gamma$, $D\widetilde{V}$ is the Jacobian matrix of $\widetilde{V}$ and $n$ is the unit outward normal to $\Omega$ on $\Gamma$. It can be shown that this quantity is independent of the choice of the extension (see also J. Sokolowski and J.P. Zolésio [37] for a more recent treatment of this notion). It turns out that for such a domain the mean curvature is given by

$$H = div_\Gamma\, n \quad \text{ on } \Gamma.$$

Here the definition of mean curvature does not use the factor $1/(N-1)$ which is often used in the literature as a normalization factor to make the radius of curvature of the unit sphere equal to one in all dimensions.

Recall from the previous section that for a domain $A$ with a boundary $\partial A$ of class $C^2$ $b_A \in C^2(U)$ and that $\nabla b_A$ is a unitary extention of the normal $n$ on $\partial A$. Necessarily

$$H = div_\Gamma \nabla b_A = div \nabla b_A - D(\nabla b_A)n \cdot n\,|_{\partial A}.$$

Now

$$|\nabla b_A|^2 = 1 \quad \text{ in } U$$

and necessarily

$$0 = \nabla|\nabla b_A|^2 = D(\nabla b_A)\nabla b_A$$

which yields the following result

$$H = \Delta b_A \text{ on } \partial A.$$

This elegant result is true for a boundary $\partial A$ of class $C^2$ since $m(\partial A) = 0$ and necessarily

$$|\nabla b_A|^2 = 1 \quad \text{everywhere in } U$$
$$\nabla|\nabla b_A|^2 \in C^0(U).$$

The question now is how could we extend this result to domains which are less smooth.

It is useful to compute the Laplacian of $b_A$ for a few examples.

*Example* 5.19. *Half plane in* $\mathbb{R}^2$ *of Example 3.15.*
Consider the domain

$$A = \{(x_1, x_2) : x_1 \leq 0\}, \quad \partial A = \{(x_1, x_2) : x_1 = 0\}.$$

It is readily seen that

$$b_A(x_1, x_2) = x_1, \quad \nabla b_A(x_1, x_2) = (1, 0), \quad \Delta b_A(x) = 0$$

and

$$b_{\partial A}(x_1, x_2) = |x_1|, \quad \nabla b_{\partial A}(x_1, x_2) = (\tfrac{x_1}{|x_1|}, 0)$$
$$< \Delta b_{\partial A}, \varphi >= 2 \textstyle\int_{\partial A} \varphi \, dx.$$

□

*Example* 5.20. *Ball of radius* $R > 0$ *in* $\mathbb{R}^2$ *of Example 3.16.*
Consider the domain

$$A = \{x \in \mathbb{R}^2 : |x| \leq R\}, \quad \partial A = \{x \in \mathbb{R}^2 : |x| = R\}.$$

Clearly

$$b_A(x) = |x| - R, \quad \nabla b_A(x) = \frac{x}{|x|}$$

and

$$< \Delta b_A, \varphi >= \int_{\mathbb{R}^2} \frac{1}{|x|} \varphi \, dx.$$

Also

$$b_{\partial A}(x) = \big|\, |x_1| - R\big|, \quad \nabla b_{\partial A}(x) = \begin{cases} \frac{x}{|x|}, & |x| > R \\ -\frac{x}{|x|}, & |x| < R \end{cases}$$

and

$$< \Delta b_{\partial A}, \varphi >= 2 \int_{\partial A} \varphi \, ds - \int_A \frac{1}{|x|} \varphi \, dx + \int_{\complement A} \frac{1}{|x|} \varphi \, dx.$$

Again we see that $\Delta b_{\partial A}$ contains twice the boundary measure on $\partial A$. □

*Example* 5.21. *Unit square in* $\mathbb{R}^2$ *of Example 3.17.*
Consider the domain

$$A = \{x = (x_1, x_2) \,:\, |x_1| \le 1,\, |x_2| \le 1\}.$$

We use the same notation as in Example 3.17. Hence

$$b_A(x) = \begin{cases} x_2 - 1 & x \in D_2 \cap Q_1 \\ |x - c_1|, & x \in D_3 \cap Q_1 \\ x_1 - 1, & x \in D_1 \cap Q_1 \end{cases}$$

$$\nabla b_A(x) = \begin{cases} (0,1) & x \in D_2 \cap Q_1 \\ \frac{x-c_1}{|x-c_1|}, & x \in D_3 \cap Q_1 \\ (1,0), & x \in D_1 \cap Q_1 \end{cases}$$

and for the whole plane

$$< \Delta b_A, \varphi >= \sum_{i=1}^{4} \int_{D_3 \cap Q_i} \frac{1}{|x - c_i|} \varphi \, dx + \sqrt{2} \int_{D_1 \cap D_2} \varphi \, dx,$$

where $D_1 \cap D_2$ is made up of the two diagonals of the square where $\nabla b_A$ has a singularity. Moreover

$$b_{\partial A}(x) = \begin{cases} |x_2 - 1|, & x \in D_2 \cap Q_1 \\ |x - c_1|, & x \in D_3 \cap Q_1 \\ |x_1 - 1|, & x \in D_1 \cap Q_1 \end{cases}$$

and

$$< \Delta b_{\partial A}, \varphi >= \sum_{i=1}^{4} \int_{D_3 \cap Q_i} \frac{1}{|x - c_i|} \varphi \, dx - \sqrt{2} \int_{D_1 \cap D_2} \varphi \, dx + 2 \int_{\partial A} \varphi \, dx.$$

Notice that the structures of the Laplacian are similar to the ones observed in the previous examples except for the presence of a singular term along the two diagonals of the square. □

The above examples are quite informative. For locally piecewise $C^2$-domains, we can conjecture that the mean curvature is the regular part of $\Delta b_A$ and that the surface measure on the boundary $\partial A$ is contained in $\Delta d_A$. It would be nice to find a formulation which would filter out the regular part of $\Delta b_A$ and extract the singular part of $\Delta d_A$ with support in $\partial A$. Recall that this surface measure was obtained from the gradient of the characteristic function $\chi_{\overline{A}} = 1 - |\nabla d_A|$ in the context of finite perimeter sets.

Another observation is that for the previous examples not only $\Delta b_A$ is a measure but also all second order derivatives of $b_A$. This suggests to use the space

$$X^{2,p}(D) = \{f \in W^{1,p}(D) \,:\, |\nabla f| = 1 \text{ and } \nabla \frac{\partial f}{\partial x_i} \in M_1(D),\ 1 \leq i \leq N\}$$

where $M_1(D) = \mathcal{D}^0(D)'$. For $D$ compact the embedding of $X^{2,p}(D)$ into $W^{1,p}(D)$ is now compact and if we could show that for a sequence $\{b_{A_n}\}$, the norms of $\{\nabla \partial_i b_{A_n}\}$ are uniformly bounded with respect to $i$ and $n$, then we could extract a subsequence which converges in $W^{1,p}(D)$-strong. Of course it would be a lot nicer to reach the same conclusion only with the uniform boundedness of the Laplacians of the sequence. In that case we could use the space

$$Y^{2,p}(D) = \{f \in W^{1,p}(D) \,:\, |\nabla f| = 1 \text{ and } \Delta f \in M_1(D)\}.$$

*Note added in the proofs.* The oriented function has also found an interesting application to the theory of thin/shallow shells. The reader is referred to Delfour and Zolésio [13], [11].

## REFERENCES


[1] R.A. ADAMS, *Sobolev Spaces*, Academic Press, New York, London 1975.

[2] J.P. AUBIN, *L'analyse Non Linéaire et Ses Motivations Économiques*, Masson, Paris, New York, Barcelona, Milan, Mexico, Sao Paulo 1984.

[3] J.P. AUBIN AND H. FRANKOWSKA, *Set-Valued Analysis*, Birkhäuser, Boston, Basel, Berlin 1990.

[4] M. BARDI AND M. FALCONE, *Discrete approximation of the minimal time function for systems with regular optimal trajectories*, in *Analysis and Optimization of Systems*, (eds., A. BENSOUSSAN AND J.L. LIONS) Springer-Verlag, Berlin, Heidelberg, New York 1990, pp. 103–112.

[5] G. BARLES, *Remarks on a flame propagation model*, Internal report, Rapport de Recherche INRIA no. 464, December 1985.

[6] M. BERGER, *Géométrie*, Vols 1, 2, 3, 4, 5, 6, CEDIC/Fernand Nathan, Paris 1977, (English Transl.) *Geometry I, II*, Springer-Verlag, Berlin, Heidelberg, New York 1987.

[7] R. CACCIOPPOLI, *Misura e integrazione sugli insiemi dimensionalmente orientati*, **VIII** (12) Rend. Accad. Naz. Lincei, Cl. Sc. fis mat nat (1952), pp. 3–11, 137–146.

[8] F.H. CLARKE, *Optimization and Nonsmooth Analysis*, John Wiley and Sons, New York, Chichester, Brisbane, Toronto, Singapore 1983.

[9] R. CORREA AND A. SEEGER, *Directional derivative of a minimax function*, Nonlinear Analysis, Theory, Methods & Applications **9** (1985), pp. 13–22.

[10] E. DE GIORGI, *Su una teoria generale della misura*, Ann. Mat. Pura Appl. **36** (serie IV) (1954), pp. 191–213.

[11] M.C. DELFOUR AND J.P. ZOLÉSIO, *A boundary differential equation for thin shells*, J. Differential Equations (to appear).

[12] M.C. DELFOUR AND J.P. ZOLÉSIO, *Functional analytic methods in shape analysis*, in *Boundary Control and Variation*, Chapter 6, (ed., J.-P. ZOLÉSIO), Marcel Dekker, New York 1994, pp. 105–139.

[13] M.C. Delfour and J.P. Zolésio, *On a variational equation for thin shell*, in *Control and Optimal Design of Distributed Parameter Systems*, (editors, J. Lagnese, D. Russell and L. White) Springer-Verlag, New York (this volume).

[14] M.C. Delfour and J.P. Zolésio, *Shape analysis via oriented distance functions*, J. Functional Analysis (in press).

[15] M.C. Delfour and J.P. Zolésio, *Shape sensitivity analysis via min max differentiability*, SIAM J. Control and Optim. **26** (1988), pp. 834–862.

[16] C. Dellacherie, *Ensembles Analytiques, Capacités, Mesures de Hausdorff*, in *Lecture Notes in Applied Mathematics*, Vol. 295, Springer-Verlag, Berlin, Heidelberg, New York 1972.

[17] J. Dieudonné, *Éléments d'analyse, 1. Fondement de l'analyse moderne*, Gauthier-Villars, Paris 1969 (Bordas 1979). (Transl. from *Foundations of Modern Analysis*, Academic Press, Inc., New York and London 1960.)

[18] J. Dugunji, *Topology*, (second printing) Allyn and Bacon, Inc., Boston 1966.

[19] I. Ekeland and R. Temam, *Analyse convexe et problèmes variationnels*, Dunod, Gauthier-Villars, Paris 1974.

[20] M. Falcone, *The minimum time problem and its applications to front propagation*, Internal report, Dipartimento di matematica "Guido Castelnuovo", Università degli Studi di Roma "La Sapienza", Roma 1992/93.

[21] M. Falcone and R. Ferretti, *High-order approximations for viscosity solutions of Hamilton-Jacobi-Bellman equations*, in sl ICOSAHOM 92, International Conference on Spectral and High-Order Methods, Montpellier, France, June 1992.

[22] M. Falcone, T. Giorgi and P. Loreti, *Level sets of viscosity solution and applications*, Internal report, (revised version) Istituto per le applicazioni del calcolo "Mauro Picone", Roma, May 1992.

[23] H. Federer, *Geometric Measure Theory*, Springer-Verlag, Berlin, Heidelberg, New York 1969.

[24] D. Gilbarg and N.S. Trudinger, *Elliptic Partial Differential Equations of Second Order*, Springer-Verlag, Berlin, Heidelberg, New York, Tokyo 1983.

[25] E. Giusti, *Minimal Surfaces and Functions of Bounded Variation*, Birkhäuser, Boston, Basel, Stuttgart 1984.

[26] J. Horváth, *Topological vector spaces and distributions*, Vol. I, Addison–Wesley, Reading, Mass. 1966.

[27] S.R. Kulbarni, S. Mitter and T.J. Richardson, *An existence theorem and lattice approximations for a variational problem arising in computer vision*, in *Signal Processing, Part I. Signal Processing Theory*, (eds., L. Auslander, T. Kailath and S. Mitter) IMA Series, Springer-Verlag, Heidelberg, Berlin, New York 1990, pp. 189–210.

[28] V.G. Maz'ja, *Sobolev Spaces*, Springer-Verlag, Berlin, Heidelberg, New York, Tokyo 1980.

[29] F. Morgan, *Geometric Measure Theory, a Beginner's Guide*, Academic Press, Inc. (Hartcourt Brace Jovanovich Publisher) Boston, San Diego 1988.

[30] C.B. Morrey, Jr., *Multiple Integrals in the Calculus of Variations*, Springer-Verlag, New York 1966.

[31] J.A.F. Plateau, *Statique expérimentale et théorique des liquides soumis aux seules forces moléculaires*, Gauthier–Villars, Paris 1873.

[32] T.J. Richardson, *Scale independent piecewise smooth segmentation of images via variational methods*, Report CICS-TH-194, Center for Intelligent Control Systems, Massachusetts Institute of Technology, Cambridge, Mass., February 1990.

[33] T.L. Rockafellar, *Convex analysis*, (second printing) Princeton University Press, Princetown, New Jersey 1972.

[34] W. Rudin, *Principles of Mathematical Analysis*, McGraw–Hill, New York 1967.

[35] L. Schwartz, *Analyse mathématique, Vol I*, Hermann, Paris 1967.

[36] N. SHIMAKURA, *Partial differential operators of elliptic type*, Transl. of Math. Monographs, Vol. **99**, American Mathematical Society, Providence, R.I. 1992.

[37] J. SOKOLOWSKI AND J.P. ZOLÉSIO, *Introduction to shape optimization: Shape sensitivity analysis*, Springer-Verlag, New York, Berlin, Heidelberg 1992.

[38] G. STAMPACCHIA, *Equations elliptiques du second order à coefficients discontinus*, Séminaire de Mathématiques Supérieures, Presses de l'Université de Montréal, Université de Montréal 1966.

[39] V. ŠVERÁK, *On optimal shape design*, Journal Math. Pures et Appl. (to appear) (report 1992).

[40] V. ŠVERÁK, *On shape optimal design*, C.R. Acad. Sc Paris Sér. I **315** (1992), pp. 545–549.

[41] F.A. VALENTINE, *Convex Sets*, McGraw–Hill 1964.

[42] E.H. ZARANTONELLO, *Projections on convex sets in hilbert spaces and spectral theory, parts I and II*, in *Contributions to Nonlinear Functional Analysis*, (ed., E.H. ZARANTONELLO) Academic Press, New York, London 1971, pp. 237–424.

[43] J.P. ZOLÉSIO, *Identification de domaines par déformation*, (Thèse de doctorat d'état) Université de Nice, France 1979.

[44] J.P. ZOLÉSIO, *The material derivative (or speed) method for shape optimization*, in *Optimization of Distributed Parameter Structures, vol II*, (eds., E.J. HAUG AND J. CÉA) Sijhofff and Nordhoff, Alphen aan den Rijn 1981, pp. 1089–1151.

[45] J.P. ZOLÉSIO, *Optimisation de Domaines*, (Thèse de spécialisation) Université de Nice, France 1973.

[46] J.P. ZOLÉSIO, *Shape optimization problems and free boundary problems*, in *Shape optimization and free boundaries*, (ed., M.C. DELFOUR) Kluwer Academic Publishers, Dordrecht, Boston, London 1992, pp. 397–457.

# ON THE DENSITY OF THE RANGE OF THE SEMIGROUP FOR SEMILINEAR HEAT EQUATIONS*

CAROLINE FABRE†, JEAN-PIERRE PUEL‡, AND ENRIQUE ZUAZUA§

**Abstract.** We consider the semilinear heat equation $u_t - \Delta u + f(u) = 0$ in a bounded domain $\Omega \subset \mathbb{R}^n$, $n \geq 1$, for $t > 0$ with Dirichlet boundary conditions $u = 0$ on $\partial\Omega \times (0, \infty)$. For $T > 0$ fixed we consider the map $S(T) : C_0(\Omega) \to C_0(\Omega)$ such that $S(T)u^0 = u(x, T)$ where $u$ is the solution of this heat equation with initial data $u(x, 0) = u^0(x)$ and $C_0(\Omega)$ is the space of uniformly continuous functions on $\Omega$ that vanish on its boundary. When $f$ is globally Lipschitz and for any $T > 0$ we prove that the range of $S(T)$ is dense in $C_0(\Omega)$. Our method of proof combines backward uniqueness results, a variational approach to the problem of the density of the range of the semigroup for linear heat equations with potentials and a fixed point technique. These methods are similar to those developed by the authors in an earlier paper in the study of the approximate controllability of semilinear heat equations.

**Key words.** Semilinear heat equation, density of semigroup-map, approximate controllability.

**1. Introduction and main results.** Let $\Omega$ be a bounded domain of $\mathbb{R}^n$, $n \geq 1$, with boundary of class $C^2$ and $f$ a globally Lipschitz function from $\mathbb{R}$ to $\mathbb{R}$.

Consider the semilinear heat equation

$$\begin{cases} u_t - \Delta u + f(u) = 0 & \text{in} \quad \Omega \times (0,\infty) \\ u = 0 \quad \text{on} \quad \partial\Omega \times (0,\infty) \\ u(x,0) = u^0(x). \end{cases} \tag{1.1}$$

It is well known that, when $1 \leq p < \infty$, and $u^0 \in L^p(\Omega)$ equation (1.1) admits a unique solution

$$u \in C([0,\infty); L^p(\Omega)).$$

The same result holds in $C_0(\Omega)$ (the space of uniformly continuous functions on $\Omega$ that vanish on $\partial\Omega$ endowed with the norm of the supremum), i.e. if $u^0 \in C_0(\Omega)$ there exists a unique solution

$$u \in C([0,\infty); C_0(\Omega)).$$

---

* Supported by DGICYT Project PB90-0245 and EEC grant SC1*-CT91-0732. This work was done while the second and third authors were visiting the IMA of the University of Minnesota. These authors acknowledge the hospitality and support of the IMA.

† Université Paris XII-Val de Marne, U.F.R. Sciences, Laboratoire de Mathématiques, Av. du Général de Gaulle, 94010 Creteil Cedex, and Centre de Mathématiques Appliquées, Ecole Polytechnique, 91128 Palaiseau Cedex, France.

‡ Université de Versailles Saint-Quentin, Département de Mathématiques, and Centre de Mathématiques Appliquées, Ecole Polytechnique, 91128 Palaiseau Cedex, France.

§ Departamento de Matemática Aplicada, Universidad Complutense, 28040 Madrid, Spain.

For every $T > 0$ we define the map

$$[S(T)]u^0 = u(x,T) \tag{1.2}$$

where $u = u(x,t)$ is the solution of (1.1).

The main result is the following:

THEOREM 1.1. *Suppose that $f : \mathbb{R} \to \mathbb{R}$ is globally Lipschitz. For every $T > 0$ the map*

$$S(T) : C_0(\Omega) \to C_0(\Omega)$$

*has dense range in $C_0(\Omega)$.*

As an immediate corollary we have the following result:

COROLLARY 1.2. *For every $T > 0$ and $1 \le p < \infty$ the range of the map $S(T) : L^p(\Omega) \to L^p(\Omega)$ is dense in $L^p(\Omega)$.*

*Remark* 1.3.

1) When $f$ is linear, or more generally, in the framework of linear heat equations with a bounded potential $a = a(x,t)$,

$$u_t - \Delta u + a(x,t)u = 0 \tag{1.3}$$

the density of the range of the semigroup-map is a consequence of the following backward uniqueness result:

$$\text{(B.U.)} \qquad \left. \begin{cases} -\psi_t - \Delta\psi + a(x,t)\psi = 0 & \text{in} \quad \Omega \times (0,T) \\ \psi = 0 \quad \text{on } \partial\Omega \times (0,T) \\ \psi(x,0) = 0 \end{cases} \right\} \Rightarrow \psi \equiv 0$$

This backward-uniqueness result holds for every solution $\psi$ with final data a bounded measure. Note that, due to the smoothing effect of the heat equation, the solution is smooth for every $0 \le t < T$. Since the potential we consider is uniformly bounded, the backward uniqueness property (B.U.) we need is classical and well known (see for instance [1], [3], [6], [8], [9] and [10]). In particular it is a direct consequence of Theorem 1.8, in S. Agmon and L. Nirenberg [1, p. 140].

Let us briefly show how (B.U.) and a simple duality argument implies the density of the range of $S_a(T) : L^2(\Omega) \to L^2(\Omega)$ in $L^2(\Omega)$ where, now, $S_a(T)$ stands for the semigroup-map associated to (1.3). Suppose that $u^1 \in L^2(\Omega)$ is orthogonal to the range of $S_a(T)$ in $L^2(\Omega)$. We solve the system

$$\begin{cases} -\psi_t - \Delta\psi + a(x,t)\psi = 0 & \text{in} \quad \Omega \times (0,T) \\ \psi = 0 \quad \text{on} \quad \partial\Omega \times (0,T) \\ \psi(T) = u^1 \end{cases}$$

which has a unique solution $\psi \in C([0,T]; L^2(\Omega))$. Let $u$ be a solution of

$$\begin{cases} u_t - \Delta u + a(x,t)u = 0 & \text{in} \quad \Omega \times (0,T) \\ u = 0 \quad \text{on} \quad \partial\Omega \times (0,T) \\ u(0) = u^0 \in L^2(\Omega) \end{cases}$$

for any $u^0 \in L^2(\Omega)$. Multiplying by $u$ the equation satisfied by $\psi$ we get

$$\int_\Omega u(x,T)u^1 dx = \int_\Omega u^0 \psi(x,0)dx, \quad \forall u^0 \in L^2(\Omega).$$

Since $u^1$ is orthogonal to $u(x,T)$ in $L^2(\Omega)$ we deduce that

$$\int_\Omega u^0 \psi(x,0)dx = 0, \quad \forall u^0 \in L^2(\Omega)$$

but then $\psi(x,0) = 0$ in $\Omega$ and by (B.U.) $\psi \equiv 0$ and $u^1 = 0$.

2) Our proof of Theorem 1.1 is based on a fixed point technique. However, in order to apply this fixed point method we need some information on how the range of $S_a(T)$ depends on the potential $a$ that the duality argument above does not provide. Therefore, we will use a different approach to the problem of the density of the range of the semigroup for linear heat equations with potentials. This approach is variational and is inspired by our recent works ([4], [5]) on the approximate controllability of semilinear heat equations.

3) Our Theorem 1.1 is sharp in the sense that it fails for some nonlinearities that, at infinity, grow in a superlinear way.

Consider for instance

$$\begin{cases} u_t - \Delta u + u^3 = 0 & \text{in} \quad \Omega \times (0,\infty) \\ u = 0 \quad \text{on} \quad \partial\Omega \times (0,\infty) \\ u(x,0) = u^0(x). \end{cases} \tag{1.4}$$

It is easy to check that for every $T > 0$ there exists $C(T) > 0$ such that for every $u^0 \in L^2(\Omega)$ the solution of (1.4) satisfies

$$||u(\cdot,T)||_{L^2(\Omega)} \le C(T).$$

Thus, the range of $S(T)$ is not dense in $L^2(\Omega)$.

On the other hand, the existence of an universal super-solution

$$w(t) = \frac{1}{\sqrt{2}} t^{-\frac{1}{2}}$$

implies that the range of $S(T)$ is not dense in any $L^p(\Omega)$ nor in $C_0(\Omega)$.

In the case of non-linearities that grow superlinearly at infinity one may expect the density of the range of $S(T)$ in some ball of $C_0(\Omega)$ around the origin. However, our methods do not seem to provide such a result. □

The method of proof of Theorem 1.1 allows us to prove some variants of it. Let us mention one of them. Let us introduce the following set of initial data:

$$\mathcal{B} = \{\varphi \in L^\infty(\Omega) : \varphi(x) \in \lambda sgn(\phi(x)) \text{ for some } \lambda \in \mathbb{R} \text{ and some measurable function } \phi = \phi(x)\}. \tag{1.5}$$

By *sgn* we denote the multi-valued sign function:

$$sgn(s) = \begin{cases} 1 & \text{if } s > 0 \\ [-1,1] & \text{if } s = 0 \\ -1 & \text{if } s < 0. \end{cases}$$

We have the following result.

THEOREM 1.4. *If $f$ is globally Lipschitz, for every $T > 0$, $S(T)\mathcal{B}$ is dense in $C_0(\Omega)$.*

*Remark* 1.5. The initial data we use in the proof of Theorem 1.4 are in fact of the form

$$\varphi \in \lambda sgn(\phi)$$

where $\phi = \psi(x,0)$ with $\psi = \psi(x,t)$ solution of the heat equation

$$\begin{cases} -\psi_t - \Delta\psi + a(x,t)\psi = 0 & \text{in} \quad \Omega \times (0,T) \\ \psi(x,T) = \psi_0 \in M(\Omega) \\ \psi = 0 \quad \text{on} \quad \partial\Omega \times (0,T) \end{cases}$$

for some potential $a \in L^\infty(\Omega \times (0,T))$. By $M(\Omega)$ we denote space of bounded measures in $\Omega$ (the dual of $C_0(\Omega)$).

In dimension $n = 1$ the results by S. Angenent [2] show that the zero set of $\psi(x,0)$ is of zero Lebesgue measure. Thus, in one space dimension, the density of the range of the semigroup map $S(T)$ holds with initial data of bang-bang form, i.e. $\varphi = \pm\lambda$ a.e. in $\Omega$ for some $\lambda \in \mathbb{R}$. □

The rest of this paper is organized as follows. In Section 2 we prove Theorem 1.1 for linear equations of the form (1.3) by the variational method mentioned above. For the sake of completeness, we first give a direct proof of Corollary 1.2 in the linear framework. In Section 3 we complete the proof of Theorem 1.1. Finally, in Section 4 we prove Theorem 1.4.

In the sequel we will denote by $Q$ the cylinder $\Omega \times (0,T)$ and by $\Sigma$ the lateral boundary $\partial\Omega \times (0,T)$. The norm in $L^p(\Omega)$ will be denoted by $||\cdot||_p$ and the norm in $M(\Omega)$ by $|\cdot|$.

**2. The linear heat equation with a potential.** Consider a potential $a = a(x,t) \in L^\infty(Q)$ with $T > 0$ fixed and the heat equation

$$(2.1) \qquad \begin{cases} u_t - \Delta u + au = 0 & \text{in} \quad Q \\ u = 0 & \text{on} \quad \Sigma \\ u(x,0) = u^0(x). \end{cases}$$

It is well known that if $u^0 \in L^p(\Omega)$ with $1 \leq p < \infty$ then (2.1) has a unique solution

$$u \in C([0,T]; L^p(\Omega)).$$

On the other hand, if $u^0 \in C_0(\Omega)$ this solution belongs to $C([0,T]; C_0(\Omega))$.

We define the linear map

$$S_a(T)u^0 = u(x,T)$$

which is continuous from $L^p(\Omega)$ into $L^p(\Omega)$ when $1 \leq p < \infty$ and from $C_0(\Omega)$ into $C_0(\Omega)$.

In this section we prove the following result

PROPOSITION 2.1. *(i) The range of $S_a(T) : L^p(\Omega) \to L^p(\Omega)$ is dense in $L^p(\Omega)$ when $1 \leq p < \infty$.*

*(ii) The range of $S_a(T) : C_0(\Omega) \to C_0(\Omega)$ is dense in $C_0(\Omega)$.*

As we mentioned in the introduction this result can be easily proved by duality as a consequence of the backward uniqueness of solutions of the heat equation.

Our goal here is to give a proof based on a variational method that, in the sequel, will allow us to extend this density result to the semilinear equation (1.1).

We distinguish the cases $L^p(\Omega)$ with $1 < p < \infty$, $p = 1$ and $C_0(\Omega)$.

**2.1. The case $1 < p < \infty$.** Let $1 < q < \infty$ be such that $\frac{1}{p} + \frac{1}{q} = 1$. Consider the functional defined for $\varphi^0 \in L^p(\Omega)$

$$(2.2) \qquad J(\varphi^0) = \frac{1}{q}||\varphi(x,0)||_q^q + \alpha||\varphi^0||_q - \int_\Omega u^1\varphi^0 dx$$

where $u^1 \in L^p(\Omega)$ and $\alpha > 0$ are fixed such that $||u^1||_p > \alpha$ and $\varphi = \varphi(x,t)$ is the solution of

$$(2.3) \qquad \begin{cases} -\varphi_t - \Delta\varphi + a\varphi = 0 & \text{in} \quad Q \\ \varphi = 0 & \text{on} \quad \Sigma \\ \varphi(T) = \varphi^0 \end{cases}$$

which belongs to $C([0,T]; L^q(\Omega))$.

We have the following result:

PROPOSITION 2.2. *The minimum of $J$ in $L^q(\Omega)$ is achieved at a unique $\hat{\varphi}^0 \in L^q(\Omega)$. If $\hat{\varphi}$ is the solution of (2.3) corresponding to this final data ($\hat{\varphi}(T) = \hat{\varphi}^0$) then the solution of*

$$\begin{cases} u_t - \Delta u + au = 0 & \text{in} \quad Q \\ u = 0 & \text{on} \quad \Sigma \\ u(x,0) = u^0(x) = |\hat{\varphi}(x,0)|^{q-2}\hat{\varphi}(x,0) & \text{in} \quad \Omega \end{cases} \tag{2.4}$$

*satisfies*

$$||u(T) - u^1||_p = \alpha. \tag{2.5}$$

*Moreover, for $\alpha > 0$ fixed, $\hat{\varphi}^0$ remains bounded in $L^q(\Omega)$ when $a$ is uniformly bounded in $L^\infty(Q)$ and $u^1$ is relatively compact in $L^p(\Omega)$.*

*In addition, if*

$$\begin{cases} a_n \to a & \text{in} \quad L^\infty(Q) \text{ weak-}* \\ u_n^1 \to u^1 & \text{in} \quad L^p(\Omega) \end{cases} \tag{2.6}$$

*then*

$$\hat{\varphi}_n^0 \to \hat{\varphi}^0 \text{ in } L^q(\Omega) \text{ weak} \tag{2.7}$$

*where $\hat{\varphi}^0$ is the minimizer associated to $a$ and $u^1$. The corresponding initial values $u_n^0 = |\hat{\varphi}_n(x,0)|^{q-2}\hat{\varphi}_n(x,0)$ satisfy*

$$u_n^0 \to u^0 = |\hat{\varphi}(x,0)|^{q-2}\hat{\varphi}(x,0) \text{ strongly in } L^\infty(\Omega). \tag{2.8}$$

*Remark* 2.3. As a consequence of this proposition we deduce that the map $S(T) : L^p(\Omega) \to L^p(\Omega)$ has dense range. □

*Proof of Proposition* 2.2. The functional $J$ is continuous in $L^q(\Omega)$ and strictly convex. Moreover $J$ is coercive. More precisely,

$$\varliminf_{||\varphi^0||_q \to \infty} \frac{J(\varphi^0)}{||\varphi^0||_q} \geq \alpha. \tag{2.9}$$

Thus, $J$ achieves its minimum at a unique $\hat{\varphi}^0 \in L^q(\Omega)$.

Let us prove (2.9). Consider a sequence $(\varphi_n^0) \subset L^q(\Omega)$ such that $||\varphi_n^0||_q \to \infty$. Define

$$I_n = \frac{J(\varphi_n^0)}{||\varphi_n^0||_q} = \frac{1}{q}||\varphi_n^0||_q^{q-1}||\tilde{\varphi}_n(x,0)||_q^q + \alpha - \int_\Omega u^1 \tilde{\varphi}_n^0 dx$$

where $\tilde{\varphi}_n^0 = \varphi_n^0/||\varphi_n^0||_q$ and $\tilde{\varphi}_n = \varphi_n/||\varphi_n^0||_q$.

If

$$\varliminf_{n\to\infty} ||\tilde{\varphi}_n(x,0)||_q > 0$$

then (2.9) holds.

If this does not hold, then for a subsequence (that we still denote by the index $n$) we have

$$||\tilde{\varphi}_n(x,0)||_q \to 0. \tag{2.10}$$

By extracting subsequences (still denoted by the index $n$) we have

$$\tilde{\varphi}_n^0 \rightharpoonup \tilde{\varphi}^0 \quad \text{in } L^q(\Omega) \text{ weakly}$$
$$\tilde{\varphi}_n \rightharpoonup \tilde{\varphi} \text{ weakly in } L^q(Q)$$

where $\tilde{\varphi}$ is the solution of

$$\begin{cases} -\tilde{\varphi}_t - \Delta\tilde{\varphi} + \tilde{\varphi} = 0 & \text{in } Q \\ \tilde{\varphi} = 0 \quad \text{on } \Sigma \\ \tilde{\varphi}(T) = \tilde{\varphi}^0. \end{cases}$$

On the other hand (2.10) implies

$$\tilde{\varphi}(x,0) = 0$$

and thus, by backward uniqueness, $\tilde{\varphi} \equiv 0$ and $\tilde{\varphi}^0 \equiv 0$. This implies that

$$\tilde{\varphi}_n^0 \rightharpoonup 0 \text{ weakly in } L^q(\Omega)$$

and therefore

$$\varliminf_{n\to\infty} I_n \geq \varliminf_{n\to\infty} \alpha - \int_\Omega u^1\tilde{\varphi}_n^0 dx = \alpha.$$

This concludes the proof of (2.9)

Let $\hat{\varphi}_0 \in L^q(\Omega)$ be the minimum of $J$ in $L^q(\Omega)$, since $||u^1||_p > \alpha$, $\hat{\varphi}^0 \neq 0$ and then the $L^q$-norm is differentiable. By calculating the first variation of $J$ at $\hat{\varphi}^0$ we deduce that

$$\int_\Omega |\hat{\varphi}(x,0)|^{q-2}\hat{\varphi}(x,0)\theta(x,0)dx + \alpha||\hat{\varphi}^0||_q^{1-q}\int_\Omega |\hat{\varphi}^0|^{q-2}\hat{\varphi}^0\theta^0 dx \tag{2.11}$$
$$-\int_\Omega u^1\theta^0 dx = 0$$

for every solution $\theta = \theta(x,t)$ of

$$\left\{\begin{array}{ll} -\theta_t - \Delta\theta + a\theta = 0 & \text{in} \quad Q \\ \theta = 0 & \text{on} \quad \Sigma \\ \theta(T) = \theta^0 \end{array}\right. \tag{2.12}$$

with $\theta^0 \in L^q(\Omega)$. Multiplying by $\theta$ in (2.4) and integrating over $Q$ we deduce that

$$\int_\Omega |\hat{\varphi}(x,0)|^{q-2}\hat{\varphi}(x,0)\theta(x,0)dx = \int_\Omega u(T)\theta^0\,dx\,.$$

Combining (2.11) and (2.12) we obtain

$$u^1 - u(T) = \alpha||\hat{\varphi}^0||_q^{1-q}|\hat{\varphi}^0|^{q-2}\hat{\varphi}^0$$

and thus (2.5).

Let us now prove the uniform bound for the minimizers when $a$ remains bounded in $L^\infty(Q)$ and $u^1$ in a compact set of $L^p(\Omega)$. We argue by contradiction. Let $\{a_n\}$ be a sequence of potentials and $u_n^1$ a sequence of final data such that

$$\begin{array}{lll} a_n \rightharpoonup a & \text{in} & L^\infty(Q) \text{ weak-}* \\ u_n^1 \to u^1 & \text{in} & L^p(\Omega) \text{ strongly} \end{array}$$

and suppose that the corresponding minimizers $\{\hat{\varphi}_n^0\}$ satisfy

$$||\hat{\varphi}_n^0||_q \to \infty. \tag{2.13}$$

Let us denote by $J_n$ the functional corresponding to the potential $a_n$ and final data $u_n^1$. It is easy to check that, since $||u_n^1||_p > \alpha$,

$$J_n(\hat{\varphi}_n^0) \le 0, \quad \forall n.$$

Thus, we will obtain a contradiction if we show that

$$\underline{\lim}_{n\to\infty} \frac{J_n(\hat{\varphi}_n^0)}{||\hat{\varphi}_n^0||_q} \ge \alpha. \tag{2.14}$$

We proceed as in the proof of (2.6). Let us normalize the minimizers

$$\psi_n^0 = \hat{\varphi}_n^0/||\hat{\varphi}_n^0||_q$$

and let $\psi_n$ be the solution of (2.6) with final data $\psi_n^0$. We have

$$I_n = \frac{J_n(\hat{\varphi}_n^0)}{||\hat{\varphi}_n^0||_q} = \frac{||\hat{\varphi}_n^0||_q^{q-1}}{q}||\psi_n(x,0)||_q^q + \alpha - \int_\Omega u_n^1\psi_n^0\,dx.$$

Clearly, in view of (2.13), (2.14) holds if

$$\underline{\lim}_{n\to\infty} \|\psi_n(x,0)\|_q > 0.$$

If this does not hold, for a subsequence (still denoted by the index $n$) we have

$$\|\psi_n(x,0)\|_q \to 0. \tag{2.15}$$

Since $\psi_n^0$ and $\psi_n$ are bounded in $L^q(\Omega)$ and $C([0,T];L^q(\Omega))$ respectively, by extracting subsequences we have

$$\begin{aligned} \psi_n^0 &\rightharpoonup \psi^0 \quad \text{weakly in} \quad L^q(\Omega) \\ \psi_n &\rightharpoonup \psi \quad \text{weakly in} \quad L^q(Q). \end{aligned}$$

But since the potentials weakly-$*$ converge in $L^\infty(Q)$ this does not suffice to pass to the limit in equation (2.3) satisfied by $\psi_n$ with potential $a_n$.

However, $\psi_n$ is also uniformly bounded in $X_q(0,T-\delta) = L^q(0,T-\delta;W_0^{1,q}(\Omega)) \cap W^{1,q}(0,T-\delta;L^q(\Omega))$ (see [7], p. 341) for every $\delta > 0$. The compactness of the embedding $X_q(0,T-\delta) \subset L^1(\Omega\times(0,T-\delta))$ allows us to pass to the limit in the equations and show that $\psi$ satisfies:

$$\begin{cases} -\psi_t - \Delta\psi + a\psi = 0 & \text{in } Q \\ \psi = 0 & \text{on } \Sigma. \end{cases}$$

Let us now prove that $\psi$ takes the final data $\psi^0$, i.e. $\psi(T) = \psi^0$. For every $\theta \in C^2(\overline{\Omega}) \cap C_0(\Omega)$ we have

$$\begin{aligned} \left| -\int_\Omega \psi_n^0 \theta dx + \int_\Omega \psi_n(t)\theta dx \right| &= \left| \int_\Omega \int_t^T \psi_n \Delta\theta dx dt - \int_\Omega \int_t^T a_n \psi_n \theta dx dt \right| \\ &\le C(T-t) \end{aligned}$$

for some $C > 0$ that does not depend on $n$. Passing to the limit as $n \to \infty$ we obtain

$$\left| -\int_\Omega \psi^0 \theta dx + \int_\Omega \psi(t)\theta dx \right| \le C(T-t)$$

which shows that $\psi(T) = \psi^0$.

On the other hand, (2.15) implies that

$$\psi(x,0) = 0$$

and therefore by backward uniqueness $\psi^0 = 0$. This shows that

$$\psi_n^0 \rightharpoonup 0 \text{ weakly in } L^q(\Omega)$$

and therefore

$$\varliminf_{n\to\infty} I_n \geq \varliminf_{n\to\infty} \left[\alpha - \int_\Omega u_n^1 \psi_n^0 dx\right] = \alpha$$

which concludes the proof of (2.14).

Since $\hat\varphi_n^0$ is bounded in $L^q(\Omega)$ we may extract a subsequence (still denoted by the index $n$) such that

$$\hat\varphi_n^0 \rightharpoonup \psi^0 \quad \text{in} \quad L^q(\Omega) \text{ weakly}\ .$$

Let us see that $\psi^0 = \hat\psi^0$ where $\hat\psi^0$ is the minimizer associated with $a$ and $u^1$. It is sufficient to check that

$$J(\psi^0) \leq J(\varphi^0), \quad \forall \varphi^0 \in L^q(\Omega). \tag{2.16}$$

By the weak lower semicontinuity of the $L^q$-norm, the strong convergence of $\{u_n^1\}$ and the weak convergence of $\hat\varphi_n^0$ to $\psi^0$, we have

$$J(\psi^0) \leq \varliminf_{n\to\infty} J_n(\hat\varphi_n^0) \leq \varliminf_{n\to\infty} J_n(\varphi^0), \quad \forall \varphi^0 \in L^q(\Omega)$$

and on the other hand

$$\lim_{n\to\infty} J_n(\varphi^0) = J(\varphi^0), \quad \forall \varphi^0 \in L^q(\Omega).$$

Thus (2.16) holds. Since the limit $\psi^0$ has been identified the whole sequence $\{\hat\varphi_n^0\}$ converges. This concludes the proof of (2.16).

Finally (2.8) is immediate from the regularizing effect of the heat equation since we have

$$\hat\varphi_n(\cdot, t) \rightharpoonup \hat\varphi(\cdot, t) \text{ in } L^\infty(\Omega)$$

for every $0 \leq t < T$. □

**2.2. The case $p = 1$.** Consider the functional

$$J : L^\infty(\Omega) \to \mathbb{R}$$

defined as follows

$$J(\varphi^0) = \frac{1}{2}||\varphi(x,0)||_2^2 + \alpha||\varphi^0||_\infty - \int_\Omega u^1 \varphi^0 dx \tag{2.17}$$

for $u^1 \in L^1(\Omega)$ and $\alpha > 0$ fixed such that $||u^1||_1 > \alpha$ and $\varphi$ being the solution of (2.3) with initial data $\varphi^0 \in L^\infty(\Omega)$. In this case $\varphi \in L^\infty(Q)$ and $\varphi \in C([0,T); C_0(\Omega))$. Thus $J$ is well defined in $L^\infty(\Omega)$.

We have the following result.

PROPOSITION 2.4. *The minimum of $J$ in $L^\infty(\Omega)$ is achieved at a unique $\hat{\varphi}^0 \in L^\infty(\Omega)$. If $\hat{\varphi}$ is the solution of (2.3) corresponding to this final data ($\hat{\varphi}(T) = \hat{\varphi}^0$) then, the solution $u$ of*

$$\begin{cases} u_t - \Delta u + au = 0 & in\ Q \\ u = 0 & on\ \Sigma \\ u(0) = u^0 = \hat{\varphi}(x,0) \end{cases} \tag{2.18}$$

*satisfies*

$$||u(T) - u^1||_1 \leq \alpha \tag{2.19}$$

*Moreover, $\hat{\varphi}^0$ is uniformly bounded in $L^\infty(\Omega)$ when $a$ is bounded in $L^\infty(Q)$ and $u^1$ remains in a compact set of $L^1(\Omega)$. In addition, if*

$$\begin{cases} a_n \rightharpoonup a\ weakly\text{-}*\ in\ L^\infty(Q) \\ u_n^1 \to u^1\ strongly\ in\ L^1(\Omega) \end{cases} \tag{2.20}$$

*then*

$$\hat{\varphi}_n^0 \rightharpoonup \hat{\varphi}^0\ weakly\text{-}*\ in\ L^\infty(\Omega) \tag{2.21}$$

*where $\hat{\varphi}^0$ is the minimizer associated with the limit potential $a$ and $u^1$ and the initial data $u_n^0 = \hat{\varphi}_n(x,0)$ satisfy*

$$u_n^0 \to u^0 = \hat{\varphi}(x,0)\ strongly\ in\ L^\infty(\Omega). \tag{2.22}$$

*Remark* 2.5. This result implies that the range of $S(T) : L^1(\Omega) \to L^1(\Omega)$ is dense in $L^1(\Omega)$. □

*Proof of Proposition 2.4*

$J$ is continuous in $L^\infty(\Omega)$ and convex. Moreover, as a consequence of the backward uniqueness, $J$ is strictly convex. On the other hand, $J$ is lower semicontinuous in the weak-$*$ topology of $L^\infty(\Omega)$.

The existence of a unique minimizer in $L^\infty(\Omega)$ is a direct consequence of the following coercivity property

$$\underline{lim}_{||\varphi^0||_\infty \to \infty} \frac{J(\varphi^0)}{||\varphi^0||_\infty} \geq \alpha. \tag{2.23}$$

The proof of (2.23) is analogous to that of (2.6). The minimizer $\hat{\varphi}^0$ is such that

$$\int_\Omega \hat{\varphi}(x,0)\theta(x,0)dx + \alpha[||\hat{\varphi}^0 + \theta^0||_\infty - ||\hat{\varphi}^0||_\infty] - \int_\Omega u^1\theta^0 dx \geq 0$$

for every $\theta^0 \in L^\infty(\Omega)$ where $\theta = \theta(x,t)$ is the solution of (2.12). Multiplying in (2.18) by $\theta$ we obtain

$$\int_\Omega \hat{\varphi}(x,0)\theta(x,0)dx = \int_\Omega u(T)\theta^0 dx.$$

Therefore

$$\int_\Omega (u(T) - u^1)\theta^0 dx \leq \alpha[||\hat{\varphi}^0 + \theta^0||_\infty - ||\hat{\varphi}^0||_\infty] \leq \alpha||\theta^0||_\infty, \quad \forall \theta^0 \in L^\infty(\Omega).$$

Thus

$$||u(T) - u^1||_1 \leq \alpha.$$

The proof of the uniform bound on $\hat{\varphi}^0$ when $a$ remains in a bounded set of $L^\infty(Q)$ and $u^1$ in a compact set of $L^1(\Omega)$ is similar to that of Proposition 2.2. The proof of (2.21) and (2.22) is also very similar. □

*Remark* 2.6. In this case we do not get $||u(T) - u^1||_1 = \alpha$ since the $L^\infty(\Omega)$-norm is not differentiable. □

**2.3. The $C_0(\Omega)$-case.** This case requires to work in $M(\Omega)$, the space of finite measures. Let us recall that $M(\Omega)$ is the dual of $C_0(\Omega)$. The norm in $M(\Omega)$ is defined as follows:

$$|\mu| = \sup_{\substack{\varphi \in C_0(\Omega) \\ ||\varphi||_\infty \leq 1}} |\langle \mu, \varphi \rangle|.$$

By $\langle \cdot, \cdot \rangle$ we denote the duality between $M(\Omega)$ and $C_0(\Omega)$.
Let us consider the functional

$$J : M(\Omega) \to \mathbb{R}$$

such that

$$J(\varphi^0) = \frac{1}{2}||\varphi(x,0)||_2^2 + \alpha|\varphi^0| - \langle \varphi^0, u^1 \rangle$$

for $u^1 \in C_0(\Omega)$ and $\alpha > 0$ fixed such that $||u^1||_\infty > \alpha$, where $\varphi$ is the solution of (2.3). Let us recall that the solution of (2.3) with a measure as final data belongs to

$$\varphi \in L^\infty(0,T; L^1(\Omega)) \cap C([0,T); L^2(\Omega))$$

and takes the final data $\varphi^0 \in M(\Omega)$ in the following sense

$$\int_\Omega \varphi(x,t)\theta(x)dx \to \langle \varphi^0, \theta \rangle, \text{ as } t \nearrow T \text{ for every } \theta \in C_0(\Omega).$$

We have the following result:

PROPOSITION 2.7. *The minimum of $J$ on $M(\Omega)$ is achieved at a unique $\hat{\varphi}^0 \in M(\Omega)$. If $\hat{\varphi}$ is the solution of (2.3) with this final data and $u$ solves (2.18) then*

$$\|u(T) - u^1\|_\infty \leq \alpha. \tag{2.24}$$

*Moreover, $\hat{\varphi}^0$ remains bounded in $M(\Omega)$ when $a$ is bounded in $L^\infty(Q)$ and $u^1$ remains in a relatively compact set of $C_0(\Omega)$. In addition, if*

$$\begin{cases} a_n \rightharpoonup a & \text{weak-}* \text{ in } L^\infty(Q) \\ u_n^1 \to u^1 & \text{in } C_0(\Omega) \end{cases} \tag{2.25}$$

*then*

$$\hat{\varphi}_n^0 \rightharpoonup \varphi^0 \quad \text{weakly-}* \text{ in } M(\Omega) \tag{2.26}$$

*and*

$$u_n^0 = \hat{\varphi}_n(x,0) \to \quad u^0 = \hat{\varphi}(x,0) \text{ in } C_0(\Omega). \tag{2.27}$$

*Remark* 2.8. As a consequence of this result we obtain Proposition 2.1 in the $C_0(\Omega)$-case. □

*Proof of Proposition 2.7*

The functional $J$ is strictly convex because of the backward-uniqueness property. On the other hand $J$ is continuous in $M(\Omega)$.

The coercivity of $J$ and the fact that it achieves its minimum in $M(\Omega)$ follow from the arguments of Proposition 2.2 and 2.4 but now one has to work with the weak-$*$ topology of $M(\Omega)$.

The uniform bound on the minimizers as well as (2.26) and (2.27) follow also by similar arguments. □

*Remark* 2.9. Note that, again, due to the non-differentiability of the $M(\Omega)$-norm we do not get $\|u(T) - u^1\|_\infty = \alpha$ but just $\|u(T) - u^1\|_\infty \leq \alpha$. □

**3. The semilinear heat equation: Theorem 1.1.** In this section we give the proof of Theorem 1.1. The same arguments allow to give a direct proof of Corollary 1.2.

Consider $u^1 \in C_0(\Omega)$ and $\alpha > 0$. Our goal is to find an initial data $u^0 \in C_0(\Omega)$ such that the solution $u$ of

$$\begin{cases} u_t - \Delta u + f(u) = 0 & \text{in } Q \\ u = 0 \quad \text{on } \Sigma \\ u(0) = u^0 \end{cases} \tag{3.1}$$

satisfies

$$\|u(T) - u^1\|_\infty \leq \alpha. \tag{3.2}$$

We consider first a nonlinearity $f$ which is $C^1$ and globally Lipschitz and introduce the continuous and bounded function:

$$g(s) = \begin{cases} \frac{f(s)-f(0)}{s} & \text{if } s \neq 0 \\ f'(0) & \text{if } s = 0. \end{cases} \tag{3.3}$$

The general case where $f$ is globally Lipschitz function (non-necessarily $C^1$) will be covered later on by a density argument.

For every $v \in L^1(Q)$ we consider the linearized equation

$$\begin{cases} u_t - \Delta u + g(v)u = -f(0) & \text{in } Q \\ u = 0 \quad \text{on } \Sigma \\ u(0) = u_0. \end{cases} \tag{3.4}$$

For each $v$, proceeding as in Section 2.3, we will find an initial data $u_0 \in C_0(\Omega)$ (defined in a unique way) such that the solution $u$ of (3.4) satisfies (3.2). Thus we will have a nonlinear map

$$K : L^1(Q) \to L^1(Q) \tag{3.5}$$

such that $Kv = u$ with $u$ solution of (3.4) satisfying (3.2). The problem will be reduced to prove the existence of a fixed point for $K$.

Let us construct the map $K$.

We decompose the solution of (3.4) as follows

$$u = U + w \tag{3.6}$$

where $U$ solves

$$\begin{cases} U_t - \Delta U + g(v)U = -f(0) & \text{in } Q \\ U = 0 \quad \text{on } \Sigma \\ U(0) = 0 \end{cases} \tag{3.7}$$

and $w$ satisfies

$$\begin{cases} w_t - \Delta w + g(v)w = 0 & \text{in } Q \\ w = 0 \quad \text{on } \Sigma \\ w(0) = u^0. \end{cases} \tag{3.8}$$

Condition (3.2) is equivalent to

$$\|w(T) - (u^1 - U(T))\|_p \leq \alpha. \tag{3.9}$$

As a consequence of Proposition 2.7 we know that given $u^1 \in C_0(\Omega)$ and $\alpha > 0$, for every $v \in L^1(Q)$ there exists $u^0 \in C_0(\Omega)$ (which of course depends on $v$) which is defined in a unique way by minimizing a suitable functional over $M(\Omega)$ such that the solution $w$ of (3.8) satisfies (3.9). Moreover, since $g \in L^\infty(\mathbb{R})$, $g(v)$ is uniformly bounded in $L^\infty(Q)$ and due to the smoothing effect of the heat equation, $U(T)$ remains in a compact set of $C_0(\Omega)$. Thus, as a consequence of Proposition 2.7, $u_0$ belongs to a compact set of $C_0(\Omega)$ and therefore $u$ remains in a compact set of $L^1(Q)$.

We have shown that $K$ maps $L^1(Q)$ into a compact set of $L^1(Q)$. If we show that $K$ is continuous, the existence of a fixed point of $K$ will be a direct consequence of Schauder's Theorem.

Let us prove that $K : L^1(Q) \to L^1(Q)$ is continuous.

Consider a sequence $\{v_n\} \subset L^1(Q)$ such that

$$v_n \to v \quad \text{in } L^1(Q) \tag{3.10}$$

and let us prove that

$$u_n = Kv_n \to u = Kv \quad \text{in } L^1(Q). \tag{3.11}$$

Since $g$ is continuous and uniformly bounded from (3.10) we deduce that

$$g(v_n) \rightharpoonup g(v) \text{ weakly-}* \text{ in } L^\infty(Q). \tag{3.12}$$

Therefore

$$U_n \to U \quad \text{in } L^1(Q). \tag{3.13}$$

Let us see that

$$w_n \to w \quad \text{in } L^1(Q). \tag{3.14}$$

In view of (3.12), the fact that $\{U_n(T)\}$ remains in a compact set of $C_0(\Omega)$ and Proposition 2.7 we deduce that the initial data $\{u_n^0\}$ satisfy

$$u_n^0 \to u^0 \quad \text{in } C_0(\Omega)$$

and then (3.14) holds. From (3.13)–(3.14) we obtain (3.11).

Schauder's fixed-point theorem ensures the existence of some $v \in L^1(Q)$ such that

$$u = Kv = v$$

but then $u = u(x,t)$ solves

$$\begin{cases} u_t - \Delta u + f(u) = 0 & \text{in } Q \\ u = 0 & \text{on } \Sigma \\ u(0) = u^0 \end{cases}$$

and by construction

$$\|u(T) - u^1\|_\infty \leq \alpha.$$

This concludes the proof of Theorem 1.1 when $f \in C^1(\mathbb{R})$.

Let us consider now a globally Lipschitz function $f$ and define the regularized sequence

$$f_n = f * \rho_n$$

where $\rho_n(s) = n\rho(ns)$ for some $\rho \in \mathcal{D}(\mathbb{R})$ with $\int_{\mathbb{R}} \rho = 1$ and $\rho \geq 0$.

It is easy to check that $f_n \in C^1(\mathbb{R})$, $f_n \to f$ uniformly on compact intervals and that every $f_n$ is globally Lipschitz with the same Lipschitz constant as $f$.

For $u^1 \in C_0(\Omega)$ and $\alpha > 0$ fixed we take, for each $n \in \mathbb{N}$, an initial data $u_n^0 \in C_0(\Omega)$ such that the solution $u_n$ of

$$\begin{cases} u_t - \Delta u + f_n(u) = 0 & \text{in } Q \\ u = 0 & \text{on } \Sigma \\ u(0) = u_n^0 \end{cases}$$

satisfies

$$\|u_n(T) - u^1\|_\infty \leq \alpha. \tag{3.15}$$

Since the nonlinearities $f_n$ are uniformly globally Lipschitz and $f_n(0)$ is a bounded sequence, in view of Proposition 2.7 we deduce that $\{u_n^0\}$ is relatively compact in $C_0(\Omega)$. By extracting subsequences (that we denote by the index $n$) we deduce that

$$u_n^0 \to u^0 \text{ in } C_0(\Omega)$$
$$u_n \to u \text{ in } L^\infty(Q)$$

where $u$ satisfies

$$\begin{cases} u_t - \Delta u + f(u) = 0 & \text{in } Q \\ u = 0 & \text{on } \Sigma \\ u(0) = u^0 & \text{in } \Omega \end{cases}$$

and by (3.15),

$$\|u(T) - u^1\|_\infty \leq \alpha.$$

This concludes the proof of Theorem 1.1.

**4. The semilinear heat equation: Theorem 1.4.** This section is devoted to the proof of Theorem 1.4. Let us fix any $u^1 \in C_0(\Omega)$ and $\alpha > 0$. Our goal is to find some $u^0 \in \mathcal{B}$ such that the solution of

$$\begin{cases} u_t - \Delta u + f(u) = 0 & \text{in } Q \\ u = 0 & \text{on } \Sigma \\ u(0) = u^0 \end{cases} \tag{4.1}$$

satisfies

$$\|u(T) - u^1\|_\infty \leq \alpha. \tag{4.2}$$

We proceed in several steps.

**Step 1.** Suppose that $f \in C^1(\mathbb{R})$ and introduce the continuous and bounded function $g$ as in (3.3).

For every $v \in L^1(Q)$ consider the linearized problem

$$\begin{cases} u_t - \Delta u + g(v)u = -f(0) & \text{in } Q \\ u = 0 & \text{on } \Sigma \\ u(0) = u^0. \end{cases} \tag{4.3}$$

Consider any $q > 1$. We claim that there exists an initial data of the form

$$u^0(x) = \left( \int_\Omega |\hat{\varphi}(x,0)|^q dx \right) |\hat{\varphi}(x,0)|^{q-2} \hat{\varphi}(x,0) \tag{4.4}$$

so that (4.2) holds for the solution of (4.3). Moreover $\hat{\varphi} = \hat{\varphi}(x,t)$ is solution of

$$\begin{cases} -\varphi_t - \Delta \varphi + g(v)\varphi = 0 & \text{in } Q \\ \varphi = 0 & \text{on } \Sigma \\ \varphi(T) = \varphi^0 \end{cases} \tag{4.5}$$

for some $\hat{\varphi}^0 \in M(\Omega)$ which is bounded in $M(\Omega)$ uniformly with respect to $v \in L^1(Q)$ and $q > 1$.

The proof of this claim is very similar to the proof of Proposition 2.7. For any $v \in L^1(Q)$ fixed we consider the functional

$$J_q : M(\Omega) \to \mathbb{R}$$

with

$$J_q(\varphi^0) = \frac{1}{2q} \left( \int_\Omega |\varphi(x,0)|^q dx \right)^2 + \alpha |\varphi^0| - \langle \varphi^0, \widetilde{u^1} \rangle$$

where $\varphi$ is solution of (4.5) and $\widetilde{u^1} = u^1 - U(T)$ where $U$ solves (3.7). Minimizing $J_q$ over $M(\Omega)$ we obtain $\hat{\varphi}^0$ such that the corresponding solution of (4.5) is such that (4.2) holds for the solution $u$ of (4.3) with initial data (4.4). The arguments of Proposition 2.7 allow us to prove that $\hat{\varphi}^0$ is uniformly bounded in $M(\Omega)$ with respect to $v \in L^1(Q)$ and $q > 1$.

**Step 2.** For $u^1 \in C_0(\Omega)$, $\alpha > 0$ and $q > 1$ fixed we may proceed as in Section 3 by applying Schauder's fixed point theorem and we deduce that there exists some $v \in L^1(Q)$ such that the solution $u$ of (4.3)–(4.4) satisfies $u = v$. Thus $u$ solves (4.1) for the initial data (4.4) and by construction satisfies also (4.2).

**Step 3.** Now, we have to pass to the limit as $q \to 1$. Since $\hat{\varphi}^0$ is uniformly bounded in $M(\Omega)$ and $f$ is globally Lipschitz we deduce that $\hat{\varphi}(x,0)$ is relatively compact in $C_0(\Omega)$. By extracting a subsequence $q_n \downarrow 1$ we deduce that

$$\begin{aligned} &\hat{\varphi}_n^0 \rightharpoonup \hat{\varphi}^0 \quad \text{in } M(\Omega) \\ &\hat{\varphi}_n \rightharpoonup \hat{\varphi} \quad \text{in } L^1(Q) \\ &\hat{\varphi}_n(x,0) \to \hat{\varphi}(x,0) \quad \text{in } C_0(\Omega) \\ &u_n \to u \quad \text{in } L^1(Q) \end{aligned}$$

where $\hat{\varphi}_n^0$, $\hat{\varphi}_n$ and $u_n$ denote respectively the minimizer of $J_{q_n}$, the corresponding solution of (4.5) with $v = u_n$ and the corresponding solution of (4.1).

It is easy to see that $u$ solves (4.1) with the initial data

$$u^0 = \left( \int_\Omega |\hat{\varphi}(x,0)| dx \right) \psi$$

for some $\psi = \psi(x) \in sgn(\hat{\varphi}(x,0))$. By construction $u$ satisfies (4.2). This concludes the proof of Theorem 1.4 when $f \in C^1(\mathbb{R})$.

**Step 4.** In the general case of a globally Lipschitz function the proof can be completed by approximation as in Section 3. □

## REFERENCES

[1] S. Agmon and L. Nirenberg, *Properties of solutions of ordinary differential equations in Banach spaces*, Comm. Pure and Applied Math., 16 (1963), pp. 121–239.

[2] S. Angenent, *The zero set of a solution of a parabolic equation*, J. reine angew. Math. 390 (1988), pp. 79–96.

[3] C. Bardos and L. Tartar, *Sur l'unicité rétrograde des équations paraboliques et quelques questions voisines*, Arch. Rational Mech. Anal. vol 50 (1973), pp. 10–25.

[4] C. Fabre, J.P. Puel and E. Zuazua, *Contrôlabilité approchée de l'équation de la chaleur semilinéaire*, C.R. Acad. Sci. Paris, t. 315, Série I (1992), pp. 807–812.

[5] C. Fabre, J.P. Puel and E. Zuazua, *Approximate controllability of the semilinear heat equation*, IMA Preprint Series, 1992.

[6] A. Friedman, *Partial Differential Equations of Parabolic Type*, Prentice-Hall Inc., 1974.

[7] O.A. Ladyzenskaja, V.A. Solonnikov and N.N. Ural'ceva, *Linear and quasilinear equations of parabolic type*, A.M.S., Rhode Island, 1968.

[8] M. Lees and M.H. Protter, *Unique continuation for parabolic differential equations and inequalities*, Duke Math. J., 28 (1961), pp. 369–382.

[9] J.L. Lions and B. Malgrange, *Sur l'unicité rétrograde des équations paraboliques*, Math. Scand. vol 8, (1960), pp. 277–286.

[10] S. Mizohata, *Le problème de Cauchy dans le passé pour quelques équations paraboliques*, Proc. Japan. Acad. 34 (1958), pp. 693–696.

# RELAXATION IN SEMILINEAR INFINITE DIMENSIONAL SYSTEMS MODELLING FLUID FLOW CONTROL PROBLEMS

H.O. FATTORINI* AND S.S. SRITHARAN†

**Abstract.** Fluid flow control problems lead to models described by semilinear abstract differential equations in Hilbert space, where relaxed controls must be incorporated to guarantee existence of optimal controls. We show that trajectories driven by relaxed controls can be uniformly approximated by trajectories driven by ordinary controls.

**Key words.** relaxed controls, optimal controls, relaxation, control of fluid flow.

**AMS(MOS) subject classifications.** 93E20, 93E25.

**1. Introduction.** Certain problems in control of fluid flow, [10], [11] can be modelled by the semilinear system

$$\partial_t y(t) + Ay(t) + B(y(t)) = N(y(t),\ u(t)),\ y(0) = \zeta \in H \tag{1.1}$$

in a Hilbert space $H$, where the operator $A$ is self adjoint and positive. For the purposes of this paper we may assume that (1.1) is the Navier–Stokes system in a domain $\Omega$ in $\mathbb{R}^2$ or $\mathbb{R}^3$ with a control term $N(y, u)$, so that $Ay = -\nu P\Delta y$ (boundary conditions included in the domain of $\Delta$) $By = P((y \cdot \nabla)y)$, $P$ the projection operator of $L^2$ into the space of divergence-free vectors that are parallel to the boundary and die down at infinity in the unbounded case and $\nu$ is the coefficient of kinematic viscosity. The control operator $N(y, u)$ satisfies assumptions specified in next section. See [9], [10] for actual flow control problems leading to this kind of model.

The system (1.1) can be equipped with **ordinary** or **relaxed** controls. Spaces $\mathcal{U}_{ad}(0, \bar{t}; \mathbf{U})$ of ordinary controls consist of functions $u(\cdot)$ taking values in a **control set U** and making the right side of (1.1) (strongly measurable and) square summable in norm for $y(\cdot)$ continuous in the norm of $D(A^{1/2})$. Most spaces of ordinary controls have a fundamental drawback: except in certain particular cases, optimal controls (solutions of control problems) do not exist. This sort of situation is familiar in calculus of variations and finite dimensional control theory, and has motivated the introduction of **Young measures** [15] [16] and their control theory counterparts, **relaxed controls** [14] (for references to other equivalent approaches to relaxed controls in finite dimensional systems see [16]). Relaxed controls

* Department of Mathematics, University of California, Los Angeles, CA 90024-1555.
Supported in part by the National Science Foundation under Grant PDQ–9801234.

† Code 574, Fluid Mechanics and Energy Research, Naval Command, Control and Ocean Surveillance Center, San Diego, California 92152–5000. Supported by the Office of Naval Research, Mathematical Sciences and the Mechanics Divisions through Grant No: N 0001493 WX2422.

for infinite dimensional systems were introduced in [1], [2] and, in a different setting in [5], [6], [7]; for flow control problems, they appear in [11]. In this version, spaces $\mathcal{V}_{bc}(0,T;\mathbf{U})$ of relaxed controls for (1.1) consist of functions $\mu(t,du)$ whose values are finitely additive probability measures in the control set $\mathbf{U}$ satisfying weak measurability conditions with respect to $t$; the relaxed equation is

$$(1.2)\quad \partial_t y(t) + Ay(t) + B(y(t)) = \int_{\mathbf{U}} N(y(t),\ u)\mu(t,\ du),\ y(0) = \zeta$$

Growth or "tightness" conditions on the measures are necessary to insure that (1.2) makes sense (see §2). Ordinary controls $u(t)$ appear in the guise of relaxed controls as $\boldsymbol{\mu}(t) = \delta_{u(t)}$, the Dirac measure concentrated at $u(t)$. The spaces $\mathcal{V}_{bc}(0,T;\mathbf{U})$ are very large; for control sets $\mathbf{U}$ of special form, spaces intermediate between $\mathcal{U}_{ad}(0,T;U)$ and $\mathcal{V}_{bc}(0,T;\mathbf{U})$ may be used.

A fundamental requirement for any notion of relaxed control is that solutions or trajectories $y(t,\boldsymbol{\mu})$ of (1.2) corresponding to a relaxed control $\mu$ should be approximable by trajectories $y(t,u)$ of (1.1); for physical reasons[1], uniform approximation in the norm of $D(A^{1/2})$,

$$(1.3)\qquad \|y(t,\boldsymbol{\mu}) - y(t,u)\|_{1/2} \le \varepsilon \qquad (0 \le t \le \bar{t})$$

is especially desirable. In fact, if this approximation property fails, one may argue that the spaces of relaxed controls are "too large", thus the relaxed problem is really unrelated to the original problem. Approximation results of the sort of (1.3) are called **relaxation theorems** and have been considered in [1], [2], [7]. Of the two approaches in [7], one extends to the present situation with some changes involving the need for a generalized version of Gronwall's inequality (Lemma 3.1), which has other applications to approximation (Corollaries 3.2 and 3.3). The main result is Theorem 4.1. Section 2 introduces the spaces of ordinary and relaxed controls with reference to [11] for proofs.

**2. Ordinary and relaxed controls. The relaxed system.** We assume that the operator $A$ in (1.1) is self adjoint and positive definite and that $0 \in \rho(A)$ (if $0 \in \sigma(A)$ we replace $A$ by $A + cI$ and $B$ by $B - cI$ for some $c > 0$). Under this last condition, the fractional powers $A^\alpha$ are bounded for $\alpha < 0$. For $\alpha \ge 0$, we write $H_\alpha = D(A^\alpha)$ and equip this space with the inner product $(y,z)_\alpha = (A^\alpha y,\ A^\alpha z)$, corresponding to the norm $\|y\|_\alpha = \|A^\alpha y\|$. For $\alpha \le 0$, $H_\alpha$ is the completion of $H$ under $\|\cdot\|_\alpha$.

The operators $B(y)$ and $N(y,u)$ satisfy Assumptions 1 and 2 below respectively. These assumptions appear in [11] as theorems for specific operators $B$ and $N$.

---

[1] For fluid flow fields in bounded domains with homogeneous boundary conditions, this norm equals the **enstrophy**.

**Hypothesis 1** $B(y)$ is defined in $H_{1/2}$ and there exists $\beta$, $0 \leq \beta < 1/2$ such that $B$ maps $H_{1/2}$ into $H_{-\beta}$. Moreover, $B : H_{1/2} \to H_{-\beta}$ is continuous, locally bounded, and locally Lipschitz continuous, i.e. for every $C > 0$ there exist constants $K_B = K_B(C)$, $L_B = L_B(C)$ such that $\|B(y)\|_{-\beta} \leq K_B$ for $y \in H_{1/2}, \|y\|_{1/2} \leq C$ and $\|B(y') - B(y)\|_{-\beta} \leq L_B\|y' - y\|_{1/2}$ for $y,\ y' \in H_{1/2},\ \|y\|_{1/2}, \|y'\|_{1/2} \leq C$. This is equivalent to requiring that $A^{-\beta}B(A^{-1/2})$, as a map from $H$ into $H$, is continuous, locally bounded, and locally Lipschitz continuous: for each $C > 0$

$$(2.1)\quad \|A^{-\beta}B(A^{-1/2}\eta)\| \leq K_B \qquad (\eta \in H, \|\eta\| \leq C)$$

$$(2.2)\quad \|A^{-\beta}B(A^{-1/2}\eta')-A^{-\beta}B(A^{-1/2}\eta)\| \leq L_B\|\eta'-\eta\| \quad (\eta, \eta' \in H, \|\eta\|, \|\eta'\| \leq C)$$

The **control set U** is an arbitrary normal topological space. The **control operator** $N(y,\ u)$ satisfies

**Hypothesis 2** $N(\cdot, \cdot)$ maps continuously $H_{1/2} \times \mathbf{U}$ into $H$, and there exists a continuous function $\kappa : \mathbf{U} \to \mathbb{R}$, $\kappa(u) \geq 1$ such that, for every $C > 0$ there exist $K_N = K_N(C)$ and $L_N = L_N(C)$ such that $\|N(y, u)\| \leq K_N\kappa(u)$ for $y \in H_{1/2},\ \|y\|_{1/2} \leq C,\ u \in \mathbf{U}$ and $\|N(y', u) - N(y, u)\| \leq L_N\|y' - y\|_{1/2}\kappa(u)$ for $y, y' \in H_{1/2}, \|y\|_{1/2}, \|y'\|_{1/2} \leq C,\ \ u \in U$. This is equivalent to the inequalities (in the norm of $H$) for the operator $N(A^{-1/2}\cdot, \cdot)$ mapping $H \times \mathbf{U}$ into $H$:

$$(2.3)\quad \|N(A^{-1/2}\eta, u)\| \leq K_N\kappa(u) \qquad (\eta \in H, \|\eta\| \leq C,\ u \in \mathbf{U})$$

$$(2.4)\quad \|N(A^{-1/2}\eta', u) - N(A^{-1/2}\eta, u)\| \leq L_N\|\eta' - \eta\|\kappa(u)$$

$$(\eta,\ \eta' \in H,\ \|\eta\|,\ \|\eta'\| \leq C, \quad u \in \mathbf{U})$$

The system (1.1) can be equipped with ordinary or relaxed controls. In the first case the space $\mathcal{U}_{ad}(0, T; \mathbf{U}, \kappa)$ of **admissible controls** is a space of **U**-valued functions defined almost everywhere and satisfying

**Control space hypothesis**: (a) $\kappa(u(\cdot)) \in L^2(0, T)$, (b) For every $y(\cdot) \in C([0, T];\ H_{1/2}), N(y(\cdot), u(\cdot))$ is strongly measurable.

By definition, **solutions** or **trajectories** of the initial value problem (1.1) in an interval $0 \leq t \leq T'$ are $H_{1/2}$-valued functions $y(\cdot)$, continuous in the norm of $H_{1/2}$ and satisfying

$$(2.5)\quad y(t) = S(t)\zeta - \int_0^t A^{\beta}S(t-r)A^{-\beta}B(y(r))dr + \int_0^t S(t-r)N(y(t), u(r))dr, \quad 0 \leq t \leq T'$$

where $S(t) = e^{-tA}$ is the semigroup generated by $A$. Instead of solving (2.5) directly, we transform it into an integral equation for $\eta(t) = A^{1/2}y(t)$,

$$\eta(t) = (\mathbf{M}\eta)(t) = A^{1/2}S(t)\zeta - \\ + \int_0^t A^{1/2+\beta}S(t-r)A^{-\beta}B(A^{-1/2}\eta(r))dr + \\ + A^{1/2}\int_0^t S(t-r)N(A^{-1/2}\eta(r), u(r))dr, \quad 0 \le t \le T'; \tag{2.6}$$

If $\eta(\cdot)$ is a continuous $H$-valued solution of (2.6) then $y(t) = A^{-1/2}\eta(t)$ is a solution of (2.5) and vice versa. Equation (2.6) is solved as follows. We recall [13] that, for any $\alpha > 0$ there exists a constant $C_\alpha$ such that

$$||A^\alpha S(t)|| \le C_\alpha t^{-\alpha}, t > 0, \tag{2.7}$$

so that the first integral on the right side of (2.6) is convergent. For the second integral, we use the fact [9] that if $g(\cdot) \in L^2(a, b; H)$ then the function

$$y(t) = \int_a^t A^{1/2}S(t-r)g(r)dr \qquad (a \le t \le b) \tag{2.8}$$

belongs to $C([a, b]; H)$ with

$$||y(\cdot)||_{C([a,b];H)} \le \frac{1}{\sqrt{2}}||g(\cdot)||_{L^2(a,b;H)}. \tag{2.9}$$

The operator $\mathbf{M}$ in (2.6) acts in $C([0, T']; H)$, with $T'$ to be determined later. Let $B(S(\cdot)A^{1/2}\zeta,\ 1)$ be the ball of center $S(\cdot)A^{1/2}\zeta \in C([0, T']; H)$ and radius 1. If $\eta(\cdot) \in B(S(\cdot)A^{1/2}\zeta, 1)$,

$$||\mathbf{M}\eta(t) - S(t)A^{1/2}\zeta|| \le \\ \le C_{1/2+\beta}K_B \int_0^t (t-r)^{-1/2-\beta}dr + \frac{K_N}{\sqrt{2}}\kappa(u(\cdot))_{L^2(0,T',H)} \tag{2.10}$$

where $C_{1/2+\beta}$ is the constant in (2.7), $K_B$ is a bound for $||A^{-\beta}B(A^{-1/2}\eta)||$ in $||\eta - S(t)A_{1/2}\zeta|| \le 1 \quad (0 \le t \le T)$ and $K_N$ is the corresponding bound for $||N(A^{-1/2}\eta)||$. Accordingly, choosing $T'$ sufficient small $\mathbf{M}$ maps $B(S(\cdot)A^{1/2}\zeta, 1)$ into $B(S(\cdot)A^{1/2}\zeta, 1)$. Using now (2.2) and (2.4) for the ball $B(S(\cdot)A^{1/2}\zeta, 1)$ and (2.9),

$$||\mathbf{M}\eta(t) - \mathbf{M}\nu(t)|| \le \tag{2.11}$$

$$\leq C_{1/2+\beta}\int_0^t (t-r)^{-1/2-\beta}\|B(A^{-1/2}\eta(r)) - B(A^{-1/2}\nu(r))\|dr +$$

$$+\frac{1}{\sqrt{2}}\left(\int_0^{T'} \|N(A^{-1/2}\eta(r)) - N(A^{-1/2}\nu(r))\|^2 dr\right)^{1/2} \leq$$

$$\leq C_{1/2+\beta}L_\beta \int_0^t (t-r)^{-1/2-\beta}\|\eta(r) - \nu(r))\|dr +$$

$$+\frac{L_N}{\sqrt{2}}\left(\int_0^{T'} \|\eta(r) - \nu(r)\|^2\kappa(u(r))^2 dr\right)^{1/2}$$

where $L_B$, $L_N$ are Lipschitz constants corresponding to the ball $B(S(\cdot)A^{1/2}\,\zeta, 1)$. It follows that $\mathbf{M}$ is a contraction map in $B(S(\cdot)A^{1/2}\zeta\ 1) \subset C([0,\ T']; H)$ for $T'$ small enough, thus a unique solution of (2.6) exists by the contraction mapping principle. We have then proved

LEMMA 2.1. *The integral equation (2.5) has a (unique) solution in some interval* $0 \leq t \leq T'$.

Local existence and uniqueness imply the existence of a maximal interval of existence $[0,\ T_m)$ for each solution. We have

LEMMA 2.2. *Let* $y(\cdot)$ *be a solution of (2.5) in the interval* $0 \leq t < T'$ *with* $\zeta \in D(A^{1/2})$. *Suppose that*

$$\|y(t)\|_{1/2} \leq C, \quad 0 \leq t < T' \ . \tag{2.12}$$

*Then, if* $[0,\ T_m)$ *is the maximal interval of existence of* $y(\cdot)$, *we have* $T_m > T'$.

See [11] for further details. We denote below by $y(t,u)$ the solution (or **trajectory**) of (1.1) (or, rather, of the integral equation (2.5)).

Optimal control problems for (1.1) are studied in [9], [10], [11]. In general, spaces of ordinary controls are too small to prove existence, and must be augmented to spaces of **relaxed controls**. We summarize the construction of one of these spaces, called $\mathcal{V}_{bc}(0,T;\mathbf{U},\kappa)$. It is based on the dual of the space $L^1(0,T;H)$ of all (equivalence classes of) strongly measurable $E$-valued functions $f(\cdot)$ such that $\|f\|_1 = \int \|f(t)\|dt < \infty$ endowed with $\|\cdot\|_1$, where $E$ is an arbitrary Banach space. If $E^*$ is the dual of $E$, an $E^*$-valued function $g(\cdot)$ is $E$-**weakly measurable** if and only if $\langle g(\cdot), y\rangle$ is measurable for each $y \in E$. The space of all such $g(\cdot)$ with

$$|\langle g(t), y\rangle| \leq C\|y\| \qquad \text{a.e. in } \ 0 \leq t \leq T \tag{2.13}$$

("a.e" depending on $y$) is named $L^\infty_w(0,T;E^*)$ and equipped with the norm $\|g(\cdot)\|$ = infimum of all $C$ such that (2.13) holds. The equivalence relation

$L^\infty_w(0,T;E^*)$ is $g(\cdot) \approx h(\cdot)$ if $\|g(\cdot)-h(\cdot)\| = 0$, that is, if $\langle g(t), y\rangle = \langle h(t), y\rangle$ a.e. in $0 \le t \le T$, "a.e." depending on $y$. A function $g(\cdot) \in L^\infty_w(0,T;E^*)$ produces an element in $L^1(0,T;E)^*$ through

$$\langle g(\cdot), f(\cdot)\rangle = \int_0^T \langle g(\sigma), f(\sigma)\rangle d\sigma \quad (f(\cdot) \in L^1(0,T;E)) \tag{2.14}$$

This functional only depends on the equivalence class of $g(\cdot)$ and its norm coincides with the $L^\infty_w(0,T;E^*)$-norm of $g(\cdot)$. Conversely, every element of $L^1(0,T;E)^*$ admits the representation (2.14) with $g(\cdot) \in L^\infty_w(0,T;E^*)$, thus $L^1(0,T;E)^* = L^\infty_w(0,T;E^*)$ algebraically and metrically. There exists a linear **lifting operator** $\mathcal{S} : L^\infty_w(0,T;E^*) \to L^\infty_w(0,T;E^*)$ such that $\mathcal{S}g$ belongs to the equivalence class of $g$, $t \to \|(\mathcal{S}g)(t)\|$ is measurable in $0 \le t \le T$ and $\sup_{0\le t\le T} \|(\mathcal{S}g)(t)\| = \|g\|$. For further details see [6], [7].

These results will be applied to the space $E = C_b(\mathbf{U})$ of all bounded continuous functions in $\mathbf{U}$ equipped with the supremum norm, whose dual $C_b(\mathbf{U})^*$ is isometrically isomorphic to the space $\Sigma_{rba}(\mathbf{U}, \Phi_c)$ of all bounded, finitely additive measures defined in the ring $\Phi_c$ generated by the closed sets of $\mathbf{U} : \Sigma_{rba}(\mathbf{U}, \Phi)$ is equipped with the total variation norm [3 p. 258]. The duality pairing is

$$\langle \boldsymbol{\mu}, f\rangle = \int_{\mathbf{U}} f(u)\mu(du); \tag{2.15}$$

The space $\mathcal{V}_{bc}(0,T;\mathbf{U},\kappa)$ of relaxed controls consists of all $\boldsymbol{\mu}(\cdot) \in L^1(0,T;C_b(\mathbf{U}))^* = L^\infty_w(0,T;\Sigma_{rba}(\mathbf{U},\Phi))$ that satisfy: (i)

$$\|\boldsymbol{\mu}(\cdot)\| \le 1 \tag{2.16}$$

($\|\cdot\|$ the norm in $L^\infty_w(0,T;\Sigma_{rba}(\mathbf{U},\Phi))$ (ii) if $f(\cdot) \in L^1(0,T;B(\mathbf{U}))$ is such that $f(t,u) \ge 0$ for $u \in \mathbf{U}$ a.e. in $0 \le t \le T$ then

$$\int_0^T \int_{\mathbf{U}} f(t,u)\mu(t,du)dt \ge 0\ . \tag{2.17}$$

(iii) if $\chi(\cdot)$ is the characteristic function of a measurable set $e \subseteq [0,T]$ and $\mathbf{1}$ is the function identically 1 in $\mathbf{U}$, then

$$\int_0^T \int_{\mathbf{U}} (\chi(t) \otimes \mathbf{1})\mu(t,du)dt = \mathrm{meas}(e)\ . \tag{2.18}$$

(iv) if $\kappa(u)$ is the function in Hypothesis 2,

$$\int_{\mathbf{U}} \kappa(u)^2 \mu(\cdot, du) \in L^1(0,T). \tag{2.19}$$

The result below shows that every equivalence class of relaxed controls contains "nice" elements.

LEMMA 2.3. *Let $\boldsymbol{\nu}(\cdot)$ be an element of $\mathcal{V}_{bc}(0,T;\mathbf{U},\kappa)$, that is, an element of $L^\infty_w(0,T;\Sigma_{rba}(\mathbf{U},\Phi_c))$ satisfying (i), (ii), (iii), (iv). Then there exists an element $\boldsymbol{\mu}(\cdot)$ in the equivalence class of $\boldsymbol{\nu}$ in $\mathcal{V}_{ba}(0,T;\mathbf{U},\kappa)$ such that*

$$\boldsymbol{\mu}(t) \geq 0, \qquad \mu(t,\mathbf{U}) = 1 \qquad (0 \leq t \leq T)$$

For the proof, see [11].

The equation (2.1) is equipped with relaxed controls as follows. Let $\Sigma_{rba}(\mathbf{U},\Phi_c,\ \kappa)$ be the subspace of $\Sigma_{rba}(\mathbf{U},\Phi_c)$ whose elements satisfy

$$\|\boldsymbol{\mu}\|_\kappa^2 = \int_{\mathbf{U}} \kappa(u)^2 \boldsymbol{\mu}(du) < \infty. \tag{2.20}$$

The **relaxed control set RU** consists of all $\boldsymbol{\mu} \in \Sigma_{rba}(\mathbf{U},\Phi_c,\kappa)$ that satisfy $\mu \geq 0$, $\mu(\mathbf{U}) = 1$. Equipped with $\|\cdot\|_\kappa$ $\Sigma_{rba}(\mathbf{U},\Phi_c,\kappa)$ is a Banach space, and **RU** inherits the topology of $\Sigma_{rba}(\mathbf{U},\Phi_c,\kappa)$. In view of Lemma 2.3 our relaxed controls take values in **RU** (modulo equivalence).

We define a **relaxed control operator N** : $H_{1/2}\times\Sigma_{rba}(\mathbf{U},\Phi_c,\kappa) \to H$ as follows: $\mathbf{N}(y)\boldsymbol{\mu}$ is the unique element of $H$ satisfying

$$(z,\mathbf{N}(y)\boldsymbol{\mu}) = \int_{\mathbf{U}} (z, N(y,u))\mu(du)$$

for all $z \in H$. In view of (2.3) and (2.20) the integral is bounded by

$$\int_{\mathbf{U}} |(z,N(y,u))|\mu(du) \leq K_N\|z\| \int_{\mathbf{U}} \kappa(u)\mu(du) \leq$$

$$\leq K_N\|z\| \left(\int_{\mathbf{U}} \mu(du)\right)^{1/2} \left(\int_{\mathbf{U}} \kappa(u)^2\mu(du)\right)^{1/2}$$

where $K_N$ is the constant in (2.3) so that $\mathbf{N}(y)\boldsymbol{\mu}$ satisfies

$$\|\mathbf{N}(y)\boldsymbol{\mu}\| \leq K_N\|\boldsymbol{\mu}\|_\kappa \quad (0 \leq t \leq T,\ \|y\|_{1/2} \leq C) \tag{2.21}$$

LEMMA 2.4. *The operator* **N** *satisfies Hypothesis 2 in* **RU**, *with* $\kappa(\boldsymbol{\mu}) = \|\boldsymbol{\mu}\|_\kappa$.

In view of (2.21), it is only necessary to prove that $\mathbf{N}(y)\boldsymbol{\mu}$ is continuous in $H_{1/2}\times$ **RU** and that

$$\|\mathbf{N}(y')\boldsymbol{\mu} - \mathbf{N}(y)\boldsymbol{\mu}\| \leq L_N\|y'-y\|_{1/2}\|\boldsymbol{\mu}\|_\kappa \tag{2.22}$$
$$(y,y' \in H_{1/2},\ \|y\|_{1/2},\ \|y'\|_{1/2} \leq C, \quad \boldsymbol{\mu} \in \mathbf{RU})$$

For a proof see [11, Lemma 4]. Once Hypothesis 2 has been established for $\mathbf{N}$, it only remains to show that the Control Space Hypothesis is satisfied for the control space $\mathcal{V}_{bc}(0,T;\mathbf{U},\kappa)$. Part (a) with $\kappa(\boldsymbol{\mu}) = \|\boldsymbol{\mu}\|_\kappa$ is (2.19). Part (b) is

LEMMA 2.5. *Let* $y(\cdot) \in C(0,T;H_{1/2}), \boldsymbol{\mu}(\cdot) \in \mathcal{V}_{bc}(0,T;\mathbf{U},\kappa)$. *Then* $\mathbf{N}(y(\cdot))\boldsymbol{\mu}(\cdot)$ *is strongly measurable.*

*Proof*. See Lemma 5 in [11].

The **relaxed** system corresponding to the control space $\mathcal{V}_{bc}(0,T;\mathbf{U},\kappa)$ is

$$\partial_t y(t) + Ay(t) + B(y(t)) = \mathbf{N}(y(t))\boldsymbol{\mu}(t). \tag{2.23}$$

Since this system satisfies Assumptions 1 and 2 and, equipped with $\mathcal{V}_{bc}$ $(0,T;\mathbf{U},\kappa)$ the Control Space Hypothesis, all results for (1.1) apply automatically.

Ordinary controls (that is, controls in $\mathcal{U}_{ad}(0,T;\mathbf{U},\kappa)$) can be replicated by relaxed controls. If $u(\cdot) \in \mathcal{U}_{ad}(0,T;\mathbf{U})$, we define a relaxed control by

$$\boldsymbol{\mu}(t) = \delta_{u(t)}.$$

It is obvious that $\boldsymbol{\mu}(\cdot)$ satisfies (i), (ii), (iii): as to (iv),

$$\|\boldsymbol{\mu}(t)\|_\kappa^2 = \int_{\mathbf{U}} \kappa(u)^2 \mu(t,du) = \kappa(u(t))^2$$

which belongs to $L^1(0,T)$ by the Control Space Hypothesis.

We note an important special case of the theory. Assume $\mathbf{U}$ is a **compact metric** space. Then, by Alexandrov's theorem [3 p. 138] each measure $\boldsymbol{\mu} \in \Sigma_{rba}(\mathbf{U},\Phi_c)$ is actually $\sigma$-additive and admits a unique, regular countably additive extension to the $\sigma$-field $\Phi_b$ generated by the closed sets. In other words, each $\mu$ can be extended to a bounded Borel measure in the Borel field $\Phi_b$. Thus, in this case, the instantaneous values of our relaxed controls are bona fide probability measures.

**3. A generalized Gronwall lemma.** An inequality of the form

$$\phi(t) \le \eta(t) + \int_a^t \mathcal{K}(t,s)\phi(s)ds \tag{3.1}$$

with arbitrary $\mathcal{K}(t,s)$ does not necessary imply

$$\phi(t) \le \eta(t) + \int_a^t \mathcal{L}(t,s)\eta(s)ds \tag{3.2}$$

for some other kernel $\mathcal{L}(t,s)$ independent of $\phi$. A simple example is $\mathcal{K}(t,s) = t^{-1}\beta(t^{-1}s)$ in $0 \le s,\ t \le 1$ where $\beta(\cdot)$ is a nonnegative test function with support in $0 \leq s \leq 1$ and integral 1. In fact, (3.1) will then hold for $\phi(t) \equiv$ arbitrary constant and $\eta(t) \equiv 0$.

The classical Gronwall Lemma corresponds to $\mathcal{K} \equiv$ constant or, in its generalized version, to $\mathcal{K}(t,s) = \mathcal{K}(s)$, where $\mathcal{K}(\cdot) \in L^1$. The case $\mathcal{K}(t,s) = (t-s)^\alpha (0 < \alpha < 1)$ is in [12]. We prove below a result "(3.1) implies (3.2)" that includes these two cases and others. This generalized Gronwall lemma is then applied to the relaxed equation in Corollaries 3.2 and 3.3. The result (which is the basis of our relaxation theorem in §4) implies also global uniqueness for (2.23).

LEMMA 3.1. *Let $\mathcal{K}(t,s) \geq 0$ be measurable in $a \leq s \leq t \leq b$. Assume that there exists $\rho < 1$ such that the interval $[a,b]$ can be partitioned in a finite number of intervals $[c,d]$ with*

$$\sup_{c \leq t \leq d} \int_c^t \mathcal{K}(t,s)ds \leq \rho\ . \tag{3.3}$$

*Then there exists $\mathcal{L}(t,s) \geq 0$ measurable in $a \leq s \leq t \leq b$ depending only on $\mathcal{K}$ with*

$$\sup_{a \leq t \leq b} \int_a^t \mathcal{L}(t,s)ds < \infty \tag{3.4}$$

*and such that, if $\phi(\cdot)$, $\eta(\cdot) \in L^\infty(0,T)$ are nonnegative and satisfy (3.1) then (3.2) holds as well.*

*Proof*. We assume first that (3.3) holds in the whole interval $a \leq t \leq b$;

$$\sup_{a \leq t \leq b} \int_a^b \mathcal{K}(t,s)ds \leq \rho < 1.$$

In this case, the integral inequation (3.1) can be solved by iteration: setting $\mathcal{K}_0(t,\sigma) = \mathcal{K}(t,\sigma), \ldots$

$$\mathcal{K}_n(t,s) = \int_s^t \mathcal{K}(t,\sigma)\mathcal{K}_{n-1}(\sigma,s)d\sigma$$

we obtain by induction

$$\int_a^t \mathcal{K}_n(t,\sigma)d\sigma \leq \rho^n \qquad (a \leq t \leq b)$$

hence

$$\mathcal{L}(t,s) = \sum_{n=0}^{\infty} \mathcal{K}_n(t,s)$$

is convergent in $L^1(a,b)$ for each $t$ and $\|\mathcal{L}(t,\cdot)\|_{L^1(a,b)} \le 1/(1-\rho)$, so that (3.2) follows. For the general result we limit ourselves to the case where the partition consists of two intervals $[a,c],[c,b]$, from which the pattern of the multi-interval case emerges clearly.

We first consider (3.1) in the interval $a \le t \le c$, where Lemma 3.1 stands proved. If $\mathcal{L}_1(t,s)$ is the kernel in (3.2) corresponding to the interval $[a,c]$ we have

$$\phi(t) \le \eta(t) + \int_a^t \mathcal{L}_1(t,s)\eta(s)ds \qquad (a \le t \le c)$$

We rewrite (3.1) for $t \ge c$ in the form

$$\phi(t) \le \eta(t) + \int_a^c \mathcal{K}(t,s)\phi(s)ds + \int_c^t \mathcal{K}(t,s)\phi(s)ds = \tag{3.5}$$

$$= \eta_0(t) + \int_c^t \mathcal{K}(t,s)\phi(s)ds \qquad (c \le t \le b).$$

Now,

$$\eta_0(t) = \eta(t) + \int_a^c \mathcal{K}(t,s)\phi(s)ds \le$$

$$\le \eta(t) + \int_a^c \mathcal{K}(t,s)\left(\eta(s) + \int_a^s \mathcal{L}_1(s,\sigma)\eta(\sigma)d\sigma\right)ds =$$

$$= \eta(t) + \int_a^c \mathcal{M}_1(t,s)\eta(s)ds \qquad (c \le t \le b),$$

where

$$\mathcal{M}_1(t,s) = \mathcal{K}(t,s) + \int_s^c \mathcal{K}(t,\sigma)\mathcal{L}_1(\sigma,s)d\sigma$$

and we check easily using Fubini's theorem that $\mathcal{M}_1(t,s)$ satisfies (3.4). Let next $\mathcal{L}_2(t,s)$ be the kernel in (3.2) corresponding to $\mathcal{K}$ in the interval $c \le t \le b$. Then it follows from (3.5) that

$$\phi(t) \le \eta_0(t) + \int_c^t \mathcal{L}_2(t,s)\eta_0(s)ds =$$

$$= \eta(t) + \int_a^c \mathcal{M}_1(t,s)\eta(s)ds +$$

$$+ \int_c^t \mathcal{L}_2(t,s)\left(\eta(s) + \int_a^c \mathcal{M}_1(s,\sigma)\eta(\sigma)d\sigma\right)ds =$$

$$= \eta(t) + \int_a^t \mathcal{M}_2(t,s)\eta(s)ds \qquad (c \le t \le b)$$

with

$$\mathcal{M}_2(t,s) = \mathcal{L}_2(t,s) + \int_c^t \mathcal{L}_2(t,\sigma)\mathcal{M}_1(\sigma,s)d\sigma$$

This ends the proof.

COROLLARY 3.2. *Let $\boldsymbol{\mu}(\cdot) \in \mathcal{V}_{bc}(0,\bar{t};\mathbf{U})$. Assume that $y(t,\mu)$ exists in $0 \le t \le \bar{t}$, and let $\{\boldsymbol{\mu}_\rho\} \subseteq \mathcal{V}_{bc}(0,\bar{t};\mathbf{U})$ be a generalized sequence with*

$$\left\| \int_0^t A^{1/2}S(t-\sigma)\{\mathbf{N}(y(\sigma,\mu))(\boldsymbol{\mu}_\rho(\sigma) - \boldsymbol{\mu}(\sigma))\}d\sigma \right\| \to 0 \tag{3.6}$$

*uniformly in $0 \le t \le \bar{t}$. Then there exists $\rho_0$ and a constant $C$ such that, if $\rho \ge \rho_0$ then $y(t,\boldsymbol{\mu}_\rho)$ exists in $0 \le t \le \bar{t}$ and*

$$\|y(t,\boldsymbol{\mu}_\rho) - y(t,\boldsymbol{\mu})\|_{1/2} \le \tag{3.7}$$

$$\le C \max_{0 \le t \le \bar{t}} \left\| \int_0^t A^{1/2}S(t-\sigma)\{\mathbf{N}(y(\sigma,\boldsymbol{\mu}))(\boldsymbol{\mu}_\rho(\sigma) - \boldsymbol{\mu}(\sigma))\}d\sigma \right\|$$

*Proof*. Let $[0,t_\rho]$ be the largest interval such that $y(t,\boldsymbol{\mu}_\rho)$ exists and satisfies $\|y(t,\boldsymbol{\mu}_\rho) - y(t,\boldsymbol{\mu})\|_{1/2} \le 1$. In this interval we have

$$A^{1/2}y(t,\boldsymbol{\mu}_\rho) - A^{1/2}y(t,\boldsymbol{\mu}) =$$

$$= -\int_0^t A^{1/2+\beta}S(t-\sigma)A^{-\beta}\{B(y(\sigma,\boldsymbol{\mu}_\rho)) - B(y(\sigma,\boldsymbol{\mu}))\}d\sigma +$$

$$+ \int_0^t A^{1/2}S(t-\sigma)\{\mathbf{N}(y(\sigma,\boldsymbol{\mu}_\rho))\boldsymbol{\mu}_\rho(\sigma) - \mathbf{N}(y(\sigma,\boldsymbol{\mu}))\boldsymbol{\mu}_\rho(\sigma)\}d\sigma +$$

$$+ \int_0^t A^{1/2}S(t-\sigma)\{\mathbf{N}(y(\sigma,\boldsymbol{\mu}))(\boldsymbol{\mu}_\rho(\sigma) - \boldsymbol{\mu}(\sigma))\}d\sigma$$

Estimating,

$$\|A^{1/2}y(t,\boldsymbol{\mu}_\rho)-A^{1/2}y(t,\boldsymbol{\mu})\|\le \tag{3.8}$$

$$C_{1/2+\beta}L_B\int_0^t(t-\sigma)^{-1/2-\beta}\|A^{1/2}y(\sigma,\boldsymbol{\mu}_\rho)-A^{1/2}y(\sigma,\boldsymbol{\mu})\|d\sigma$$

$$+\frac{L_N}{\sqrt{2}}\left(\int_0^t\|\boldsymbol{\mu}_\rho(\sigma)\|_\kappa^2\|A^{1/2}y(\sigma,\boldsymbol{\mu}_\rho)-A^{1/2}y(\sigma,\boldsymbol{\mu})\|^2d\sigma\right)^{1/2}$$

$$+\left\|\int_0^t A^{1/2}S(t-\sigma)\{\mathbf{N}(y(\sigma,\boldsymbol{\mu}))(\boldsymbol{\mu}_\rho(\sigma)-\boldsymbol{\mu}(\sigma))\}d\sigma\right\|$$

where $L_B=L_B(C), L_N=L_N(C)$ with $C=\max_{0\le t\le\bar{t}}\|A^{1/2}y(t,\boldsymbol{\mu})\|+1$. Since $(a+b+c)^2\le 3(a^2+b^2+c^2)$, we have

$$\|A^{1/2}y(t,\boldsymbol{\mu}_\rho)-A^{1/2}y(t,\boldsymbol{\mu})\|^2\le \tag{3.9}$$

$$\le 3(C_{1/2+\beta}L_B)^2\left(\int_0^t(t-\sigma)^{-1/2-\beta}\|A^{1/2}y(\sigma,\boldsymbol{\mu}_\rho)-A^{1/2}y(\sigma,\boldsymbol{\mu})\|d\sigma\right)^2$$

$$+\frac{3L_N^2}{2}\int_0^t\|\boldsymbol{\mu}_\rho(\sigma)\|_\kappa^2\|A^{1/2}y(\sigma,\boldsymbol{\mu}_\rho)-A^{1/2}y(\sigma,\boldsymbol{\mu})\|^2d\sigma+$$

$$+3\left\|\int_0^t A^{1/2}S(t-\sigma)\{\mathbf{N}(y(\sigma,\boldsymbol{\mu}))(\boldsymbol{\mu}_\rho(\sigma)-\boldsymbol{\mu}(\sigma))\}d\sigma\right\|^2\le$$

$$\le 3(C_{1/2+\beta}L_B)^2\bar{t}^{1/2-\beta}\int_0^t(t-\sigma)^{-1/2-\beta}\|A^{1/2}y(\sigma,\boldsymbol{\mu}_\rho)-A^{1/2}y(\sigma,\boldsymbol{\mu})\|^2d\sigma$$

$$+\frac{3L_N^2}{2}\int_0^t\|\boldsymbol{\mu}_\rho(\sigma)\|_\kappa^2\|A^{1/2}y(\sigma,\boldsymbol{\mu}_\rho)-A^{1/2}y(\sigma,\boldsymbol{\mu})\|^2d\sigma$$

$$+3\left\|\int_0^t A^{1/2}S(t-\sigma)\{\mathbf{N}(y(\sigma,\boldsymbol{\mu}))(\boldsymbol{\mu}_\rho(\sigma)-\boldsymbol{\mu}(\sigma))\}d\sigma\right\|^2$$

applying in the first integral the Schwarz inequality to the factors $(t-\sigma)^{-1/4-\beta/2}$ and $(t-\sigma)^{-1/4-\beta/2}\|A^{1/2}y(\sigma,\boldsymbol{\mu}_\rho)-A^{1/2}y(\sigma,\boldsymbol{\mu})\|$. Calling $\eta_\rho(t)$ the last integral, $\phi_\rho(t)=\|A^{1/2}y(t,\boldsymbol{\mu}_\rho)-A^{1/2}y(\sigma,\boldsymbol{\mu})\|^2, \mathcal{K}_\rho(t,s)=3$ $(C_{1/2+\beta}L_B)^2\bar{t}^{1/2-\beta}(t-s)^{-1/2-\beta}+(3L_N^2/2)\|\boldsymbol{\mu}_\rho(s)\|_\kappa^2$, (3.9) can be rewritten as

$$\phi_\rho(t) \le \eta_\rho(t) + \int_0^t \mathcal{K}_\rho(t,s)\phi_\rho(s)ds,$$

where $\mathcal{K}_\rho(t,s)$ is easily seen to satisfy the assumptions of Lemma 3.1. Accordingly, if $\mathcal{L}_\rho(t,s)$ is the kernel for $\mathcal{K}_\rho(t,s)$ in $0 \le t \le \bar{t}$,

$$\phi_\rho(t) \le \eta_\rho(t) + \int_0^t \mathcal{L}_\rho(t,s)\eta_\rho(s)ds.$$

Inequality (3.7) then follows in the interval $[0, t_\rho]$. In view of condition (3.6), there exists $\rho_0$ such that, if $\rho \ge \rho_0$ then $\|A^{1/2}y(t_\rho, \boldsymbol{\mu}_\rho) - A^{1/2}y(t_\rho, \boldsymbol{\mu})\| < 1$, which means $t_\rho = \bar{t}$. This ends the proof.

COROLLARY 3.3. *Let $\boldsymbol{\mu}(\cdot) \in \mathcal{V}_{bc}(0,\bar{t};\mathbf{U},\mathcal{K})$. Assume that $y(t,\boldsymbol{\mu})$ exists in $0 \le t \le \bar{t}$, and let $\{\boldsymbol{\mu}_\rho\} \subseteq \mathcal{V}_{bc}(0,\bar{t};\mathbf{U},\mathcal{K})$ be a generalized sequence with*

$$\int_0^{\bar{t}} \|\boldsymbol{\mu}_\rho(t) - \boldsymbol{\mu}(t)\|_\kappa^2 dt \to 0 \tag{3.10}$$

*Then there exists $\rho_0$ and a constant $C$ such that, if $\rho \ge \rho_0$ then $y(t, \boldsymbol{\mu}_\rho)$ exists in $0 \le t \le \bar{t}$ and*

$$\|y(t,\boldsymbol{\mu}_\rho) - y(t,\boldsymbol{\mu})\|_{1/2} \le C \left( \int_0^{\bar{t}} \|\boldsymbol{\mu}_\rho(\sigma) - \boldsymbol{\mu}(\sigma)\|_\kappa^2 d\sigma \right)^{1/2} \tag{3.11}$$

*Proof*. Use (2.21), (2.9) and Corollary 3.2.

**4. Approximation of relaxed trajectories.** In this section $E$ is a reflexive separable Banach space, $\mathbf{U}$ a normal topological space, $\kappa : \mathbf{U} \to \mathbb{R}$ a continuous function with $\kappa(u) \ge 1$ ($u \in \mathbf{U}$). We denote by $\Sigma^+_{rba}(\mathbf{U}, \Phi_c)$ (resp. $\Pi_{rba}(\mathbf{U}, \Phi_c)$) the subspace of $\Sigma_{rba}(\mathbf{U}, \Phi_c) = C_b(\mathbf{U})^*$ consisting of all positive measures (resp. all probability measures). We denote by $\mathbf{D} \subseteq \mathbf{RU} \subseteq \Pi_{rba}(\mathbf{U}, \Phi_c)$ the set of all Dirac measures $\{\delta_u,\ u \in \mathbf{U}\}$. The space $\mathcal{V}_{bc}(0,T;\mathbf{U})$ consists of all elements of $L^\infty_w(0,T;\Sigma_{rba}(\mathbf{U},\Phi_c)) = L^1(0,T;C_b(\mathbf{U}))^*$ such that (i), (ii) and (iii) in §3 is satisfied; equivalently (Lemma 2.3) $\mathcal{V}_{bc}(0,T;\mathbf{U})$ consists of all $\boldsymbol{\mu}(\cdot) \in L^\infty_w(0,T;\Sigma_{rba}(\mathbf{U},\Phi_c))$ whose equivalence class contains an element whose values are probability measures for all $t$. The space $\mathcal{V}^0_{bc}(0,T;\mathbf{U})$ consists of all **countably valued** elements of $\mathcal{V}_{bc}(0,T;\mathbf{U})$ (that is, all relaxed controls $\boldsymbol{\mu}(\cdot)$ constant in each of the sets of a finite or countable partition of $[0,T]$ in measurable sets). Finally, we denote by $\mathcal{V}_{bc}^{\infty}(0,T;\mathbf{U},\kappa)$ the subspace of $\mathcal{V}_{bc}(0,T;\mathbf{U},\kappa)$ consisting of all $\boldsymbol{\mu}(\cdot)$ with

$$\| \, \|\boldsymbol{\mu}(\cdot)\|_\kappa \|_\infty = \text{ess. sup. } \|\boldsymbol{\mu}(t)\|_\kappa < \infty \tag{4.1}$$

(recall that the condition for membership of $\boldsymbol{\mu}(\cdot) \in \mathcal{V}_{bc}(0,T;\mathbf{U})$ in $\mathcal{V}_{bc}(0,T;\mathbf{U},\kappa)$ is $\|\boldsymbol{\mu}(\cdot)\|_\kappa \in L^2(0,T)$) and the space $\mathcal{V}_{bc}^{\infty,0}(0,T;\mathbf{U},\kappa)$ consists of all countably valued elements of $\mathcal{V}_{bc}^{\infty}(0,T;\mathbf{U},\kappa)$.

Corresponding definitions are made for the admissible (ordinary) control space: $\mathcal{U}_{ad}^{\infty}(0,T,\mathbf{U};\kappa)$ is the subspace of $\mathcal{U}_{ad}(0,T,\mathbf{U};\kappa))$ whose elements satisfy (4.1), which for ordinary controls takes the form

$$\|\kappa(u(\cdot))\|_\infty = \text{ess. sup. } \kappa(u(t)) < \infty \tag{4.2}$$

and $\mathcal{U}_{ad}^{\infty,0}(0,T,\mathbf{U};\kappa)$ is the subspace of $\mathcal{U}_{ad}^{\infty}(0,T,\mathbf{U};\kappa)$ whose elements are countably valued. Obviously, we have $\mathcal{U}_{ad}^{\infty,0}(0,T,\mathbf{U};\kappa) \subseteq \mathcal{V}_{bc}^{\infty,0}(0,T,\mathbf{U};\kappa) \subseteq \mathcal{V}_{bc}^{0}(0,T,\mathbf{U};\kappa)$. It is also plain that (4.2) implies (with excess) part (a) of the Control Space Hypothesis for $\mathcal{U}_{ad}^{\infty,0}(0,T,\mathbf{U};\kappa)$; part (b) is automatic. We shall then assume that (1.1) is equipped with the admissible control space $\mathcal{U}_{ad}(0,T,\mathbf{U};\kappa) = \mathcal{U}_{ad}^{\infty,0}(0,T,\mathbf{U};\kappa)$.

THEOREM 4.1. *Let $\boldsymbol{\mu}(\cdot) \in \mathcal{V}_{bc}(0,\bar{t};\mathbf{U},\kappa)$ be such that the trajectory $y(t,\boldsymbol{\mu})$ exists in $0 \le t \le \bar{t}$, and let $\varepsilon > 0$. Then there exists $u(\cdot) \in \mathcal{U}_{ad}^{\infty,0}(0,T,\mathbf{U};\kappa)$ such that the trajectory $y(t,u)$ exists in $0 \le t \le \bar{t}$ and*

$$\|y(\cdot,u) - y(\cdot,\boldsymbol{\mu})\|_{1/2} \le \varepsilon \qquad (0 \le t \le \bar{t}) \tag{4.3}$$

The proof of Theorem 4.1 is the consequence of several auxiliary results. We do **not** assume that the operator $A^{-1}$ is compact, thus the result applies to fluid flow in unbounded domains.

Note first that we may assume the control $\boldsymbol{\mu}(\cdot)$ in Theorem 4.1 belongs to $\mathcal{V}_{bc}^{\infty}(0,T,\mathbf{U};\kappa)$. To see this, take $\boldsymbol{\mu}(\cdot) \in \mathcal{V}_{bc}^{\infty}(0,T,\mathbf{U};\kappa)$, and define for $N$ large enough, the $N$-**truncation** $\boldsymbol{\mu}_N$ of $\boldsymbol{\mu}(\cdot)$ by $\boldsymbol{\mu}_N(t) = \boldsymbol{\mu}(t)$ if $\|\boldsymbol{\mu}(t)\|_\kappa \le N$, $\boldsymbol{\mu}_N(t) = \boldsymbol{\mu}_0$ otherwise, where $\boldsymbol{\mu}_0$ is a fixed element of $\mathbf{RU}$ with $\|\boldsymbol{\mu}_0\|_\kappa \le N$ (say, the Dirac delta at some fixed $u_0 \in \mathbf{U}$). Obviously, if $N$ is restricted to (sufficiently large) integers, the sequence $\{\boldsymbol{\mu}_N\}$ satisfies (3.10) and, by Corollary 3.3 the trajectory $y(t,\boldsymbol{\mu}_N)$ exists in $0 \le t \le \bar{t}$ for $N$ large enough and approximates uniformly $y(t,\boldsymbol{\mu})$ in the norm of $C([0,\bar{t}];H_{1/2})$.

We denote below by $C_b(\mathbf{U},E)$ the space of all bounded and continuous functions $\psi : \mathbf{U} \to E$, endowed with the supremum norm. Likewise, $C_{b,\kappa}(\mathbf{U},E)$ is the space of all continuous $\phi : \mathbf{U} \to E$ obeying the bound $\|\phi(u)\| \le C\kappa(u)^2$, endowed with the supremum norm of $\kappa^{-2}\phi$. Given $\psi \in C_b(\mathbf{U},E)$ define $\boldsymbol{\psi} : \Sigma_{rba}(\mathbf{U},\Phi_c) \to E$ by: $\boldsymbol{\psi\mu}$ is the only element of $E$ satisfying

$$\langle y^*, \boldsymbol{\psi\mu}\rangle = \int_{\mathbf{U}} \langle y^*, \psi(u)\rangle \mu(du) \tag{4.4}$$

for $y^* \in E^*$. We have

$$\int_{\mathbf{U}} |\langle y^*, \psi(u)\rangle| |\mu(du) \le \|y^*\| \int_{\mathbf{U}} \|\psi(u)\| |\mu|(du),$$

so that $\boldsymbol{\psi\mu}$ exists and $\|\boldsymbol{\psi\mu}\| \le \|\psi\| \, \|\boldsymbol{\mu}\|(\|\mu\|$ the norm in $\Sigma_{rba}(\mathbf{U}, \Phi_c))$. Likewise, if $\phi \in C_{b,\kappa}(\mathbf{U}, E)$ we may define $\boldsymbol{\phi} : \mathbf{RU} \to E$ by: $\boldsymbol{\phi\mu}$ is the only element of $E$ satisfying (4.4). We show in the same way that $\boldsymbol{\phi\mu}$ is well defined and $\|\boldsymbol{\phi\mu}\| \le \|\phi\| \, \|\boldsymbol{\mu}\|_\kappa^2$; the corresponding inequality is

$$\int_{\mathbf{U}} |\langle y^*, \phi(u)\rangle| |\mu(du) \le \|y^*\| \int_{\mathbf{U}} \kappa(u)^{-2} \|\phi(u)\| \kappa(u)^2 \mu(du). \tag{4.5}$$

Lemma 4.2. *Let $\psi : [0,T] \times \mathbf{U} \to E$ be such that (a) $\psi(t,\cdot) \in C_b(\mathbf{U}; E)$ for every $t$ fixed, (b) The function $t \to \psi(t,\cdot)$ belongs to $L^1(0,T; C_b(\mathbf{U}; E))$. Let $\boldsymbol{\mu}(\cdot) \in \mathcal{V}_{bc}(0,T;\mathbf{U}), \varepsilon > 0$. Then there exists $\boldsymbol{\nu}(\cdot) \in \mathcal{V}_{bc}^0(0,T;\mathbf{U})$ taking values in conv $(\mathbf{D})$ such that*

$$\|\boldsymbol{\psi}(t)\boldsymbol{\nu}(t) - \boldsymbol{\psi}(t)\boldsymbol{\mu}(t)\| \le \varepsilon \quad a.e. \text{ in } \ 0 \le t \le T. \tag{4.6}$$

For a proof, see [7, Lemma 4.5, and Lemma 4.2].

Corollary 4.3. *Let $\phi : [0,T] \times \mathbf{U} \to E$ be such that (a) $\phi(t,\cdot) \in C_{b,\kappa}(\mathbf{U}; E)$ for every $t$ fixed (b) the function $t \to \kappa(u)^{-2}\phi(t,\cdot)$ belongs to $L^1(0,T; C_b\ (\mathbf{U}; E))$. Let $\boldsymbol{\mu}(\cdot) \in \mathcal{V}_{bc}^\infty(0,T;\mathbf{U},\kappa), \varepsilon > 0$. Then there exists $\boldsymbol{\lambda}(\cdot) \in \mathcal{V}_{bc}^{\infty,0}(0,T;\mathbf{U},\kappa)$ taking values in conv $(D)$ such that*

$$\|\boldsymbol{\phi}(t)\boldsymbol{\lambda}(t) - \boldsymbol{\phi}(t)\boldsymbol{\mu}(t)\| \le \varepsilon \quad a.e. \text{ in } \ 0 \le t \le T \tag{4.7}$$

$$\| \, \|\boldsymbol{\lambda}(\cdot)\|_\kappa \|_\infty \le (1+\varepsilon) \| \, \|\boldsymbol{\mu}(\cdot)\|_\kappa \|_\infty. \tag{4.8}$$

*Proof*. For each $t$ $\|\boldsymbol{\mu}(t)\|_\kappa^{-2}(\kappa^2\boldsymbol{\mu})(t)$ is a probability measure, and $\langle f, \|\boldsymbol{\mu}(t)\|_\kappa^{-2}(\kappa^2\boldsymbol{\mu})(t)\rangle$ is measurable for each $f \in C_b(\mathbf{U})$, so that $\|\boldsymbol{\mu}(\cdot)\|_\kappa^{-2}$ $(\kappa^2\boldsymbol{\mu})(\cdot) \in \mathcal{V}_{ab}(0,T;\mathbf{U})$. We apply Lemma 4.2 to this measure and to the function $\psi(u) = (\kappa(u)^{-2}, \kappa(u)^{-2}\phi(t,u))$ with values in $\mathbb{E} = \mathbb{R} \times E$ (norm $\|(r,y)\| = |r| + \|y\|$), and obtain $\nu(\cdot) \in \mathcal{V}_{bc}^0(0,T;\mathbf{U})$ with values in conv $(\mathbf{D})$ satisfying

$$\begin{aligned}
&\|\boldsymbol{\psi}(t)\boldsymbol{\nu}(t) - \boldsymbol{\psi}(t)(\|\boldsymbol{\mu}(t)\|_\kappa^{-2}(\kappa^2\boldsymbol{\mu})(t))\| = \\
&= \left| \int_{\mathbf{U}} \kappa(u)^{-2}\nu(t,du) - \|\boldsymbol{\mu}(t)\|_\kappa^{-2} \int_{\mathbf{U}} \mu(t,du) \right| + \\
&+ \|(\boldsymbol{\kappa}^{-2}\boldsymbol{\phi}(t))\boldsymbol{\nu}(t) - (\boldsymbol{\kappa}^{-2}\boldsymbol{\phi}(t))(\|\boldsymbol{\mu}(t)\|_\kappa^{-2}(\kappa^2\boldsymbol{\mu})(t))\| \le \delta,
\end{aligned} \tag{4.9}$$

in $0 \le t \le T$. Multiplying by $\|\boldsymbol{\mu}(t)\|_\kappa^2$,

$$\begin{aligned}
&\left| \|\boldsymbol{\mu}(t)\|_\kappa^2 \int_{\mathbf{U}} \kappa(u)^{-2}\nu(t,du) - \int_{\mathbf{U}} \mu(t,du) \right| + \\
&+ \|\boldsymbol{\phi}(t)(\|\boldsymbol{\mu}(t)\|_\kappa^2\boldsymbol{\nu})(t)) - \boldsymbol{\phi}(t)\boldsymbol{\mu}(t)\| \le \delta \| \, \|\boldsymbol{\mu}(t)\|_\kappa^2 \|
\end{aligned} \tag{4.10}$$

in $0 \le t \le T$, which implies (4.7); however, Corollary 3.3 is not yet proved, as $\|\boldsymbol{\mu}(t)\|_\kappa^2(\kappa^{-2}\boldsymbol{\nu})(t)$ may not be a probability measure, that is, $\rho(t) = \|\boldsymbol{\mu}(t)\|_\kappa^2 \int_{\mathbf{U}} \kappa(u)^{-2}\boldsymbol{\nu}(t,du)$ need not equal 1. However (4.10) implies that $|\rho(t)-1| \le \delta' = \delta\| \, \|\boldsymbol{\mu}(\cdot)\|_\kappa^2\|_\infty$, so that if we define $\boldsymbol{\lambda}(t) = \rho(t)^{-1}\|\boldsymbol{\mu}(t)\|_\kappa^2(\kappa^{-2}\boldsymbol{\nu})(t)$, then $\boldsymbol{\lambda}(t)$ is a probability measure and

$$
\begin{aligned}
&\|\phi(t)\boldsymbol{\lambda}(t) - \phi(t)\boldsymbol{\mu}(t)\| \le \\
&\le \|\phi(t)(\rho(t)^{-1}\|\boldsymbol{\mu}(t)\|_\kappa^2(\kappa^{-2}\boldsymbol{\nu})(t)) - \Phi(\|\boldsymbol{\mu}(t)\|_\kappa^2\boldsymbol{\nu})(t)\| + \\
&+\|\phi(t)(\|\boldsymbol{\mu}(t)\|_\kappa^2(\kappa^{-2}\boldsymbol{\nu})(t)) - \phi(t)\boldsymbol{\mu}(t)\| \le \\
&\le |\rho(t)-1| \; \|\phi(t)(\|\boldsymbol{\mu}(t)\|_\kappa^2(\kappa^{-2}\boldsymbol{\nu})(t))\| + \\
&+\|\phi(t)(\|\boldsymbol{\mu}(t)\|_\kappa^2(\kappa^{-2}\boldsymbol{\nu})(t)) - \phi(t)\boldsymbol{\mu}(t)\|.
\end{aligned}
$$

To prove (4.8), we note that

$$
\begin{aligned}
&\|\boldsymbol{\lambda}(t)\|_\kappa^2 = \|\rho(t)^{-1}\|\boldsymbol{\mu}(t)\|_\kappa^2(\kappa^{-2}\boldsymbol{\nu}(t))\|_\kappa^2 = \\
&= \rho(t)^{-1}\|\boldsymbol{\mu}(t)\|_\kappa^2 \int_{\mathbf{U}} \nu(t,du) = \rho(t)^{-1}\|\boldsymbol{\mu}(t)\|_\kappa^2.
\end{aligned}
$$

We note finally that $\rho(t)^{-1}\|\boldsymbol{\mu}(t)\|_\kappa^2$ must be constant in any set where $v(t)$ is constant, so that $\boldsymbol{\lambda}(\cdot) \in \mathcal{V}_{bc}^{\infty,0}(0,T;U,\kappa)$. Also, in any such set $\boldsymbol{\lambda}(t) = \Sigma\alpha_\kappa\delta_{u_k}$ with $\alpha_k \ge 0$; since $\lambda(t)$ is a probability measure $\Sigma\alpha_k = 1$ and $\boldsymbol{\lambda}(t) \in \operatorname{conv}(\mathbf{D})$ as claimed.

LEMMA 4.4. *Let $\phi : [0,T]\times\mathbf{U} \to E$ be such that (a) $\phi(t,\cdot) \in C_{b,\kappa}(\mathbf{U};E)$ for every $t$ fixed (b) The function $t \to \kappa(\cdot)^{-2}\phi(t,\cdot)$ belongs to $L^1(0,T;C_b(\mathbf{U};E))$. Let $t \to S(t) \in L(E,X)$ be strongly continuous in $0 \le t \le T$ ($L(E,X)$ the space of bounded linear operators from the space $E$ into another Banach space $X$). Finally, let $\boldsymbol{\mu}(\cdot) \in \mathcal{V}_{bc}^\infty(0,T;\mathbf{U},\kappa)$ and $\varepsilon > 0$. Then there exists $u(\cdot) \in \mathcal{U}_{\mathrm{ad}}^{\infty,0}(0,T;\mathbf{U},\kappa)$ such that*

$$
\left\| \int_{\mathbf{U}} S(t)\{\phi(t,u(t)) - \phi(t)\boldsymbol{\mu}(t)\}dt \right\|_X \le \varepsilon \tag{4.11}
$$

$$
\|\kappa(u(\cdot))\|_\infty \le (1+\varepsilon)\| \, \|\boldsymbol{\mu}(\cdot)\|_\kappa\|_\infty \tag{4.12}
$$

Lemma 4.4 generalizes Lemma 5.4 in [7] and is proved as follows. By Corollary 4.3 there exists $\boldsymbol{\lambda}(\cdot) \in \mathcal{V}_{bc}^{\infty,0}(0,T;\mathbf{U},\kappa)$ such that $\boldsymbol{\lambda}(t) \in \operatorname{conv}(\mathbf{D})$ and (4.7) and (4.8) hold. Select a countably $C_b(\mathbf{U};E)$-valued function $t \to \psi_\varepsilon(t,\cdot)$ such that, after eventual modification in a null set we have $\|\psi_\varepsilon(t) - (\kappa^{-2}\phi)(t)\|_{C_b(\mathbf{U})} \le \delta$ $(0 \le t \le T)$. Then, if $\phi_\varepsilon(t,u) = \kappa(u)^2\psi_\varepsilon(t,u)$ we have

$$
\begin{aligned}
&\|\phi_\varepsilon(t)\boldsymbol{\mu}(t) - \phi(t)\boldsymbol{\mu}(t)\| = \\
&= \|(\psi_\varepsilon(t) - \kappa^{-2}\phi(t))(\kappa^2\boldsymbol{\mu}(t))\| \le \delta\|\boldsymbol{\mu}(t)\|_\kappa^2
\end{aligned}
$$

a.e. in $0 \leq t \leq T$, with a similar estimate for $\|\phi_\varepsilon(t)\boldsymbol{\lambda}(t) - \phi(t)\boldsymbol{\lambda}(t)\|$. Accordingly, we only have to show (4.11) for $\phi = \phi_\varepsilon$.

Let $\Xi_\varepsilon$ be a collection of pairwise disjoint measurable sets covering $[0, T]$ and such that both $\phi_\varepsilon(\cdot)$ and $\boldsymbol{\lambda}(\cdot)$ are constant in each $e \in \Xi_\varepsilon$. If $t \in e \in \Xi_\varepsilon$, then $\phi_\varepsilon(t) = \phi_\varepsilon$, $\boldsymbol{\lambda}(t) = \boldsymbol{\lambda} = \Sigma\alpha_j(e)\delta(\cdot - u_j(e))$ for $t \in e$, where $\alpha_j(e) \geq 0$, $\alpha_1(e) + \cdots + \alpha_n(e) = 1$; moreover,

$$\|\boldsymbol{\lambda}\|_\kappa^2 = \sum \alpha_j(e)\kappa(u_j(e))^2 \leq (1+\varepsilon)\| \, \|\boldsymbol{\mu}(\cdot)\|_\kappa^2\|_\infty \tag{4.13}$$

(with a different $\varepsilon$). Let $\Sigma[e]$ be the subspace of $\Sigma_{rba}(\mathbf{U}, \Phi_c)$ generated by the measures $\delta_{u_j(e)}$, $j = 1, \ldots n$, $E[e]$ the image of $\Sigma[e]$ by $\boldsymbol{\phi}_\varepsilon$, that is, the subspace of $E$ generated by the elements $\boldsymbol{\phi}_\varepsilon\delta_{u_j(e)}$, $j = 1, \ldots, n$. The $L(E[e],\ X)$-valued function $S(\cdot)$ is strongly continuous, thus (since $E[e]$ is finite dimensional) continuous in the norm of $L(E[e], X)$ and can be approximated uniformly in the $L(E[e], X)$ norm by a countably valued $L(E[e], X)$-valued function $S_\varepsilon(t)$:

$$\|S_\varepsilon(t) - S(t)\|_{L(E[e];X)} \leq \varepsilon \qquad (t \in e) \tag{4.14}$$

Let, finally $d \subseteq e$ be a set where $S_\varepsilon(t)$ is constant. Divide $d$ into $n$ disjoint measurable sets with meas $(d_j) = \alpha_j(e)$ meas $(d)$, and define $\boldsymbol{\nu}(t) = \delta(\cdot - u_j(e))$ for $t \in d_j$. Then

$$\begin{aligned}
\int_d S_\varepsilon(t)\boldsymbol{\phi}_\varepsilon(t)\boldsymbol{\nu}(t)dt &= \sum_{j=1}^n \int_{d_j} S_\varepsilon(t)\phi_\varepsilon(t, u_j(e))dt \\
&= \sum_{j=1}^n \text{meas}(d_j).\ S_\varepsilon(t)\phi_\varepsilon(t, u_j(e)) = \sum_{j=1}^n \text{meas}(d)S_\varepsilon(t)\phi_\varepsilon(t, u_j(e))\alpha_j(e) \\
&= \sum_{j=1}^n \text{meas}(d)S_\varepsilon(t)\boldsymbol{\phi}_\varepsilon(t)\boldsymbol{\lambda}(t)) = \int_d S_\varepsilon(t)\boldsymbol{\phi}_\varepsilon(t)\boldsymbol{\lambda}(t)dt\ .
\end{aligned}$$

The control $u(\cdot) = \delta_{u(\cdot)} \in \mathcal{V}_{bc}^0(0, T; \mathbf{U})$ obtained piecing together these constructions in all $d \subseteq e$ in all $e \in \Xi_\varepsilon$ satisfies (4.11) (with a different $\varepsilon$); condition (4.12) follows from (4.13). This ends the proof.

COROLLARY 4.5. *Under the same assumptions on $E$ and $\phi$ in Lemma 4.4, let the $L(E, E)$-valued function $S(t, s)$ be strongly continuous in $0 \leq s < t \leq T$ and*

$$\|S(t, s)\| \leq \alpha(t-s) \qquad (0 \leq s < t \leq T)$$

*with $\alpha(\cdot)$ a nonincreasing function in $\in L^1(0, T)$. Then, if $\boldsymbol{\mu}(\cdot) \in \mathcal{V}_{bc}^\infty(0, T; \mathbf{U}, \kappa)$ and $\varepsilon > 0$ there exists $u(\cdot) \in \mathbf{U}_{\text{ad}}^{\infty,0}(0, T; \mathbf{U}, \kappa)$ such that*

$$\left\| \int_0^t S(t, \sigma)\{\phi(\sigma,\ u(\sigma)) - \boldsymbol{\phi}(\sigma)\boldsymbol{\mu}(\sigma)\}d\sigma \right\|_E \leq \varepsilon \ (0 \leq t \leq T) \tag{4.15}$$

*Proof.* Construct a strongly continuous $L(E, E)$-valued function $S_\varepsilon(t,s)$ in $0 \le s, t \le T$ such that $S_\varepsilon(t,s) = S(t,s)$ in $0 \le s \le t-\varepsilon \le T, S_\varepsilon(t,s) = 0$ in $s > t - \varepsilon$ and such that

$$||S_\varepsilon(t,s)|| \le \alpha(t-s) \qquad (0 \le s,\ t \le T),$$

where $\alpha(t) = 0$ for $t < 0$. Applying Lemma 4.4 with $X = C(0,T;E)$, $S_\varepsilon(t) = S_\varepsilon(\cdot, t)$, (4.15) results for $S_\varepsilon(t,s)$. On the other hand,

$$\left\| \int_0^t \{S(t,\sigma) - S_\varepsilon(t,\sigma)\}\phi(\sigma)\boldsymbol{\mu}(\sigma)d\sigma \right\|_E \le \tag{4.16}$$

$$\le C|||\boldsymbol{\mu}(\cdot)||_\kappa^2||_\infty \int_{t-\varepsilon}^t \alpha(t-\sigma)d\sigma$$

with a similar inequality for $u(\cdot)$, so that (4.15) extends to $S(t,s)$.

*Proof of Theorem 4.1.* Apply Corollary 3.2 with $S(t,\sigma) = A^{1/2}S(t-\sigma)$ and $\boldsymbol{\mu}_\rho = \boldsymbol{\mu}_\varepsilon$.

*Remark 4.6.* All the results in this paper extend to the control spaces based on the dual of the space $B(\mathbf{U})$ of all bounded functions in $\mathbf{U}$ equipped with the supremum norm; $B(\mathbf{U})^*$ is isometrically isomorphic to the space $\Sigma_{ba}(\mathbf{U}, \Phi)$ of all bounded, finitely additive measures defined in the ring $\Phi$ of all subsets of $\mathbf{U}$ equipped with the total variation norm. For details on these spaces see [11].

## REFERENCES


[1] N.U. Ahmed, *Properties of relaxed trajectories for a class of nonlinear evolution equations in a Banach space*, SIAM J. Control and Optimization 21 (1983), 953–967.

[2] N.U. Ahmed, *Existence of optimal controls for a class of systems governed by differential inclusions in Banach space*, Jour. Optimization Theory Appl. 50 (1986), 213–237.

[3] N. Dunford and J.T. Schwartz, *Linear Operators, part 1*, Interscience, New York 1958.

[4] H.O. Fattorini, *A remark on the "bang-bang" principle for linear control systems in infinite dimensional spaces*, SIAM J. Control 6 (1968), 109–113.

[5] H.O. Fattorini, *Relaxed controls in infinite dimensional systems*, Int. Series Numer. Math. vol. 100, Birkhäuser, Basel (1991), 115–128.

[6] H.O. Fattorini, *Existence theory and the maximum principle for relaxed infinite dimensional optimal control problems*, to appear in SIAM J. Control Optimization.

[7] H.O. Fattorini, *Relaxation theorems, differential inclusions and Filippov's theorem for relaxed controls in semilinear infinite dimensional systems*, to appear in J. Differential Equations.

[8] H.O. Fattorini, *Relaxation in semilinear infinite dimensional control systems*, to appear in L. Markus Festschrift.

[9] H.O. Fattorini and S.S. Sritharan, *Existence of optimal controls for viscous flow problems*, Proc. Royal Soc. London Series A 439 (1992), 81–102.

[10] H.O. Fattorini and S.S. Sritharan, *Necessary and sufficient conditions for optimal controls in viscous flow problems*, to appear in Proceedings Royal Soc. Edinburgh series A.
[11] H.O. Fattorini and S.S. Sritharan, *Optimal chattering controls for viscous flow*, to appear.
[12] D. Henry, *Geometric Theory of Semilinear Parabolic Equations*, Springer–Verlag, Berlin 1981.
[13] E. Hille and R.S. Phillips, *Functional Analysis and Semi-Groups*, Amer. Math. Soc., Providence, 1957.
[14] J. Warga, *Necessary conditions for minimum in relaxed variational problems*, J. Math. Anal. Appl. 4 (1962), 129–145.
[15] L.C. Young, *Generalized curves and the existence of an attained absolute minimum in the calculus of variations*, Comptes Rendus Soc. des Sciences et des Lettres de Varsovie, classe III, 30 (1937), 212–234.
[16] L.C. Young, *Lectures on the Calculus of Variations and Optimal Control Theory*, W.B. Saunders, Philadelphia 1969.

# MULTIDIMENSIONAL INVERSE SCATTERING PROBLEMS IN DETERMINISTIC AND RANDOM MEDIA*

SEMION GUTMAN† AND MICHAEL KLIBANOV‡

**Abstract.** Suppose that a medium with slowly changing spatial properties is enclosed in a bounded 3-dimensional domain and it is subjected to a scattering by plane waves of a fixed frequency. Let measurements of the wave scattering field induced by this medium be available in the region outside of this domain. We study how to extract the properties of the medium from the information contained in the measurements. We are concerned with the weak scattering case of the above inverse scattering problem (ISP). That is, the unknown spatial variations of the medium are assumed to be close to a constant. Examples can be found in studies of wave propagation in oceans, in the atmosphere and in some biological media.

In this paper we study the Inverse Scattering Problems in both deterministic and random media. Since the problems are nonlinear, the methods for their linearization (the Born approximation) have been developed. However, such an approach often does not produce good results. In our method, the Born approximation is just the first iteration and further iterations improve the identification by an order of magnitude.

The iterative sequence is defined in the framework of a Quasi-Newton method. Using the measurements of the scattering field from a carefully chosen set of directions we are able to recover (finitely many) Fourier coefficients of the sought parameters of the model. Numerical experiments for the scattering from coaxial circular cylinders as well as for simulated data are presented, which show a high precision of the identification. We also discuss some results based on Carleman's estimates, applicable to the strong scattering case.

**1. Introduction.** In this paper we present a survey of our recent studies on numerical methods for 3-Dimensional Inverse Scattering Problems (ISP) in inhomogeneous media. The models under consideration are of wave type, rather than diffusion or transport type (see below). The goal of Inverse Scattering is, in particular, the development of methods capable of the determination of spatially varying PDE's coefficients from the information contained in measurements made outside of a certain region, see e.g. Lavrent'ev [22]. In the problems that we consider an unknown coefficient is to be determined from the measurements of the amplitude and the phase of scattered electromagnetic or acoustic waves made outside of the region containing media's inhomogenuities.

Numerical methods for 3-Dimensional ISPs currently attract considerable interest, cf. Baylis et al. [3], Cheney [8], Colton and Monk [9],[10], Gutman and Klibanov [14],[15], Klibanov et al. [19]–[21], W. Symes [27]. Papers [8],[9] contain surveys.

Among the numerous applications of the multidimensional ISPs, first of all we can refer to such traditional areas as ocean acoustics, geophysics and

* This research was supported by ONR grant N0014-92-J-1008.

† Department of Mathematics, University of Oklahoma, Norman, OK 73071 USA.

‡ Department of Mathematics, University of North Carolina at Charlotte, Charlotte, NC 28223 USA.

studies of the terrestrial atmosphere. A rather new and very promising area is the imaging of complicated internal structures of inhomogeneous media, such as a biological tissue. In this instance the imaging should display the subtle structures of abnormal inclusions (e.g. a cancerous tumor) in the media. The imaging is reported to be used in studies of a 3-D structure of the AIDS virus [2], non-invasive quality control of industrial devices, etc.

Perhaps the most widely known wave scattering imaging process is X-ray tomography. However, there are significant deficiencies of this method. Among them is a considerable risk of a prolonged X-ray exposure for a patient. It can be reduced if longer (e.g. laser light) electromagnetic waves are used for the diagnostic purposes. The main difficulty in this later approach is the need to take into the account the diffusive properties of these waves.

In this paper we describe our numerical algorithms designed to work for waves with nonnegligible diffusion. The 3-D ISPs under consideration are of the *weak* scattering type. That is, the spatial variations of the media are assumed to be close to a constant. Both deterministic and random media are considered. Of course, the *strong* scattering case is of major importance. In section 8 we describe our recent results [19], which, in our opinion, can be applied to the study of strong scattering.

Nevertheless, *weak* scattering is a widespread phenomenon in nature. One example is given by ocean acoustics, since the sound speed in an ocean changes very slowly, see Tappert [28]. Furthermore, it was shown in Flatte and Tappert [13] that the dispersion of the sound wave profile, due to internal waves (which are *random*), is sufficiently small. Another example is given by the *terrestrial atmosphere* which is characterized by the refractive index $n(x, y, z) = 1+\delta_n(x, y, z)$, with $|\delta_n| \leq 10^{-6}$, see Crosignani and di Porto [11].

Regarding biology, according to Ishimaru [16, p. 62] the dielectric constants of such biological media as fat, muscles and brain change very slowly with respect to the frequency of electromagnetic waves in the range from 1 MHz to 100 GHz. Furthermore, according to Table 3.2 in [16, p. 62] the index of refraction of blood plasma has the form $n = n_1 + in_2$, where $n_1 = 1.036$ and $|n_2| \leq 2.3 \cdot 10^{-4}$, in a very broad range of frequencies. *Whole blood may be viewed as a homogeneous hemoglobin medium with scattering particles made of the plasma between red blood cells*, see [16, p. 66].

Summarizing, we conclude that in all these situations one would deal with the weak scattering case. Nevertheless, we would like to point out that it would not be a simple Born-approximation case. In our numerical examples below we clearly demonstrate that our method provides an improvement of an order of magnitude compared with the Born-approximation case. Also we would like to outline why we prefer to deal with wave-type PDEs rather than with transport-type equations. In our opinion, wave-type equations could, in principle, provide much better resolution and accuracy compared with transport-type equations. There are, basically, two reasons: (i). The transport equation can be derived from a Helmholtz-like equation, but not vice versa; (ii). According to the principles of Theoretical Physics, the transport equation is valid only in the case when a relatively *small* number of photons penetrates the media.

As to the mathematical essence of our methods, we note that in the bulk of research on this subject the main object of study is integral equations (e.g. the Lippman-Schwinger equation). We, however, work directly with the differential equations. As a result, we need to numerically pre-calculate only a relatively small number of "basic" integrals and it is done only once. All other necessary integrals can be computed explicitly.

The paper is organized as follows. In Section 2 we discuss a model for scattering in deterministic media. Two versions of an iterative (Quasi-Newton) method for the ISP are described in Section 3. Numerical experiments for scattering with simulated data as well as for scattering for coaxial cylinders are presented in Section 4. Section 5 contains a mathematical framework and main results on convergence and the coefficient identification with exact data. Discussion of noisy data and finite dimensional approximations are presented in Section 6. A model for inverse scattering in random media and the resulting algorithm are given in Section 7. Finally, in Section 8 we present our global convergence results.

**2. Inverse scattering in deterministic media.** We consider the following model for deterministic media. Let $\Omega$ be a bounded domain in $\mathbb{R}^3$. Properties of the media are described by a complex valued function $\epsilon(x)$, such that $\epsilon \in L^\infty(\Omega)$, $\epsilon(x) = 0$ for $x \in \mathbb{R}^3 \backslash \Omega$. Without loss of generality we can assume $\Omega \subset W$, where $W$ is the ball of radius $\frac{\pi}{2}$ with the center at the origin. Let $\partial W$ be the spherical boundary of the ball $W$. Let $S^2$ be the unit sphere in $\mathbb{R}^3$. If $\nu \in S^2$, then $e^{i\kappa\langle x,\nu\rangle}$ is the scalar incident planar wave of the frequency $\kappa > 0$ propagating in the direction of the vector $\nu$. Here $\langle \cdot, \cdot \rangle$ is the inner product in $\mathbb{R}^3$. Let $u(x,\nu)$ be the solution of the forward scattering problem

$$\Delta u + \kappa^2(1+\epsilon)u = 0 \tag{2.1}$$

$$u(x,\nu) = e^{i\kappa\langle x,\nu\rangle} + \tilde{u}(x,\nu) \tag{2.2}$$

$$\tilde{u} = O(\frac{1}{|x|}), \quad \frac{\partial \tilde{u}}{\partial |x|} - i\kappa\tilde{u} = o\left(\frac{1}{|x|}\right), \quad |x| \to \infty, \tag{2.3}$$

where $\tilde{u} \in H^2_{\text{loc}}(\mathbb{R}^3)$.

We study how to extract the properties of the media from the information contained in the measurements, assuming that the frequency $\kappa$ is known and it is a constant. This constitutes the Inverse Scattering Problem (ISP). We are concerned with the weak scattering case, that is, the unknown spatial variations of the medium, represented by the coefficient $\epsilon$ are assumed to be close to zero.

*The ISP statement.* Let $\kappa = \text{const} > 0$ be fixed. Determine the coefficient function $\epsilon(x)$ in (2.1) assuming that the measurement function $\varphi(x,\nu)$, $\nu \in S^2$, $x \in \partial W$ is given and the solution $u(x,\nu)$ of (2.1)–(2.3) satisfies

$$u\big|_{\partial W} = \varphi(x,\nu)\,, \quad \forall \nu \in S^2\,. \tag{2.4}$$

Let $\Omega_1 = [-\pi,\pi]^3$ be a cube in $\mathbb{R}^3$ and let $\Gamma$ be its boundary. In our treatment the ISP (2.1)–(2.4) is considered as a particular case of the following more general inverse problem (IP) on the *bounded* domain $\Omega_1$.

*The IP statement.* Determine the coefficient function $\epsilon(x)$, $\epsilon \in L^\infty(\Omega_1)$ given the measurement function $\Phi(x,\nu)$, $\nu \in S^2$, $x \in \Omega_1$ such that for every fixed $\nu \in S^2$ the system

$$\Delta u + \kappa^2(1+\epsilon)u = 0\,, \quad x \in \Omega_1 \tag{2.5}$$

$$u(x,\nu)\big|_\Gamma = \Phi(x,\nu)\big|_\Gamma \tag{2.6}$$

$$\frac{\partial u}{\partial n}(x,\nu)\big|_\Gamma = \frac{\partial \Phi}{\partial n}(x,\nu)\big|_\Gamma \tag{2.7}$$

is satisfied.

Besides the fact that the above inverse problem is stated on the bounded domain $\Omega_1$, it has an additional advantage of the data $\Phi$ being given in the whole domain $\Omega_1$, and not just on its boundary as in (2.4). Also, a construction below shows that the data $\Phi$ can be chosen to be sufficiently smooth.

To see that the ISP (2.1)–(2.4) can be reduced to the form (2.5)–(2.7) we can argue as follows. Given function $\varphi(x,\nu)$, $\nu \in S^2$ one can obtain the solution $u(x,\nu)$ of (2.1)–(2.4) in the *exterior* of $W$ using Green's functions of the Helmholtz equation

$$\Delta u + \kappa^2 u = 0\,, \quad x \in \mathbb{R}^3 \backslash W\,.$$

Let $\chi \in C^3(\mathbb{R}^3)$ be a cut-off function, defined by

$$\chi(x) = \begin{cases} 1 & \text{in a neighborhood of } \Gamma \\ 0 & \text{in a neighborhood of } W \end{cases} \tag{2.8}$$

In particular, $\chi(x) = 0$ for $x \in W$. Define

$$\Phi(x,\nu) = \chi(x)\cdot u(x,\nu)\,. \tag{2.9}$$

Then

$$u|_\Gamma = \Phi|_\Gamma \quad \text{and} \quad \frac{\partial u}{\partial n}|_\Gamma = \frac{\partial \Phi}{\partial n}|_\Gamma .$$

Thus the form of (2.5)–(2.7) is obtained.

Introducing $V = u - \Phi$, system (2.5)–(2.7) becomes

$$\Delta V + \kappa^2(1+\epsilon)V = -(\Delta\Phi + \kappa^2\Phi) - \kappa^2\epsilon\Phi \tag{2.10}$$

$$V|_\Gamma = 0 \tag{2.11}$$

$$\frac{\partial V}{\partial n}|_\Gamma = 0\,, \tag{2.12}$$

which is a more convenient form for our considerations.

DEFINITION 2.1. Let $\epsilon \in L^\infty(\Omega_1)$ and let $u \in H^2_{\text{loc}}(\mathbb{R}^3)$ be the solution of (2.1)–(2.4). Function $\Phi(x,\nu)$, $x \in \Omega_1$, $\nu \in S^2$ is called an **exact data** (generated by the $\epsilon$) for the IP (2.5)–(2.7) if $\Phi(x,\nu) = u(x,\nu)$, $\nu \in S^2$ in a neighborhood of $\Gamma$.

An exact data $\Phi$ for the IP can be obtained from the measurements $\varphi(x,\nu)$ with the help of a cut-off function $\chi(x)$ as described above in (2.8)–(2.9).

Some results for *noisy* data $\Phi$ are discussed in Section 6.

**3. The Quasi-Newton method.** The Quasi-Newton method (also known as Newton-Kantorovich method) is used to solve a nonlinear equation $Ax = 0$ in a Banach space $X$. The solution is sought as the limit of the sequence $\{x_n\}_{n=1}^\infty$ defined iteratively by

$$x_n = x_{n-1} - [A'(x_0)]^{-1}Ax_{n-1}\,, \quad n \geq 1 \tag{3.1}$$

provided an initial guess $x_0 \in X$.

Writing (3.1) as

$$x_n = Tx_{n-1} \tag{3.2}$$

with $T = I - [A'(x_0)]^{-1}A$, implies that the convergence of the sequence $\{x_n\}_{n=1}^\infty$ can be analyzed in the general framework of the fixed point type arguments for the equation

$$x = Tx \tag{3.3}$$

We postpone the precise mathematical framework till Section 5 and present here the resulting algorithm with its motivation.

Fix a positive integer $N$ such that $\sqrt{3}N \le 2\kappa$ ($\sqrt{2}N \le 2\kappa$ for a 2-D case). Let $L_N^\infty$ be a subspace of functions in $L^\infty(\Omega_1)$ having a finite Fourier series expansion; more precisely

$$L_N^\infty = \{b \in L^\infty(\Omega_1) : b(x) = \sum_{|z_i| \le N} a_n e^{i\langle x, Z\rangle}\,,$$
$$Z = (z_1, z_2, z_3)\,, \quad z_i - \text{integer}\}\,.$$

Thus a function $b \in L_N^\infty$ has at most $(2N+1)^3$ nonzero Fourier coefficients.

Let $Q^0 \subset L^2(\Omega_1 \times S^2)$ be the Banach space of functions satisfying

$$\|f\|_{Q^0} = \text{ess-sup}\{\|f(\cdot,\nu)\|_{L^2(\Omega_1)} : \nu \in S^2\} < \infty\,.$$

Let

$$\begin{aligned} X &= L^\infty(\Omega_1) \times Q^0\,, \\ X_N &= L_N^\infty \times Q^0\,. \end{aligned} \tag{3.4}$$

Assuming that $\Phi$ and $\Delta\Phi \in Q^0$ the operator $T : X \to X$, corresponding to our ISP (2.1)–(2.3) can be described as follows: For $V \in Q^0$ with $\Delta V \in Q^0$ let

$$A(\epsilon, V) = \Delta V + \kappa^2 V + \kappa^2 \epsilon (V + \Phi) + (\Delta\Phi + \kappa^2\Phi)\,, \tag{3.5}$$

Then (2.10)–(2.12) is just

$$A(\epsilon, V) = 0$$

plus the boundary conditions.

According to (3.1), $T = I - [A'(x_0)]^{-1}A$. We take

$$x_0 = (0, u_0 - \Phi)\,, \tag{3.6}$$

where

$$u_0 = e^{i\kappa\langle x, \nu\rangle}\,. \tag{3.7}$$

It can be shown that the Frechet's derivative of $A$ is given by

$$A'(x_0)(h, H) = \Delta H + \kappa^2 H + \kappa^2 h u_0\,. \tag{3.8}$$

Writing (3.1) as

$$A'(x_0)x_n = [A'(x_0) - A]x_{n-1} \tag{3.9}$$

gives for $x_n = (\epsilon_n, V_n)$, $x_{n-1} = (\epsilon_{n-1}, V_{n-1})$

$$\Delta V_n + \kappa^2 V_n + \kappa^2 \epsilon_n u_0 = \kappa^2 \epsilon_{n-1} u_0 - \kappa^2 \epsilon_{n-1}(V_{n-1} + \Phi) - (\Delta\Phi + \kappa^2\Phi) \tag{3.10}$$

Thus at every iterative step, given functions $\epsilon_{n-1}$ and $V_{n-1}$ ($\Phi$ is a data), we need to solve (3.10) to obtain $\epsilon_n$ and $V_n$.

Equation (3.10) can be solved for $\epsilon_n$, $V_n$ assuming that $V_n$ satisfies the zero boundary conditions (2.11)–(2.12) and $\epsilon_n \in L_N^\infty$. Indeed, let $Z = (z_1, z_2, z_3) \in \mathbb{R}^3$ be a point with the integer coordinates $z_1, z_2, z_3$ such that $|z_i| \le N$, $i = 1, 2, 3$. Since $\sqrt{3}N \le 2\kappa$ we have $|Z| \le 2\kappa$ and therefore there exist two unit vectors $\nu_Z, \mu_Z \in S^2$ such that

$$\kappa(\mu_Z + \nu_Z) = Z\,. \tag{3.11}$$

Assuming that $\nu = \nu_Z$ in (3.10) (therefore $u_0 = e^{i\kappa\langle x,\nu_Z\rangle}$) we multiply (3.10) by $e^{i\kappa\langle x,\mu_Z\rangle}$ and integrate it over $\Omega_1$. Applying the integration by parts twice to the term with $\Delta V_n$ and using the zero boundary conditions we finally obtain

$$\int_{\Omega_1} \epsilon_n e^{i\langle x,Z\rangle}\,\mathrm{dx} = \frac{1}{\kappa^2}\int f e^{i\kappa\langle x,\mu_Z\rangle}\,\mathrm{dx}\,, \tag{3.12}$$

where $f$ denotes the right (known) side of (3.10).

Therefore all the Fourier coefficients of $\epsilon_n \in L_N^\infty$ can be computed and then the function $\epsilon_n$ itself. At this point we can proceed to find all the Fourier coefficients of the function $V_n(x, \nu_Z)$ by multiplying (3.10) by $e^{i\langle x,Y\rangle}$, where $Y \in \mathbb{R}^3$ is a point with integer coordinates. Integrating over $\Omega_1$ and applying the integration by parts (twice) to the term with $\Delta V_n$ we obtain

$$\int_{\Omega_1} V_n e^{i\langle x,Y\rangle}\,\mathrm{dx} = \frac{1}{-|Y|^2+\kappa^2}\Big[-\kappa^2\int_{\Omega_1} \epsilon_n e^{i\langle x,\kappa\nu_Z+Y\rangle}\,\mathrm{dx} + \int_{\Omega_1} f e^{i\langle x,Y\rangle}\,\mathrm{dx}\Big]\,. \tag{3.13}$$

This gives all the Fourier coefficients of $V_n(x, \nu_Z)$ and therefore the function $V_n(x, \nu_Z)$ itself.

Strictly speaking, the **zero** boundary conditions (2.11)–(2.12) are **not** satisfied for functions $\{V_n\}_{n=0}^\infty$. Nevertheless, we can use formulas (3.12)–(3.13) to define an iterative process. Its convergence to the sought solution $\epsilon$ and $V$ can be proved (see Sections 5 and 6) under the assumption of the smallness of the original coefficient function $\epsilon$ in (2.1).

To implement this method numerically, we must restrict ourselves to the computation of finitely many Fourier coefficients of $V_n$. For example, one can compute only the coefficients $V_n^Y$ with $Y = (y_1, y_2, y_3)$ satisfying $|y_i| \le M$ , $i = 1, 2, 3$ for a prescribed positive number $M$.

We summarize this (first) version of the Quasi-Newton method as follows:

(i) *Fix a positive integer $N$ such that $\sqrt{3}N \le 2\kappa$ ($\sqrt{2}N \le 2\kappa$ for a 2-D case). Let $P$ be the set of points $Z \in \mathbb{R}^3$ such that the coordinates $z_1, z_2, z_3$ of $Z$ be integer and satisfy $|z_i| \le N$, $i = 1, 2, 3$. Also, fix a positive number $M$.*

*(ii)* *For each* $Z \in P$ *find two unit vectors* $\nu_Z, \mu_Z \in S^2$ *such that* $\kappa(\nu_Z + \mu_Z) = Z$.

*(iii)* *Let* $\epsilon_0(x) = 0$, $V_0(x, \nu_Z) = u_0(x, \nu_Z) - \Phi(x, \nu_Z)$, *where* $u_0(x, \nu) = e^{i\kappa\langle x, \nu\rangle}$.

*(iv)* *Construct the sequence* $\{\epsilon_n(x), \{V_n(x, \nu_Z)\}_{Z\in P}\}_{n=1}^{\infty}$ *iteratively as follows*

a). *For every* $Z \in P$ *let*

$$\begin{aligned} f(x, \nu_Z) &= \kappa^2 \epsilon_{n-1} u_0(x, \nu_Z) \\ &\quad -\kappa^2 \epsilon_{n-1}(V_{n-1}(x, \nu_Z) + \Phi(x, \nu_Z)) \\ &\quad -(\Delta\Phi(x, \nu_Z) + \kappa^2 \Phi(x, \nu_Z)) \end{aligned}$$

b). *For every* $Z \in P$ *compute*

$$\epsilon_n^Z = \frac{1}{\kappa^2} \int_{\Omega_1} f(x, \nu_Z) e^{i\kappa\langle x, \mu_Z\rangle}\, dx$$

c). *Let*

$$\epsilon_n(x) = \frac{1}{(2\pi)^3} \sum_{Z\in P} \epsilon_n^Z e^{-i\langle x, Z\rangle}$$

d). *For every* $Z \in P$ *and* $Y \in \mathbb{R}^3$ *with the integer coordinates* $y_1, y_2, y_3$ *such that* $|y_i| \le M$, $i = 1, 2, 3$ *compute*

$$V_{n,Z}^Y = \frac{1}{\kappa^2 - |Y|^2}\Big[-\int_{\Omega_1} \kappa^2 \epsilon_n e^{i\langle x, \kappa\nu_Z + Y\rangle}\, dx + \int_{\Omega_1} f e^{i\langle x, Y\rangle}\, dx\Big]$$

e). *Let*

$$V_n(x, \nu_Z) = \frac{1}{(2\pi)^3} \sum_Y V_{n,Z}^Y e^{-i\langle x, Y\rangle}$$

*(v)* $\epsilon = \lim_{n\to\infty} \epsilon_n$ *is the sought coefficient function* $\epsilon(x)$ *in (2.1).*

*Remark* 3.1. The normalization constants $\frac{1}{(2\pi)^3}$ are replaced by $\frac{1}{(2\pi)^2}$ in the 2-D case.

While equation (3.10) can be attempted to be solved using the above method (see [14]) the resulting numerical algorithm is not very efficient. The problem is the presence of the highly oscillatory function $e^{i\kappa\langle x,\nu\rangle}$ in functions $u$, $V$ and $\Phi$ which negatively affects our method since we need to compute more Fourier coefficients of the functions $V_n$ than would probably be needed for the functions $e^{-i\kappa\langle x,\nu\rangle} V_n$. A certain justification of this conclusion can be seen from Figures 1a,b. Figure 1a shows $Re[u(x, \nu)]$ for $\nu = (1, 0)$, in the case of a scattering from the circular cylinder with the radius $a = \frac{\pi}{2}$. Value of $\epsilon$ in this circle is $E = 0.03$, and the frequency is

$\kappa = 5.3$. Thus this is a 2-D problem. Figure 1b, on the other hand, shows $Re[e^{-i\kappa\langle x,\nu\rangle}u(x,\nu)]$ with the same function $u(x,\nu)$, and it presents a much clearer view of the essential features of the scattering field. In fact, some theoretical results to this effect can be established as well.

Thus, we expect that solution of (3.10) in terms of the function $W = e^{-i\kappa\langle x,\nu\rangle}u(x,\nu)$ rather than $u$ would bring significant computational savings, which is indeed the case.

Let the set $P$ and the vectors $\nu_Z$ be defined as above

Multiplying (3.10) by $e^{i\langle x,Z-\kappa\nu_Z\rangle}$ we get

$$(\Delta V_n + \kappa^2 V_n + \kappa^2\epsilon_n u_0)e^{i\langle x,Z-\kappa\nu_Z\rangle} = f(x,\nu_Z)e^{i\langle x,Z-\kappa\nu_Z\rangle}, \tag{3.14}$$

where

$$f = \kappa^2\epsilon_{n-1}u_0 - \kappa^2\epsilon_{n-1}(V_{n-1}+\Phi) - (\Delta\Phi + \kappa^2\Phi). \tag{3.15}$$

Integrating (3.14) over $\Omega_1$ and assuming that the boundary conditions (2.11)–(2.12) are satisfied for $V_n$ we obtain (after an integration by parts)

$$\begin{aligned}(-|Z-\kappa\nu_Z|^2+\kappa^2)\textstyle\int_{\Omega_1} V_n e^{i\langle x,Z-\kappa\nu_Z\rangle}\,\mathrm{dx} + \kappa^2\int_{\Omega_1}\epsilon_n e^{i\langle x,Z\rangle}\,\mathrm{dx}\\ = \textstyle\int_{\Omega_1} f e^{i\langle x,Z-\kappa\nu_Z\rangle}\,\mathrm{dx}.\end{aligned} \tag{3.16}$$

Since $2\kappa\langle\nu_Z,Z\rangle = |Z|^2$ and $|\nu_Z| = 1$ we have

$$\epsilon_n^Z = \int_{\Omega_1}\epsilon_n e^{i\langle x,Z\rangle}\,\mathrm{dx} = \frac{1}{\kappa^2}\int_{\Omega_1} f e^{i\langle x,Z-\kappa\nu_Z\rangle}\,\mathrm{dx}, \tag{3.17}$$

and the function $\epsilon$ can be restored by its Fourier coefficients.

To extract the function $V_n$ from (3.10) we assume that it is given in the form

$$V_n(x,\nu_Z) = e^{(\gamma+i\kappa)\langle x,\nu_Z\rangle}W_n(x,\nu_Z), \tag{3.18}$$

where $\gamma$ is a positive constant, satisfying $\gamma^2 + 2\kappa^2$ is a noninteger. The Fourier coefficients of $W_n$ are obtained as follows. Multiply (3.10) by $e^{-(\gamma+i\kappa)\langle x,\nu_Z\rangle+i\langle x,Y\rangle}$, where $Y \in \mathbb{R}^3$ is an integer point. Since $\epsilon_n$ is already known from (3.17) we get

(3.19)

$$\begin{aligned}\int_{\Omega_1}\Delta V_n e^{-(\gamma+i\kappa)\langle x,\nu_Z\rangle+i\langle x,Y\rangle}\,\mathrm{dx} + \kappa^2\int_{\Omega_1} V_n e^{-(\gamma+i\kappa)\langle x,\nu_Z\rangle+i\langle x,Y\rangle}\,\mathrm{dx}\\ = [-\kappa^2\int_{\Omega_1}\epsilon_n e^{-\gamma\langle x,\nu_Z\rangle+i\langle x,Y\rangle}\,\mathrm{dx} + \int_{\Omega_1} f e^{-(\gamma+i\kappa)\langle x,\nu_Z\rangle+i\langle x,Y\rangle}\,\mathrm{dx}].\end{aligned}$$

Integrating the first term in (3.19) by parts twice and assuming the zero boundary conditions (2.11)–(2.12) for $V_n$ results in

$$p(\kappa,Y,\nu_Z)\int_{\Omega_1} V_n e^{-(\gamma+i\kappa)\langle x,\nu_Z\rangle+i\langle x,Y\rangle}\,\mathrm{dx} = g \tag{3.20}$$

where $g$ is the right side of (3.19) and

$$p(\kappa, Y, \nu_Z) = 2\kappa\langle \nu_Z, Y\rangle - 2i\gamma\langle \nu_Z, Y\rangle - |Y|^2 + \gamma^2 + 2i\gamma\kappa\,.$$

Condition $\gamma^2 + 2\kappa^2$ is a noninteger implies that $p(\kappa, Y, \nu_Z) \neq 0$ for any $Y$ and therefore all the Fourier coefficients

$$W_{n,Z}^Y = \int_{\Omega_1} W_n e^{i\langle x,Y\rangle}\,\mathrm{dx} = \int_{\Omega_1} V_n e^{-(\gamma+i\kappa)\langle x,\nu_Z\rangle + i\langle x,Y\rangle}\,\mathrm{dx}$$

can be found from (3.20).

The computational experiments resulting from this algorithm are presented in Section 4. Note that in numerical computations we can find just finitely many Fourier coefficients $W_{n,Z}^Y$. We do it by assuming that the integer points $Y = (y_1, y_2, y_3)$ satisfy $|y_i| \leq M$, $i = 1, 2, 3$ for a given number $M$.

We summarize this second version of the Quasi-Newton Method as follows.

*(i)* *Fix a positive integer $N$ such that $\sqrt{3}N \leq 2\kappa$ ($\sqrt{2}N \leq 2\kappa$ for a 2-D case). Let $P$ be the set of points $Z \in \mathbb{R}^3$ such that the coordinates $z_1, z_2, z_3$ of $Z$ be integer and satisfy $|z_i| \leq N$, $i = 1, 2, 3$. Also, fix positive numbers $M$ and $\gamma$ such that $\gamma^2 + 2\kappa^2$ be a noninteger.*

*(ii)* *For each $Z \in P$ find a unit vector $v_Z \in S^2$ such that*

$$2\kappa\langle \nu_Z, Z\rangle = |Z|^2\,.$$

*(iii)* *Let $\epsilon_0(x) = 0$, $V_0(x, \nu_Z) = u_0(x, \nu_Z) - \Phi(x, \nu_Z)$, where $u_0(x, \nu) = e^{i\kappa\langle x,\nu\rangle}$.*

*(iv)* *Construct the sequence*

$$\{\epsilon_n(x), \{V_n(x, \nu_Z)\}_{Z\in P}\}_{n=1}^{\infty}$$

*iteratively as follows.*

*a).* *For every $Z \in P$ let*

$$f(x, \nu_Z) = \kappa^2\epsilon_{n-1}u_0(x, \nu_Z) - \kappa^2\epsilon_{n-1}(V_{n-1}(x, \nu_Z) + \Phi(x, \nu_Z)) - (\Delta\Phi(x, \nu_Z) + \kappa^2\Phi(x, \nu_Z))\,.$$

*b).* *For every $Z \in P$ compute*

$$\epsilon_n^Z = \frac{1}{\kappa^2}\int_{\Omega_1} f(x, \nu_Z)e^{i\langle x, Z-\kappa\nu_Z\rangle}\,\mathrm{dx}\,.$$

*c).* *Let*

$$\epsilon_n(x) = \frac{1}{(2\pi)^3}\sum_{Z\in P} \epsilon_n^Z e^{-i\langle x,Z\rangle}\,.$$

*d).* *For every $Z \in P$ and $Y = (y_1, y_2, y_3) \in \mathbb{R}^3$ with the integer coordinates*
*$y_1, y_2, y_3$ with $|y_i| \leq M$, $i = 1,2,3$ compute*

$$W_{n,Z}^{Y} = \frac{1}{p(\kappa, Y, \nu_Z)}[-\kappa^2 \int\limits_{\Omega_1} \epsilon_n e^{-\gamma\langle x,\nu_Z\rangle + i\langle x,Y\rangle}\, dx +$$

$$+ \int\limits_{\Omega_1} f e^{-(\gamma+i\kappa)\langle x,\nu_Z\rangle + i\langle x,Y\rangle}\, dx,$$

*where*

$$p(\kappa, Y, \nu_Z) = 2\kappa\langle \nu_Z, Y\rangle - 2i\gamma\langle \nu_Z, Y\rangle - |Y|^2 + \gamma^2 + 2i\gamma\kappa$$

*e).* *Let*

$$V_n(x, \nu_Z) = \frac{1}{(2\pi)^3} e^{(\gamma+i\kappa)\langle x,\nu_Z\rangle} \sum_Y W_{n,Z}^{Y} e^{-i\langle x,Y\rangle}.$$

*(v) $\epsilon = \lim_{n\to\infty} \epsilon_n$ is the sought coefficient function $\epsilon(x)$ in (2.1).*

*Remark* 3.2. As in the first version, the boundary conditions (2.11)–(2.12) are not necessarily satisfied for functions $V_n$. Still, it can be shown that the above algorithm produces a sequence $\{\epsilon_n\}_{n=1}^{\infty}$ convergent to the sought coefficient $\epsilon$ in (2.1), provided that $\epsilon$ be sufficiently small and $\epsilon \in L_N^{\infty}$, see Theorems 5.1 and 5.2. For a detailed treatment see [14], [15].

**4. Numerical experiments.** The above methods can easily be reformulated for a 2-D case and we have conducted several 2-D numerical experiments. The computations were carried out on CRAY-Y/MP at the POPS NAVY Supercomputer Center. First we show how to simulate the data for the ISP (2.5)–(2.7). We assume that

$$\Omega = \{x_1, x_2 : -\frac{\pi}{2} < x_1, x_2 < \frac{\pi}{2}\}.$$

Let function $F(x,\nu) \in C^2(\overline{\Omega}_1)$ for every $\nu \in S^2$. Consider the boundary value problem

$$\Delta u + \kappa^2(1 + \epsilon(x))u = 0, \quad \text{in} \quad \Omega_1$$

$$\frac{\partial u}{\partial n}\Big|_{\partial\Omega_1} = \frac{\partial F}{\partial n}\Big|_{\partial\Omega_1}$$

Let $w = u - F$. Then

$$\Delta w + \kappa^2(1 + \epsilon(x))w = -[\Delta F + \kappa^2(1 + \epsilon(x))F] \tag{4.1}$$

(4.2)
$$\frac{\partial w}{\partial n}\Big|_{\partial\Omega_1} = 0 .$$

Let $t_1, t_2$ be two arbitrary integers and $t = (t_1, t_2)$. Multiply both sides of (4.1) by $e^{i\langle x,t\rangle}$ and integrate over $\Omega_1$ using (4.2). Therefore the variational form of (4.1), (4.2) is

(4.3)
$$-i\int_{\Omega_1}(\nabla w, t)e^{i\langle x,t\rangle}\,\mathrm{dx} + \kappa^2\int_{\Omega_1}(1+\epsilon)we^{i\langle x,t\rangle}\,\mathrm{dx} = -\int_{\Omega_1}[\Delta F + \kappa^2(1+\epsilon)F]e^{i\langle x,t\rangle}\,\mathrm{dx} .$$

In order to obtain an approximate solution of the system (4.3) we represent the function $\epsilon$ in the form

(4.4)
$$\epsilon(x) = \sum_{t_1,t_2=-mh}^{mh} \epsilon_t e^{-i\langle x,t\rangle}$$

and the functions $w, F$ as

(4.5)
$$w(x,\nu) = \sum_{t_1,t_2=-\ell h}^{\ell h} w_t(\nu)e^{-i\langle x,t\rangle}, \quad F(x,\nu) = \sum_{t_1,t_2=-\ell h}^{\ell h} F_t(\nu)e^{-i\langle x,t\rangle},$$

where $mh$, $\ell h$ are integers corresponding to $N$ and $M$ respectively in the above summary of the algorithm , and $\epsilon_t$, $w_t(\nu)$, and $F_t(\nu)$ are the Fourier coefficients. Hence the system (4.3) becomes a linear system with respect to $w_t(\nu)$.

To avoid using the data inside the support of $\epsilon$ the function $w(x,\nu)$ is multiplied by the cut-off function $\chi(x) = 1 - \zeta(x_1)\zeta(x_2)$, where

$$\zeta(y) = \begin{cases} 1, & \text{for} \quad |y| < \frac{\pi}{2} \\ \sin^4 y, & \text{for} \quad |y| > \frac{\pi}{2} . \end{cases}$$

Let $G(x) = w(x)\chi(x)$. Then

$$u\big|_{\partial\Omega_1} = (F+G)\big|_{\partial\Omega_1}, \quad \frac{\partial u}{\partial n}\Big|_{\partial\Omega_1} = \frac{\partial}{\partial n}(F+G)\big|_{\partial\Omega_1}$$

Then we define $\Phi(x,\nu) = F(x,\nu) + G(x,\nu)$. Thus we have the setting of (2.5)–(2.7).

In our numerical experiments we represented the function $\epsilon(x)$ in the form (4.4), and functions $\epsilon_n$, $V_n$ as

$$\epsilon_n(x) = \sum_{t_1,t_2=-mh}^{mh} \epsilon_t^{(n)} e^{-i\langle x,t\rangle}$$

$$V_n(x,\nu_Z) = \sum_{t_1,t_2=-\ell h}^{\ell h} V_n(\nu_Z) e^{-i\langle x,t\rangle}$$

Results of our numerical experiments for simulated data are presented on Figure 2a,b. *But only the real parts are plotted for the test and the estimated coefficient*. We have taken $\ell h = 12$, $mh = 5$, $\kappa = 5.5$. For test #1 (Figures 2a and 2b) we have chosen the test coefficient $\epsilon_{\text{test}}$ and the function $F$ as follows. Let

$$F(x,\nu) = e^{i\kappa\langle x,\nu\rangle}[1 + 0.02\sin(x_1\nu_1^2 + x_2\nu_2)]$$

and let the test coefficient $\epsilon_{\text{test}}$ be the finite Fourier approximation (4.4) of the function

$$\epsilon(x) = \begin{cases} 0.02 & \text{for } \frac{\pi}{4} < |x_1| < \frac{\pi}{2}, \quad |x_2| < \frac{\pi}{2} \\ 0 & \text{otherwise}. \end{cases}$$

Figure 2b represents the real part of the estimated coefficient, which has been obtained after *three* iterations. Note, that our goal is to recover the coefficient $\epsilon_{\text{test}}$, not the function $\epsilon$ itself. For this reason we did not take into the account error due to the finite Fourier approximation (4.4). Therefore we defined the relative error as

$$\frac{||\epsilon_{\text{test}} - \epsilon_{\text{est}}||}{||\epsilon_{\text{test}}||}. \tag{4.6}$$

where the $L_2(\Omega_1)$-norm is taken, and $\epsilon_{\text{est}}$ denotes the function $\epsilon$ obtained after a final iteration.

In this experiments we used 121 vectors $\nu$. The forward problem takes about 5 seconds per vector (637 seconds total). The inverse problem takes about 220 seconds per iteration (11 minutes total). The error (4.6) is about 14%.

Better results have been obtained for the scattering from a *circular cylinder*. Since the forward scattering problem admits an explicit solution in this case, there is no need to simulate the data as in the above experiments. Let $a > 0$ and

$$\epsilon(r) = \begin{cases} E, & \text{for } r \le a \\ 0, & \text{for } r > a \end{cases}$$

where $r = \sqrt{x_1^2 + x_2^2}$ for $(x_1, x_2) \in \mathbb{R}^2$. Given vector $\nu = (\cos\alpha, \sin\alpha)$ equations (2.1)–(2.3) in polar coordinates become (in 2-D case)

$$\frac{1}{r}\frac{\partial}{\partial r}\left(r\frac{\partial u}{\partial r}\right) + \frac{1}{r^2}\frac{\partial^2 u}{\partial \theta^2} + \kappa^2(1+\epsilon)u = 0 \tag{4.7}$$

$$u = e^{i\kappa\langle x,\nu\rangle} + v\,, \quad \lim_{r\to\infty} r\left(\frac{\partial v}{\partial \nu} - i\kappa v\right) = 0\,. \tag{4.8}$$

Let $\kappa_1 = \kappa \cdot \sqrt{1+E}$. If $(x,y) = (r\cos\varphi, r\sin\varphi)$ then the solution of (4.7)–(4.8) is

$$u(x,y,\nu) = \begin{cases} e^{i\kappa(x,\nu)} + \sum\limits_{m=-\infty}^{\infty} a_m H_m^{(1)}(\kappa r) e^{im(\varphi-\alpha)} & r > a \\ \sum\limits_{m=-\infty}^{\infty} b_m J_m(\kappa_1 r) e^{im(\varphi-\alpha)} & r \le a \end{cases}$$

where $J_m, H_m^{(1)}$ are Bessel and Hankel functions. The coefficients $a_m, b_m$, $-\infty < m < \infty$ are found from the compatibility (continuity) conditions for the solution $u$ and its normal derivative at $r = a$. They are

$$\begin{aligned} a_m &= \frac{i^m \kappa_1 J_m(\kappa a) J_m'(\kappa_1 a) - \kappa i^m J_m'(\kappa a) J_m(\kappa_1 a)}{-\kappa_1 H_m(\kappa a) J_m'(\kappa, a) + \kappa H_m'(\kappa a) J_m(\kappa_1 a)} \\ b_m &= i^m \frac{J_m(\kappa a)}{J_m(\kappa_1 a)} + a_m \frac{H_m(\kappa a)}{J_m(\kappa_1 a)}\,, \quad -\infty < m < \infty\,, \end{aligned}$$

$m$ is an integer.

Let $\chi(x) = 1 - \zeta(x_1)\zeta(x_2)$, where the function $\zeta$ was defined above. We set

$$\Phi(x,\nu) = u(x,\nu)\chi(x)\,. \tag{4.9}$$

Thus the setting of (2.5)–(2.7) is obtained in this way.

In test #2 we used a cylinder with $a = \frac{\pi}{2}$ and $E = 0.1$. We have taken $\kappa = 5.5$, $mh = 5$, and $\ell h = 12$. $L_2(\Omega_1)$ relative error in the representation of $\epsilon$ by (4.4) is 17% in this case. The errors (4.6) obtained after different iterations are shown in Table 1. Note that the first iteration corresponds to the Born approximation. Therefore the quasi-Newton method dramatically improves the reconstruction comparing with the Born approximation even in the case of a small $\epsilon$.

In this test we used 121 vectors $\nu$. Each iteration for the inverse problem required 7 sec. of CPU time for total of 84 sec. for 12 iterations. In

*Table 1. Relative $L_2(\Omega_1)$ error (percent)*

| Iteration | 1 | 2 | 3 | 4 | 5 | 6 | 7 | 8 | 9 | 10 | 11 | 12 |
|---|---|---|---|---|---|---|---|---|---|---|---|---|
| Test #2 | 39.0 | 20.0 | 10.5 | 5.54 | 3.27 | 1.94 | 1.88 | 1.89 | 1.90 | 1.90 | 1.90 | 1.90 |
| Test #3 | 40.2 | 20.1 | 11.1 | 6.96 | 5.43 | 4.80 | 4.78 | 4.78 | 4.78 | 4.78 | 4.78 | 4.78 |

test #3, we have chosen $\kappa = 5.5$, $mh = 7$, $\ell h = 12$, $a = \frac{\pi}{2}$, and $E = 0.1$. Accordingly, 225 vectors $\nu$ were used here. In this case the relative $L_2(\Omega_1)$ error of the approximation $\epsilon$ by Fourier series (4.4) is 13%. The relative errors (4.6) are shown in Table 1.

To test the second version of the algorithm (see Section 3) we let $\gamma = 0.3$, $\kappa = 5.3$ and consider scattering from two coaxial cylinders. That is

$$\epsilon(r) = \begin{cases} \alpha\,, & \text{for } b < r < a \\ \beta\,, & \text{for } 0 \le r < b \\ 0\,, & \text{otherwise}\,. \end{cases} \tag{4.10}$$

In this case the forward problem (2.1)–(2.3) admits an explicit solution, which can be constructed as above. This solution was used to obtain an exact data $\Phi(x,\nu)$ as it was described in Section 2. The estimated coefficient $\epsilon_{\text{est}}$ was compared with the cylinder (4.10) to judge the performance of the algorithm.

First, we considered $m = N = M = 5$, (here $m = mh = lh$). The relative error (4.6) was between 1.22% and 1.94% after 12 iterations, depending on $\alpha$, $\beta$, $a$, and $b$ in (4.10). The inverse problem required about 8 sec. of CPU time for 12 iterations, which is about *10 times faster* than for the version I of our method. For $m = N = M = 7$ the relative error (4.6) was between 1.31% and 3.86% after 12 iterations. The inverse problem required about 41 sec. of CPU time for 12 iterations. Results of our numerical experiments in this case are shown on Figures 3a–3c and 4. Figures 3a–3c correspond to test #4, where $m = N = M = 7$, $a = \frac{\pi}{2}$, $b = 0.2\pi$, $\alpha = 0.03$, and $\beta = 0.05$. Figure 4 corresponds to test #5, where $m = N = M = 7$, $a = \frac{\pi}{2}$, $b = 0.2\pi$, $\alpha = 0.03$, and $\beta = 0.00$. In this case only the crossection through the central axis of the cylinders is shown.

**5. Convergence and identification.** Recall from Section 3 that $Q^0$ is the subspace of $L^2(\Omega_1 \times S^2)$ of functions $f(x,\nu)$ having a finite norm $\|f\|_{Q^0}$, defined by

$$\|f\|_{Q^0} = \text{ess-sup}\{\|f(\cdot,\nu)\|_{L^2(\Omega_1)} : \nu \in S^2\}\,.$$

Let

$$\begin{aligned} X &= L^\infty \times Q^0 \\ X_N &= L^\infty_N \times Q^0\,, \end{aligned}$$

where $L^\infty = L^\infty(\Omega_1)$ and $X$ and $X_N$ are equipped with the norm

$$\|x\|_X = \|(\epsilon, V)\|_X = \|\epsilon\|_{L^\infty} + \|V\|_{Q^0}\,. \tag{5.1}$$

We assume that $\Phi$ belongs to a linear (not necessarily closed) subspace $Q^2$ of $Q^0$ such that both $\Phi$, $\Delta\Phi \in Q^0$. Then the operator $A$ defined in (3.5) is acting from the subspace $L^\infty(\Omega_1) \times Q^2$ of $X$ into $Q^0$. Let the operator $B_\Phi : X \to Q^0$ be the extension of $A'(p) - A$ to all of $X$, that is

$$B_\Phi(\epsilon, V) = \kappa^2 \epsilon u_0 - \kappa^2 \epsilon (V + \Phi) - (\Delta\Phi + \kappa^2 \Phi)\,, \tag{5.2}$$

where $p = (0, u_0 - \Phi)$ and $u_0 = e^{i\kappa\langle x,\nu\rangle}$.

According with (3.1) and (3.2) we would like to define

$$T = [A'(p)]^{-1}(A'(p) - A) = [A'(p)]^{-1} B_\Phi \,. \tag{5.3}$$

However, typically for ill-posed problems, an operator, defined as above, would lack the necessary continuity properties. In our case, the operator $[A'(p)]^{-1}$, generally speaking, cannot be guaranteed to exist for every $f \in Q^0$. According with Tikhonov's treatment of ill-posed problems [29], equation

$$x = Tx \tag{5.4}$$

is replaced with a sequence of "regularized" equations

$$x = T_s x \,. \tag{5.5}$$

The stability analysis would address the relation between the solutions $\{x_s\}_{s=1}^{\infty}$ of (5.5) and the "true" solution $q$ of (5.4). In our case, the regularizing parameter is the constant $N$, which makes the approximations $\epsilon_n$ to be found in the finite-dimensional space $L_N^\infty$. Note that this constant $N$ cannot be made arbitrarily large in this algorithm. It is, in fact, determined by the given choice of the fixed frequency $\kappa$, since $\sqrt{3}N \le 2\kappa$.

The iterative algorithms presented in this paper are all regularized realizations of the Quasi-Newton method. We restate the summary of the first version of the algorithm presented in Section 3 in the form more suitable for a mathematical analysis (see (6.16)-(6.19) for a finite dimensional version). The second version of the algorithm, also presented in Section 3, can be restated similarly. The results (Theorems 5.1, 5.2, 6.2, 6.3) have the same form for the second version as for the first one. See [15] for details.

Let $\kappa^2$ be a noninteger. Fix a positive integer $N$ such that $\sqrt{3}N \le 2\kappa$ ($\sqrt{2}N \le 2\kappa$ for the 2-D case). Let $P$ be the set of points $Z \in \mathbb{R}^3$ such that the coordinates $z_1, z_2, z_3$ of $Z$ be integer and satisfy $|z_i| \le N$, $i = 1, 2, 3$.

For each $Z \in P$ find two unit vectors $\nu_Z, \mu_Z \in S^2$ such that $\kappa(\nu_Z + \mu_Z) = Z$. Let $u_0(x, \nu) = e^{i\kappa\langle x, \nu\rangle}$ for $x \in \mathbb{R}^3$, $\nu \in S^2$. Let $\Phi \in Q^2$.

Define the operator $B_\Phi : X \to Q^0$ by

$$B_\Phi(\epsilon, V) = \kappa^2 \epsilon u_0 - \kappa^2 \epsilon (V + \Phi) - (\Delta\Phi + \kappa^2\Phi) \tag{5.6}$$

Let $f \in Q^0$. For every point $Z \in P$ let

$$\tilde{f}^Z = \frac{1}{\kappa^2} \int_{\Omega_1} f(x, \nu) e^{i\kappa\langle x, \mu_Z\rangle} \, \mathrm{dx} \,. \tag{5.7}$$

Define the operator $K_\epsilon : Q^0 \to L_N^\infty$ by

$$K_\epsilon f = \frac{1}{(2\pi)^3} \sum_{Z \in P} \tilde{f}^Z e^{-i\langle x, Z\rangle} \,. \tag{5.8}$$

Define the operator $K_e : L^\infty \to Q^0$ by

$$K_e \epsilon = \kappa^2 \epsilon e^{i\kappa\langle x,\nu\rangle}, \quad \nu \in S^2. \tag{5.9}$$

Let $f \in Q^0$ and $Y = (y_1, y_2, y_3) \in \mathbb{R}^3$ be a point with the integer coordinates $y_1, y_2, y_3$. Let

$$\hat{f}^Y = \frac{1}{\kappa^2 - |Y|^2} \int_{\Omega_1} f e^{i\langle x,Y\rangle}\, \mathrm{dx}. \tag{5.10}$$

Define the operator $K_V : Q^0 \to Q^0$ by

$$K_V f = \frac{1}{(2\pi)^3} \sum_Y \hat{f}^Y e^{-i\langle x,Y\rangle}. \tag{5.11}$$

It can be shown that $K_V$ is a continuous linear operator from $Q^0$ into $Q^0$.

Define the operator $K : Q^0 \to X_N \subset X$ by

$$Kf = (K_\epsilon f, K_V(f - K_e K_\epsilon f)). \tag{5.12}$$

Finally, define the (regularized) operator $T : X \to X_N$ by

$$T(\epsilon, V) = K B_\Phi(\epsilon, V). \tag{5.13}$$

We attempt to solve the equation $Tx = x$ iteratively, that is

$$(\epsilon_n, V_n) = T(\epsilon_{n-1}, V_{n-1}), \quad n \geq 1 \tag{5.14}$$

for an initial guess $(\epsilon_0, V_0)$. Let

$$U_\delta(x) = \{y \in X : \|y - x\|_X < \delta\}. \tag{5.15}$$

Our main result concerning the convergence of this iterative algorithm with an *exact* data $\Phi$ (see Section 2) is the following.

THEOREM 5.1. *Let the operator $T$ be defined by (5.13) and let $u_0(x,\nu) = e^{i\kappa\langle x,\nu\rangle}$, $\nu \in S^2$. Let $\delta = \frac{1}{4\|K\|\kappa^2}$. Then there exists a positive constant $\hat{\delta} = \hat{\delta}(\kappa, N)$ such that*

*(i) $\hat{\delta} < \frac{\delta}{2}$*

*(ii) $\|\epsilon\|_{L^\infty} < \hat{\delta}$ implies $\|\tilde{u}\|_{L^2(\Omega_1)} < \frac{\delta}{2}$, where $\epsilon$, $\tilde{u}$ are as in (2.1)–(2.3).*

*(iii) Given an* **exact** *data $\Phi$, generated by an $\epsilon \in L^\infty(\Omega_1)$ with $\|\epsilon\|_{L^\infty} < \hat{\delta}$, the sequence $\{(\epsilon_n, V_n)\}_{n=1}^\infty$ defined by (5.14) converges for any initial guess $(\epsilon_0, V_0) \in U_\delta(0, u_0 - \Phi)$.*

*(iv) The limiting point $q = (\epsilon_q, V_q)$ of $\{(\epsilon_n, V_n)\}_{n=1}^\infty$ is the unique fixed point of the operator $T$ in $U_\delta(0, u_0 - \Phi)$.*

The proof of Theorem 5.1 involves establishing two crucial facts:

(5.16) (i) $\|Tx - Ty\|_X \leq \frac{1}{4}\|x-y\|_X$ , for $x, y \in U_\delta(p)$ ,
where $p = (0, u_0 - \Phi)$, and

(5.17) (ii) $\|T(p) - p\|_X < \frac{\delta}{4}$ ,

which implies that $T$ is a contraction in $U_\delta$, and $U_\delta$ is invariant under $T$. In turn, (5.16), (5.17) are proved by showing that the operators $K_\epsilon$, $K_e$, $K_V$ and $K$ are continuous linear operators on their domains of definition and the operator $B_\Phi$ satisfies

$$\|B_\Phi x - B_\Phi y\|_{Q^0} \leq \frac{1}{4\|K\|}\|x-y\|_X$$

for $x, y \in U_\delta(p)$.

Theorem 5.1 is used to show that the algorithm identifies the sought coefficient $\hat{\epsilon} \in L_N^\infty$. More precisely, we have

THEOREM 5.2. *Suppose $\delta, \hat{\delta}$ are chosen according with Theorem 5.1. Suppose that $\hat{\epsilon} \in L_N^\infty$ is such that $\|\hat{\epsilon}\|_X \leq \hat{\delta}$ and $\Phi \in Q^2$ is an exact data for IP (2.5)–(2.7) generated by $\hat{\epsilon}$. Let the sequence $\{(\epsilon_n, V_n)\}_{n=1}^\infty$ be defined by*

$$(\epsilon_n, V_n) = T(\epsilon_{n-1}, V_{n-1}), \quad n \geq 1$$

*with $(\epsilon_0, V_0) \in U_\delta(0, u_0 - \Phi)$. Then $\lim_{n\to\infty} \|\epsilon_n - \hat{\epsilon}\|_\infty = 0$.*

For details see [14],[15].

**6. Noisy data and finite dimensional approximations.** In practice, the measurements $\varphi(x, \nu)$ of the scattered field are known only approximately. In terms of IP (2.5)–(2.7) it means that the measurement function $\Phi(x, \nu)$ is *not* necessarily an *exact* data. In the case of such a noisy data we assume that $\Phi(x, \nu)$ can be represented as the sum of two components: an exact data $\Phi_e$ and a noise $\Phi_s$. The noisy part $\Phi_s$ should be small to achieve a reasonably good identification of the unknown coefficient function $\epsilon(x)$.

DEFINITION 6.1. Function $\Phi \in Q^2$ is called a *data* with the noise of level $\delta_s$ for the IP (2.5)–(2.7) if

$$\Phi = \Phi_e + \Phi_s \tag{6.1}$$

where $\Phi_e$ is an *exact data* for (2.5)–(2.7) and $\Phi_s \in Q^2$ with $\|\Phi_s\|_{Q^0}$, $\|\Delta\Phi_s\|_{Q^0} < \delta_s$.

Since the operators $K$, $B_\Phi$ and $T$ can still be defined as in Section 5, we can iteratively define sequence $\{x_n\}_{n=1}^\infty$ by $x_n = Tx_{n-1}$, $n \geq 1$, for an initial guess $x_0 \in X$. The convergence of this sequence and an estimate of the precision of the identification with inexact data are established in the next theorem.

THEOREM 6.2. *Let operator* $T = KB_\Phi$ *be defined by (5.13) and let* $u_0(x,\nu) = e^{i\kappa\langle x,\nu\rangle}$ *for* $x \in \Omega_1$, $\nu \in S^2$. *Then there exist positive constants* $\delta(\kappa, N)$ *and* $\hat{\delta}(\kappa, N)$ *such that*

(i) $\delta = \frac{1}{4\|K\|\kappa^2}$

(ii) *Given a noisy data* $\Phi = \Phi_e + \Phi_s$ *satisfying*

$$\text{a).} \qquad \Phi_e \text{ is an exact data generated by an } \epsilon \in L^\infty(\Omega_1),$$

$$\|\epsilon\|_{L^\infty} < \hat{\delta}\,,$$

$$\text{b).} \qquad \|\Phi_s\|\,,\ \|\Delta\Phi_s\|_{Q^0} < \delta_s\,,\ 0 < \delta_s < \tfrac{1}{8}\delta^2\,,$$

*the sequence* $\{x_n\}_{n=1}^\infty$ *defined by* $x_n = Tx_{n-1}$, $n \geq 1$, $x_0 \in U_{\frac{\delta}{2}}(0, u_0 - \Phi)$ *converges in* $X$.

(iii) *The limiting point* $q = (\epsilon_q, V_q)$ *of the sequence* $\{x_n\}_{n=1}^\infty$ *is the unique fixed point of the operator* $T$ *in* $U_\delta(0, u_0 - \Phi)$.

(iv) *If* $\epsilon \in L_N^\infty$, *then the following estimate relating the coefficient* $\epsilon$ *and its identification* $\epsilon_q$ *is valid*

$$\|\epsilon - \epsilon_q\|_{L^\infty} < 4\kappa^2\|K\|\delta_s = \frac{\delta_s}{\delta}\,. \tag{6.2}$$

Thus, according with the above theorem, the identified coefficients $\epsilon_q$ approach the "true" coefficient $\epsilon$ as the level of noise decreases. In fact, the method seems to be quite stable in this regard.

In test #6, we have considered the scattering (version I) from the cylinder with $a = 0.1\pi$, $E = 0.3$ and with the *noise* contaminated data $\Phi$.

Denote

$$\Phi_1(x,\nu) = e^{-i\kappa\langle x,\nu\rangle}\Phi$$

$$\Phi_2(x,\nu) = \Delta\Phi_1 + 2i\kappa\langle\nu, \nabla\Phi_1\rangle\,.$$

Hence

$$\Phi_2(x,\nu) = (\Delta\Phi + \kappa^2\Phi)e^{-i\kappa\langle x,\nu\rangle}\,.$$

Let

$$M = \max_{x,\nu_n}|V(x,\nu_n)| = \max_{x,\nu_n}|u(x,\nu_n) - \Phi(x,\nu_n)|\,.$$

Let $\eta_1$, $\eta_2$ be random variables uniformly distributed on $[0,1]$. For each pair of vectors $x, \nu_n$ we choose its own $\eta_1(x,\nu_n)$, $\eta_2(x,\nu_n)$ and set

$$\widehat{\Phi}_2(x,\nu_n) = \Phi_2(x,\nu_n) + \sqrt{2}pM[(\eta_1(x,\nu_n) - 0.5) + i(\eta_2(x,\nu_n) - 0.5)]$$

$$\widetilde{\Phi}_2(x,\nu_n) = \widehat{\Phi}_2(x,\nu_n)e^{i\kappa\langle x,\nu_n\rangle},$$

where $p \geq 0$ is the level of noise. Then we substitute $\widetilde{\Phi}_2$ into (2.5)–(2.7) and solve the corresponding inverse problem. In test #6 we have taken $\kappa = 5.5$, $mh = 7$, $\ell h = 12$, and 225 vectors $\nu$. The function $\epsilon_{\text{test}} + 1$ is shown on Figure 5a. The functions $\epsilon_{\text{est}} + 1$ obtained after 12 iterations, for different levels of noise are shown on Figures 5b,c. The relative error (4.6) is shown in Table 2. One can conclude from this table that our method is quite stable.

*Table 2. Relative $L_2(\Omega_1)$ error for different levels of noise p.*

| noise p% | 0 | 1 | 5 | 10 | 25 | 50 | 100 | 200 |
|---|---|---|---|---|---|---|---|---|
| relative error (4.6) after 12 iterations | 4.33 | 4.36 | 4.68 | 5.45 | 9.97 | 16.8 | 33.2 | 68.9 |

Finally in this Section we consider a finite dimensional version of the iterative algorithms described in Section 3, having in mind their applications to numerical implementations. In fact, it is this version of the algorithm for which the summary was presented in Section 3. Only the first version is presented here. The second version can be treated similarly, see [15].

It follows from the definitions of the operators $K_\epsilon$, $K_e$, $K_V$, $K$, $B_\Phi$ and $T$ (see (5.6)–(5.13)) that the only definition involving infinite summation is that of the operator $K_V$ in (5.11).

Let $M > 0$ and let $D$ denote the subset of $\mathbb{R}^3$ of the points $Y = (y_1, y_2, y_3)$ with the integer coordinates satisfying $|y_i| \leq M$, $i = 1, 2, 3$.

Define $K_V^M : Q^0 \to Q^0$ by

$$K_V^M f = \frac{1}{(2\pi)^3} \sum_{Y \in D} \hat{f}^Y e^{-i\langle x,Y\rangle}, \tag{6.3}$$

where

$$\hat{f}^Y = \frac{1}{\kappa^2 - |Y|^2} \int_{\Omega_1} f e^{i\langle x,Y\rangle}\, \mathrm{dx}\,. \tag{6.4}$$

Define $T^M : X \to X_N$ by

$$T^M x = K^M B_\Phi x\,, \tag{6.5}$$

where

$$K^M = (K_\epsilon f, K_V^M(f - K_e K_\epsilon f))\,, \tag{6.6}$$

(compare with (5.10)–(5.13)).

If $\Phi \in Q^2$ is a data for IP (2.5)–(2.7) and, say, $x_0^M = (0, u_0 - \Phi)$, then a sequence $\{x_n^M\}_{n=1}^\infty$ can be iteratively defined by

$$x_n^M = T^M x_{n-1}^M, \quad n \geq 1. \tag{6.7}$$

The convergence of this sequence, the existence of a fixed point $q^M$ and the fact that $q^M \to q$ as $M \to \infty$, where $q = Tq$ are established in the next theorem. We state it in the case of an *exact* data for (2.5)–(2.7). A result similar to Theorem 6.2 can be established as well.

THEOREM 6.3. *Let the conditions of Theorem 5.1 be satisfied. Let $q = \lim_{n\to\infty} x_n$, where $x_n = Tx_{n-1}$, $n \geq 1$ and $x_0 \in U_\delta(0, u_0 - \Phi)$. Let $M > 0$ and the sequence $\{x_n^M\}_{n=1}^\infty$ be defined by (6.7). Then*

*(i) The sequence $\{x_n^M\}_{n=1}^\infty$ converges in $X$ for a sufficiently large $M > 0$.*
*(ii) Its limiting point $q^M$ is the unique fixed point of the operator $T^M$ in $U_\delta(0, u_0 - \Phi)$.*
*(iii) $\lim_{M\to\infty} q^M = q$.*

**7. Inverse scattering in random media.** Under certain conditions waves in a medium propagate mainly in "propagation channels", see e.g. [13], [28]. In such cases the scattered field, described by system (2.1)–(2.3) involving the Helmholtz equation, is replaced by a non-stationary Schrödinger equation.

As before, we assume $\operatorname{supp}\epsilon \subset W = \{x \in \mathbb{R}^3 : |x| \leq \pi/2\}$. Suppose that a plane wave $e^{i\kappa\langle x,\nu\rangle}$ propagates in the $z$ direction. That is $\nu = (0,0,1)$ and $u = e^{i\kappa z} + \tilde{u}$. Let $v = ue^{-i\kappa z}$. Then

$$\Delta_{x,y} v + v_{zz} + 2i\kappa v_z + \kappa^2 \epsilon v = 0. \tag{7.1}$$

Physical assumptions are that $|v_{zz}| \ll |2i\kappa v_z|$ and that the backscattering is small. That is, the scattered field $\tilde{u}$ is assumed to be zero on any plane $z = C$ with $C < -\frac{\pi}{2}$. Therefore $v = ue^{-i\kappa z} = (e^{i\kappa z} + \tilde{u})e^{-i\kappa z} = 1$ on such a plane. Allowing the plane waves $e^{i\kappa\langle x,\nu\rangle}$ to come from any direction $\nu \in S^2$, we obtain the following non-stationary Schrödinger equation

$$2i\kappa\nu\nabla v + \Delta_\perp v + \kappa^2 \epsilon v = 0, \tag{7.2}$$

where $\Delta_\perp$ denotes the Laplace operator with respect to the two variables orthogonal to the vector $\nu$.

Concerning the initial condition, let $\nu \in S^2$ and $\ell_\nu$ be any plane orthogonal to the vector $\nu$, located sufficiently far from the ball $W$ and before the wave reaches $W$. For definiteness let

$$\ell_\nu = \{y \in \mathbb{R}^3 : \langle y, \nu\rangle = -2\pi\}.$$

Thus, $\ell_\nu$ is located at the distance $2\pi$ from the origin, and according with the above discussion, the appropriate initial condition for $v(x,\nu)$ is

$$v(x,\nu)|_{x\in\ell_\nu} = 1. \tag{7.3}$$

The unknown coefficient function $\epsilon(x)$ in (7.2) is assumed to be a real valued random field satisfying

A1. $\epsilon(x) = 0$ for $x \in \mathbb{R}^3 \backslash W$

A2. Let $\tilde{\epsilon}(x) = \langle \epsilon(x) \rangle$ be the ensemble average of $\epsilon(x)$ at the point $x$. Thus $\epsilon(x) = \tilde{\epsilon}(x) + \epsilon_0(x)$, where $\langle \epsilon_0(x) \rangle = 0$. We assume that $\tilde{\epsilon} \in L^\infty(W)$ and the random field $\epsilon_0$ has a normal distribution for each $x \in W$ with the density function $\mu(x,t)$ given by

$$\mu(x,t) = \frac{\exp(-\frac{t^2}{2\sigma^2(x)})}{\sqrt{2\pi\sigma^2(x)}},$$

where $\sigma(x) > 0$ in $W$, $\sigma(x) = 0$ in $\mathbb{R}^3 \backslash W$ and $\sigma \in L^\infty(W)$.

A3. The random field $\epsilon_0(x)$ is delta-correlated, that is, for any two points $x_1, x_2 \in W$

$$\langle \epsilon_0(x_1)\epsilon_0(x_2) \rangle = \delta(|x_1 - x_2|)\sigma(x_1) \cdot \sigma(x_2),$$

where $\delta$ is the Dirac function.

Let $G$ be the cube

$$G = \{x \in \mathbb{R}^3 : -\pi < x_1, x_2, x_3 < \pi\},$$

and

$$U = G \backslash W.$$

Now we can state the Inverse Scattering Problem (ISP) in random media:

*Let $\kappa > 0$ be a fixed number and $v(x,\nu)$ be the solution of the Cauchy problem (7.2), (7.3) for the non-stationary Schrödinger equation. Let $w(x,\nu) = \langle v(x,\nu) \rangle$.*

*Determine functions $\tilde{\epsilon}(x)$ and $\sigma^2(x)$ assuming that*

$$w|_{x \in U} = \varphi(x,\nu), \tag{7.4}$$

*where the function $\varphi(x,\nu)$ is given for all $x \in \partial W$ and $\nu \in S^2$.*

We do not know how to investigate Cauchy problem (7.2), (7.3) because of the complicated nature of the function $\epsilon$. However, physicists accept the following *deterministic* model for these phenomena [25].

$$2i\kappa\nu\nabla w + \Delta_\perp w + \kappa^2 b(x) w = 0 \tag{7.5}$$

$$w|_{\ell_\nu} = 1, \tag{7.6}$$

where

$$b(x) = [\tilde{\epsilon}(x) + \frac{i\kappa}{4}\sigma^2(x)]. \tag{7.7}$$

Thus we consider only the following:

**Deterministic ISP**

*Determine function $b(x)$ from (7.4)–(7.7).*

Here and in the sequel the solution $w$ of the Cauchy problem (7.5)–(7.7) is understood in a weak sense. It can be shown, see [21] for details, that such a solution is well defined, exists and is also unique.

**Description of the algorithm**

We solve the above Deterministic ISP by a method similar to the one used for the deterministic media, that is by a regularized Quasi-Newton method. Again, only the weak scattering case ($b(x)$ is small) is considered. We start with a more convenient formulation of the ISP which can be obtained as follows: Let

$$\Phi(x,\nu) = w(x,\nu)\cdot\chi(x)\,,$$

where $\chi$ is 0 on $W$ and $\chi$ is 1 on $G$.

Let $V = w - \Phi$, then

$$L_\nu(V+\Phi) + \kappa^2 b(x)(V+\Phi) = 0 \quad \text{in} \quad R_\nu\,, \tag{7.8}$$

where $L_\nu u = 2i\kappa\nu\nabla u + \Delta_\perp u$. Also note that $V = 0$ in the domain $U$.

Writing (7.8) as

$$A(b,V) = 0 \tag{7.9}$$

we can use the Quasi-Newton method to obtain a solution. Accordingly, we would like to form the sequence

$$x_{n+1} = [A'(x_0)]^{-1}(A'(x_0)x_n - Ax_n)\,, \tag{7.10}$$

for $n \geq 0$, being provided $x_0 \in X$. In other words

$$A'(x_0)x_{n+1} = A'(x_0)x_n - Ax_n\,. \tag{7.11}$$

If $x = (b,V)$, then it can be shown that

$$A'(x_0)(b,V) = L_\nu V + \kappa^2 b\,,$$

provided $x_0 = (0, 1-\Phi(x,\nu))$. Thus, at each iterative step, given function $f$, we solve the equation

$$L_\nu V + \kappa^2 b = f \tag{7.12}$$

for $V$ and $b$.

This can be done as follows: Let $\xi(x) \in C^3(\mathbb{R}^2)$ be such that

$$\xi(x) = \begin{cases} 1 & x \in G \\ 0 & |x| \geq 2\pi\,. \end{cases}$$

Now multiply both sides of (7.12) by $e^{i\langle x,Z\rangle}$, where $Z$ is a point in $\mathbb{R}^3$ with the integer coordinates, and integrate over $R_\nu$.

If $V(x) = 0$ for $x \notin G$, we obtain

$$\int_G V L_\nu^*(\xi(x)e^{i\langle x,Z\rangle})\,\mathrm{dx} + \kappa^2 \int_G b(x)e^{i\langle x,Z\rangle}\,\mathrm{dx} = \int_{|x|\leq 2\pi} f(x,\nu)\xi(x)e^{i\langle x,Z\rangle}\,\mathrm{dx}\,, \tag{7.13}$$

where $L_\nu^*(U) = -2i\kappa\nu \cdot \nabla U + \Delta_\perp U$. Note that

$$L_\nu^*(e^{i\langle x,Z\rangle}) = 2\kappa\langle \nu, Z\rangle - |Z_\perp|^2\,,$$

where $Z_\perp = Z - (Z,\nu)\nu$. Thus

$$\begin{aligned} L_\nu^*(e^{i\langle x,Z\rangle}) &= 2\kappa\langle \nu, Z\rangle + \langle \nu, Z\rangle^2 - |Z|^2 \\ &= (2\kappa\cos\langle \nu, Z\rangle + |Z|\cos^2\langle \nu, Z\rangle - |Z|)|Z|\,. \end{aligned}$$

Therefore $L_\nu^*(e^{i\langle x,Z\rangle}) = 0$ if the vector $\nu_Z$ is chosen to satisfy

$$\cos(\nu_Z, Z) = \frac{-\kappa + \sqrt{\kappa^2 + |Z|^2}}{|Z|}\,.$$

Therefore all the Fourier coefficients of $b(x)$ can be recovered in this manner from (7.13) and, consequently, the function $b(x)$ itself. Now, since $b(x)$ is already known we are left with finding $V$ from

$$\begin{cases} L_\nu(V) = f(x,\nu) - \kappa^2 b(x) \\ V|_{\ell_\nu} = 0\,. \end{cases} \tag{7.14}$$

These Cauchy problems can be solved for every $\nu \in S^2$.

Again, we remark that the required zero boundary conditions for functions $V$ are assumed but not necessarily satisfied. Nevertheless, as in the case of the deterministic scattering, the above algorithm is well defined and is, in fact, a regularized Quasi-Newton method. Its convergence and the identification properties can be established along the lines similar to the deterministic case.

**8. Strong scattering.** In this section we describe our results from [19], which, we believe, can be useful for the study of the strong scattering and the global convergence.

Let $T = \text{const} > 0$, $R_T^3 = \mathbb{R}^3 \times (0,T)$ and $r_0$ be another positive constant. Introduce a cylinder $\Omega_{r_0} \subset \mathbb{R}^3$,

$$\Omega_{r_0} = \{x \in \mathbb{R}^3 : \sqrt{x_1^2 + x_2^2} < r_0\}\,.$$

Let $a(x) \in C^2(\mathbb{R}^3)$ be a nonnegative function such that $a(x) = 0$ for $x \in \Omega_{r_0}$. Consider solution of the Cauchy problem

$$u_{tt} = \Delta u + a(x)u\,, \quad x \in R^3_T \tag{8.1}$$

$$u|_{t=0} = 0\,, \quad u_t|_{t=0} = \delta(x)\,,$$

where $\delta(x)$ is the delta function.

*The ISP statement.* Determine the function $a(x)$ for $x \in (\mathbb{R}^3 \backslash \Omega_{r_0}) \bigcap \{|x| < R\}$ assuming that the following function $\psi$ is given

$$u|_\omega = \psi(x,t)\,, \quad t \in (0,T)\,, \tag{8.2}$$

where $R$ is a fixed number, $r_0 < R < \frac{1}{3}T$ and $\omega$ is the surface of the cylinder $\Omega_{r_0} \subset \mathbb{R}^3$.

We note that since function $a(x) = 0$ in $\Omega_{r_0}$, then we can uniquely determine function $u(x,t)$ for $(x,t) \in \Omega_{r_0} \times (0,T)$ as a solution of the following boundary value problem

$$u_{tt} = \Delta u\,, \quad x \in \Omega_{r_0} \times (0,T)$$

$$u|_{t=0} = 0\,, \quad u_t|_{t=0} = \delta(x) \tag{8.3}$$

$$u|_\omega = \psi(x,t)\,.$$

*Remark* 8.1. The only "useful" part of the signal $u$ measured on the surface $\omega$, (see (8.2)) is the backscattering information. Indeed, the waves propagate from the source $x = 0$, which is located inside the cylinder $\Omega_{r_0}$, but the support of the unknown function $a(x)$ is outside $\Omega_{r_0}$. If the measurement data would not contain the backscattering signal, we would not be able to prove the uniqueness result (see Theorem 8.2 below).

Introduce the spherical coordinates

$$x_1 = r\cos\theta\sin\varphi\,, \quad x_2 = r\sin\theta\sin\varphi\,, \quad x_3 = r\cos\varphi\,,$$

$$\theta \in [0,2\pi)\,, \quad \varphi \in [0,\pi)\,.$$

Let $T_1 = T - R$. Denote

$$R^3_{T_1} = \mathbb{R}^3 \times (0,T_1)\,,$$

$$D_{T_1} = \{(x,t)\,:\, r_0 < |x| < R\,,\ \sin\varphi > \frac{r_0}{R}\,,\ t \in (0,T_1)\}\,.$$

Hence, for all $(x,t) \in D_{T_1}$

$$r \sin\varphi = \sqrt{x_1^2 + x_2^2} > \frac{r_0^2}{R}$$

and the conic surface

$$\partial_1(D_{T_1}) = \{\sin\varphi = \frac{r_0}{R},\ r_0 < |x| < R,\ t \in (0, T_1)\}$$

is a part of the boundary of the domain $D_{T_1}$. Furthermore, $\partial_1(D_{T_1}) \subset \Omega_{r_0}$ because for $(x,t) \in \partial_1(D_{T_1})$ we have $\sqrt{x_1^2 + x_2^2} = r\sin\varphi = \frac{r}{R} r_0 < r_0$. Also, consider another part $\partial_2(D_{T_1})$ of the boundary $D_{T_1}$ defined by

$$\partial_2(D_{T_1}) = \{|x| = r_0\,,\ \sin\varphi > \frac{r_0}{R},\ t \in (0, T_1)\}\,.$$

Denote

$$\eta = \partial_1(D_{T_1}) \bigcup \partial_2(D_{T_1})\,.$$

Hence, by (8.2) and (8.3) the following two functions $\psi_1$ and $\psi_2$ are known from

$$u|_\eta = \psi_1\,, \quad \frac{\partial u}{\partial n}|_\eta = \psi_2\,, \tag{8.4}$$

where $n$ is the outward normal vector of the surface $\eta$.

We should also introduce some function spaces. Let $\varphi_1 = arcsin(\frac{r_0}{R})$ and $\varphi_2 = \pi - \varphi_1$. Let $L_{2,\varphi}$ be the space of functions $f(\theta, \varphi)$, $\theta \in (0, 2\pi)$, $\varphi \in (\varphi_1, \varphi_2)$ having a finite norm

$$||f||_{L_{2,\varphi}} = \int_0^{2\pi} d\theta \int_{\varphi_1}^{\varphi_2} |f(\theta,\varphi)|^2 \sin\varphi \, d\varphi\,.$$

Let $\tilde{H}_0^2(D_{T_1})$ be the Sobolev space of functions $f$ having a finite norm

$$||f||_{\tilde{H}_0^2(D_{T_1})} = \{\int_0^{2\pi} d\theta \int_{\varphi_1}^{\varphi_2} \sin\varphi \, d\varphi \int_{r_0}^{R} r^2 \, dr \int_0^{T_1} dt[|f|^2 +$$

$$+ |f_r'|^2 + |f_t'|^2 + |f_{rt}''|^2]\}^{1/2}$$

and such that $f|_{r=r_0} = 0$.

Finally, let $H_0^2(r_0, R)$ be the space of functions $f(r)$ having a finite norm

$$||f||_{H_0^1(r_0,R)} = \{\int_{r_0}^{R} (|f|^2 + |f'|^2) r^2 \, dr\}^{1/2}$$

and such that $f(r_0) = f'(r_0) = 0$.

Returning to our ISP (8.1)–(8.3), consider the function

$$v(x,t) = \frac{u(x,t+|x|)}{\int_0^t u(x,\tau+|x|)\,d\tau}\,, \qquad t>0\,. \tag{8.5}$$

Note that since $a(x) \geq 0$ we have

$$\int_0^t u(x,\tau+|x|)\,d\tau = \frac{1}{4\pi|x|} + u_1(x,t)\,,$$

where the function $u_1(x,t) \in C^2(R^3_{T_1})$ is non–negative and $u_1(x,0) = 0$. Thus the characteristic conic surface $\{t = |x|\}$ is converted into the surface $\{t = 0\}$.

We can show that the function $v(x,t)$ is a solution of the following non-linear Cauchy problem

$$L(v) = 0\,, \tag{8.6}$$

$$v|_\eta = \hat{\psi}_1\,, \qquad \frac{\partial v}{\partial n}|_\eta = \hat{\psi}_2\,,$$

where the functions $\hat{\psi}_1, \hat{\psi}_2$ are generated by the functions $\psi_1, \psi_2$ from (8.4) under the transformation (8.5) and the operator $L$ has the form

$$Lv = L_0 v + 4\nabla v \int_0^t \nabla v(x,\tau)\,d\tau - 2v_t \int_0^t v_r(x,\tau)d\tau$$

$$-2v_r v + 2\nabla v \nabla g - 2v_t g'(r)\,, \tag{8.7}$$

where

$$L_0 v = \Delta v - 2v_{rt}\,, \quad g(r) = ln(\frac{1}{4\pi r})\,, \tag{8.8}$$

and

$$\frac{\partial}{\partial r} = \sum_{i=1}^{3} \frac{x_i}{|x|}\frac{\partial}{\partial x_i}\,.$$

Furthermore, the Cauchy problem (8.6)–(8.8) is equivalent to the original ISP.

We seek the solution of this Cauchy problem in the form

$$v(x,t) = p(x,t) + w(x,t)\,,$$

where the function $p \in H^2(D_{T_1})$ is given, the function $w \in H^2(D_{T_1}) \bigcap H^2_0(D_{T_1})$ is unknown. Also we assume that the function

$w$ has a fixed finite number of possibly nonzero Fourier coefficients with respect to an orthonormal basis described below. If the function $p$ is a starting point (the first guess) for an iterative process, then the function $w$ is a finite-dimensional perturbation of the first guess.

Now we describe the orthonormal basis in $L_{2,\varphi}$ mentioned above. The basis $\{\Phi_n(\theta,\varphi)\}_{n=1}^{\infty}$ consists of functions $\Phi_n \in L_{2,\varphi}$ such that

$$\Phi_n \in C^2([0,2\pi]\times[\varphi_1,\varphi_2])$$

$$\Phi_n(\theta,\varphi_1) = \frac{\partial\Phi_n}{\partial\varphi}(\theta,\varphi_1) = \Phi_n(\theta,\varphi_2) = \frac{\partial\Phi_n}{\partial\varphi}(\theta,\varphi_2) = 0$$

Further, choose an orthonormal basis $\{Q_n(r)\}_{n=1}^{\infty}$ in $H_0^1(r_0,R)$ such that all the functions $Q_n(r)$ are piecewise analytic in $(r_0,R)$, $Q_n \in C^2[r_0,R]$ and $Q_n(r_0) = Q_n'(r_0) = 0$. Finally, let $\{S_n(t)\}_{n=1}^{\infty}$ be an orthonormal basis in $H^1(0,T_1)$ such that all the functions $S_n$ are analytic in $(0,T_1)$ and $S_n \in C^2[0,T_1]$. Hence the products

$$\{\Phi_{n_1}(\theta,\varphi)Q_{n_2}(r)S_{n_3}(t)\}_{n_1,n_2,n_3=1}^{\infty}$$

form an orthonormal basis in $\tilde{H}_0^2(D_{T_1})$.

Choose an arbitrary integer $N>1$ and denote $m = N^3$,

$$P_N = \{f \in H_0^2(D_{T_1}) : f(x,t) = \sum_{n_1,n_2,n_3=1}^{N} a_n\Phi_{n_1}(\theta,\varphi)Q_{n_2}(r)S_{n_3}(t)\,,\quad n=(n_1,n_2,n_3)\}\,.$$

*Therefore we will assume that*

$$w \in P_N\,.$$

Now, let $\alpha = \text{constant} \in (\frac{R}{T_1},\frac{1}{2})$. Introduce a function $\beta$ by

$$\beta(x,t) = \beta(r,t) = -(r+\alpha t)\,.$$

Likewise, introduce the domain $G$ by

$$G = \{(x,t) : \beta(r,t) > -R\,,\ r>r_0\,,\ \sin\varphi > \frac{r_0}{R}\,,\ t>0\,\}\,.$$

Hence $G \subset D_{T_1}$.

Finally, it is the time to consider a cost functional. Let $\lambda$ be a positive parameter. Consider the cost functional $J_\lambda(w)$,

$$J_\lambda(w) = \int_G [L(w+p)]^2 e^{2\lambda\beta}\,dxdt\,,\quad w\in P_N\,.$$

Hence, our ISP can be reformulated as follows:

*Minimize the functional* $J_\lambda(w)$.

The main result of this section is the following theorem.

THEOREM 8.2. *Let* $M$ *be an arbitrary positive constant and*

$$B_M = \{w \in H_0^2(D_{T_1}) \bigcap P_N \, : \, \|w\|_{H_0^2(D_{T_1})} < M\} \, .$$

*Then there exists a positive constant* $\tilde{\lambda} = \tilde{\lambda}(D_{T_1}, P_N, B_M, \alpha)$ *such that for all* $\lambda \geq \tilde{\lambda}$ *the functional* $J_\lambda(w)$ *is uniformly strictly convex on* $B_M$. *That is*

$$J_\lambda(w+h) - J_\lambda(w) - J_\lambda'(w)(h) \geq \frac{k}{2}\|h\|^2_{H^2(D_{T_1})}$$

*for all* $w, h \in B_M$, *where the positive constant* $k = k(D_{T_1}, P_N, B_M, \lambda, \alpha)$ *does not depend on* $w, h$.

*Remark* 8.3.

(i). The main point of this theorem is that beginning with a function, which is a finite-dimensional perturbation of a true solution of the Cauchy problem (8.6)–(8.8), one can recover this true solution by a convex minimization technique. We remind that the Cauchy problem (8.6)–(8.8) is equivalent to the original ISP (8.1)–(8.3).

(ii). Thus, this theorem quarantees a global convergence on $B_M$ of a number of traditional minimization methods, provided that the true solution $w$ is in $B_{M/2}$ and the starting point is in the center of the ball $B_M$. Note, that by Tikhonov's principle one can a priori assume that the solution belongs to the given compact set $B_{M/2}$.

(iii). This theorem is proved by the method of Carleman estimates, which are applied to inverse problems beginning from the work of Bukheim and Klibanov [5], written in 1981. For a survey see Klibanov [18].

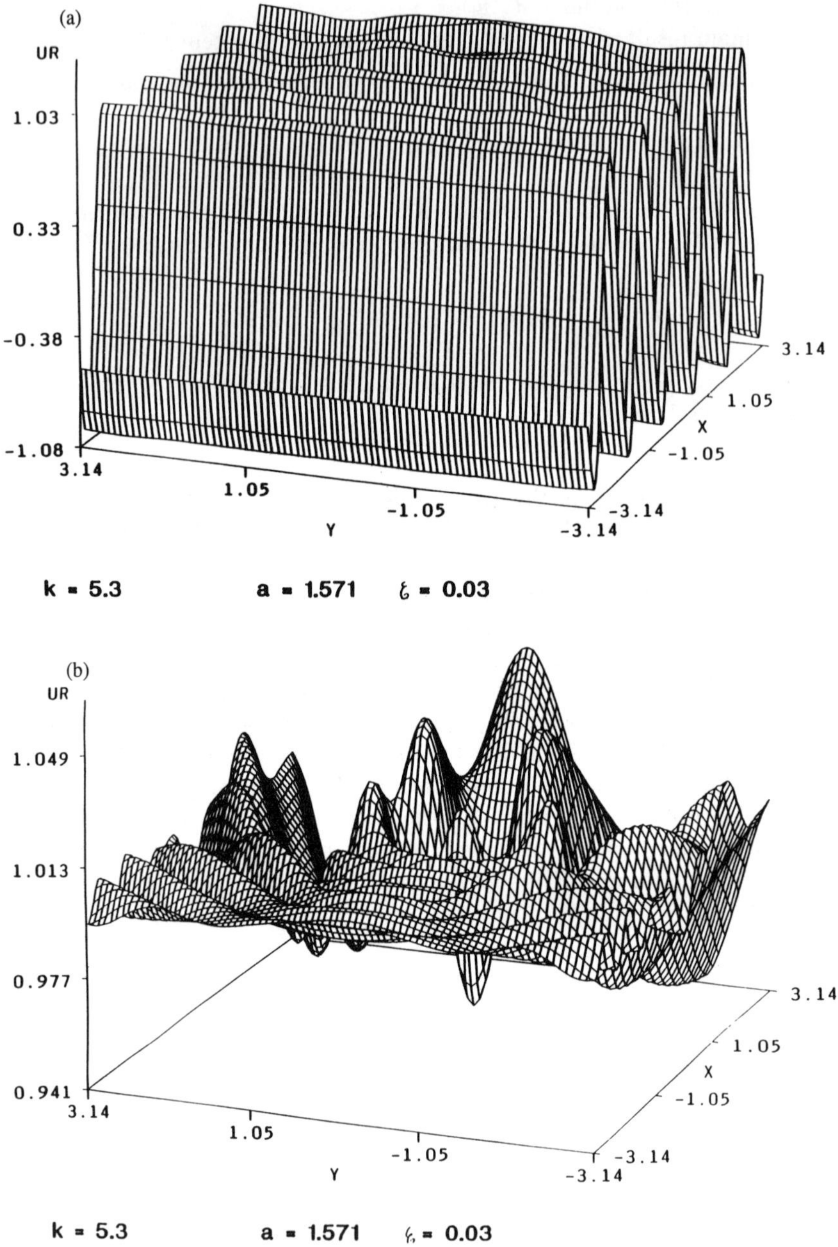


FIGURE 1a,b. *$Re[u(x,\nu)]$ and $Re[u(x,\nu)e^{-i\kappa\langle x,\nu\rangle}]$ in the case of scattering from a circular cylinder with $a = \frac{\pi}{2}$, $E = 0.03$, $\kappa = 5.3$, and $\nu = (1,0)$. The function $Re[u(x,\nu)e^{-i\kappa\langle x,\nu\rangle}]$ is much less oscillatory than the function $Re[u(x,\nu)]$.*

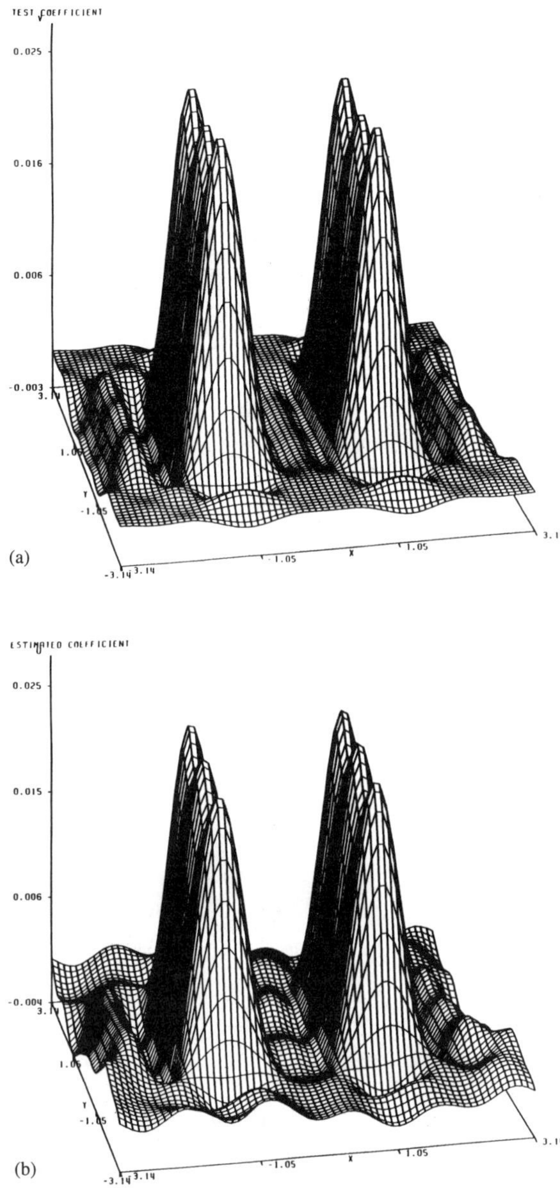


K = 5.5 mh = 5 lh = 12

FIGURE 2a,b. *Test #1.* $\epsilon_{test}$ *is the finite Fourier approximation (4.4) of*

$$E(x) = \begin{cases} 0.02\,, & \text{for } \frac{\pi}{4} < |x_1| < \frac{\pi}{2}\,, \quad |x_2| < \frac{\pi}{2} \\ 0\,, & \text{otherwise} \end{cases}$$

$\epsilon_{est}$ *represents the estimate coefficient. We have taken* $mh = 5$, $\ell h = 12$, *and* $\kappa = 5.5$. *The relative error (4.6) is 14%.*

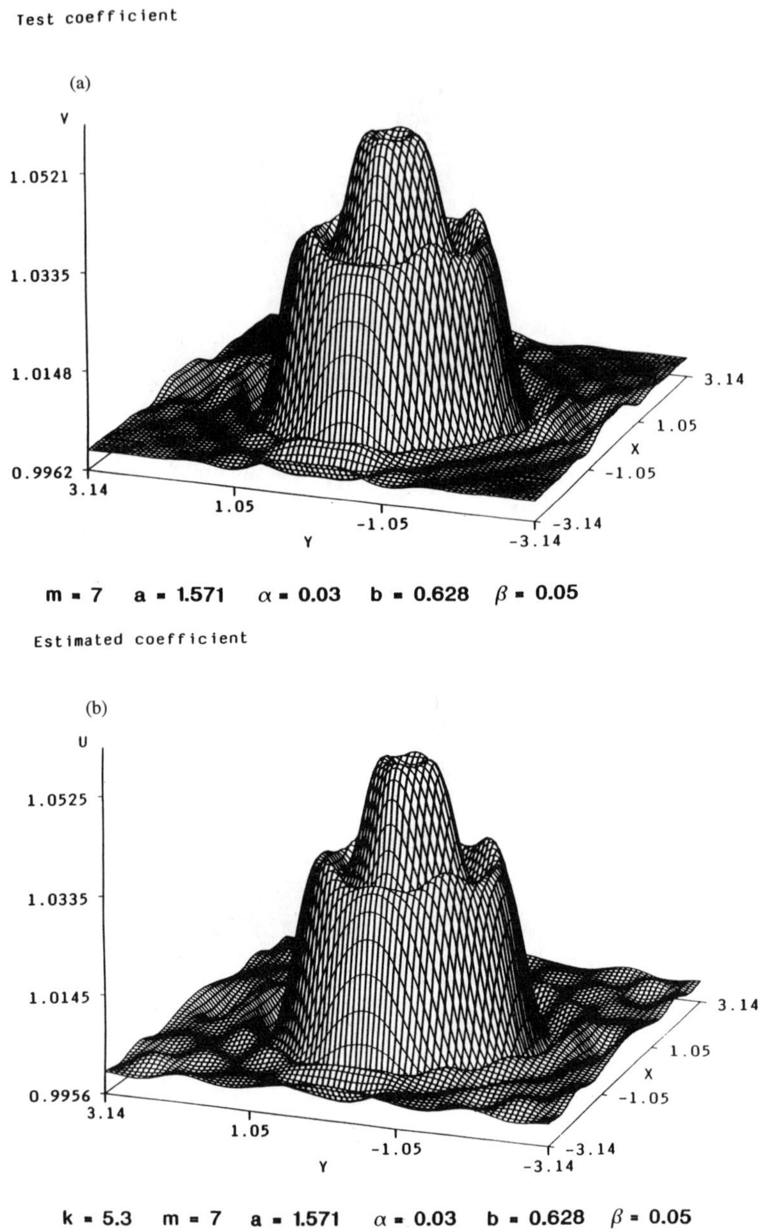


FIGURE 3a,b. *Test #4.* $\epsilon_{test} + 1$ *and* $\epsilon_{est} + 1$. $\epsilon_{test}$ *is the finite Fourier approximation of the function* $\epsilon(r)$ *in (4.10) with* $a = \frac{\pi}{2}$, $b = 0.2\pi$, *and* $\alpha = 0.03, \beta = 0.05$. *We have taken* $m = N = M = 7$. *The inverse problem required about 41 sec. of CPU time for 12 iterations. The relative error (4.6) is 3.45%.*

(c)

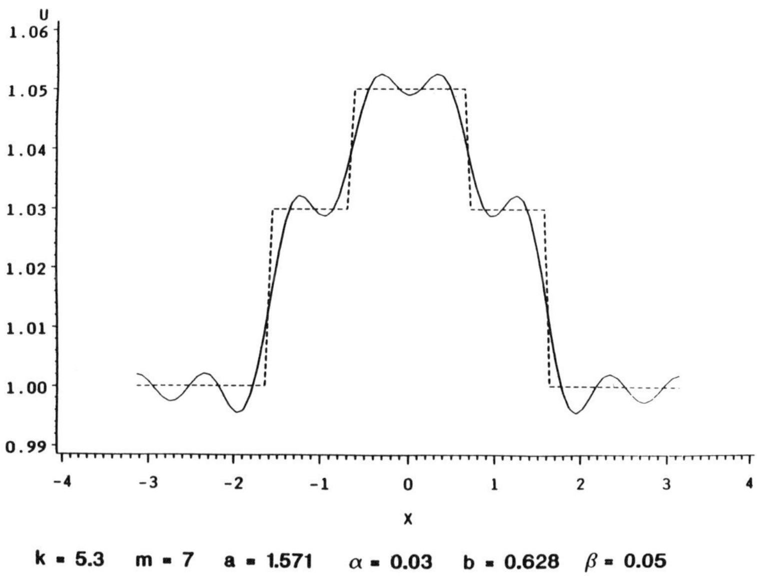


FIGURE 3c. *Test #4. Cross-section of the estimated coefficient and the coaxial cylinders (dotted lines) through the center of these cylinders.*

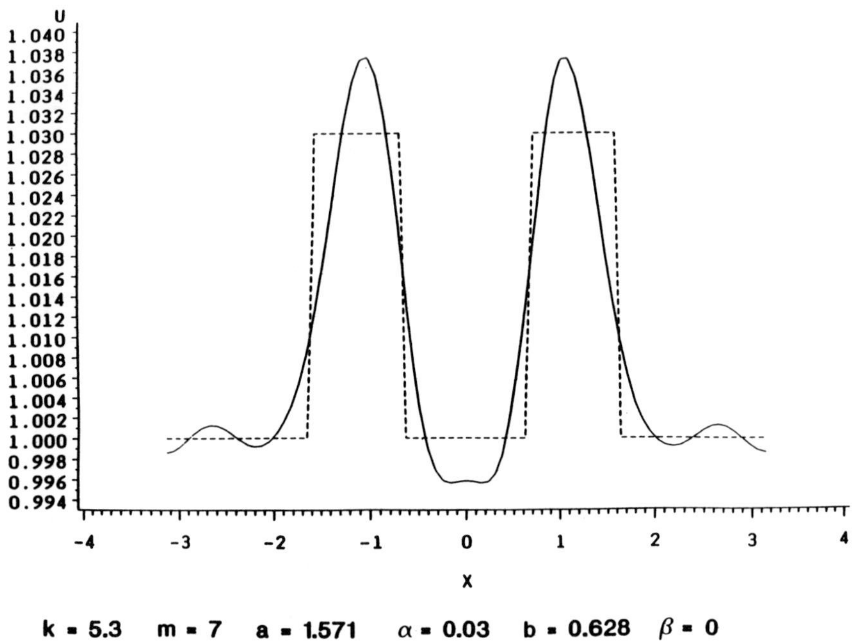


FIGURE 4. *Test #5. Cross-section of the estimated coefficient and the coaxial cylinders (dotted lines) through the center of these cylinders.* $\epsilon_{test}$ *is the Fourier approximation of the function* $\epsilon(r)$ *in (4.10) with* $a = \frac{\pi}{2}$, $b = 0.2\pi$, $\alpha = 0.03$, *and* $\beta = 0.00$. *Here* $m = N = M = 7$. *The relative error (4.6) obtained after 12 iterations is 3.86%.*

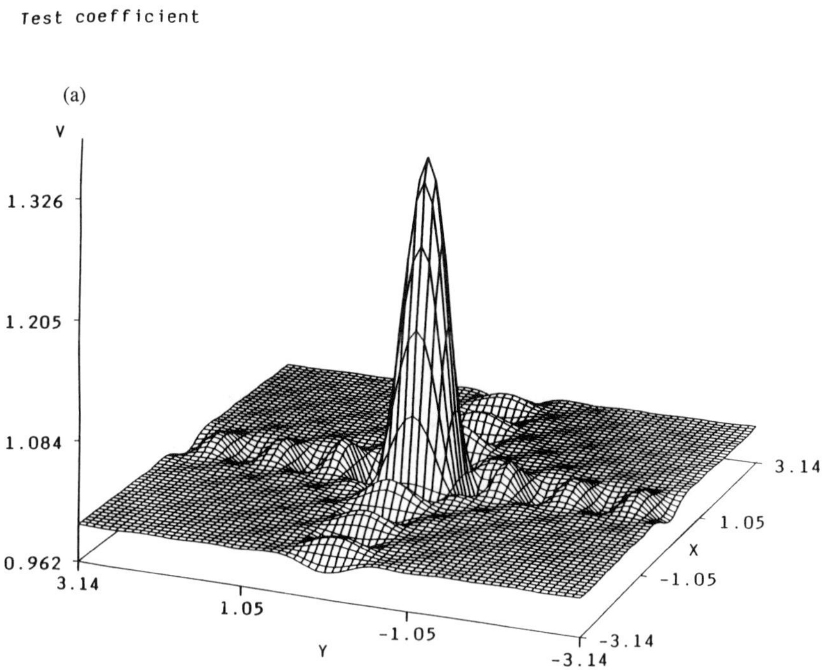


FIGURE 5a. *Test #6.* $\epsilon_{test} + 1$. $\epsilon_{test}$ *is the finite Fourier approximation (4.4) of*

$$\epsilon(r) = \begin{cases} 0.3\,, & \text{for } r \leq 0.1\pi \\ 0\,, & \text{otherwise.} \end{cases}$$

*We have taken* $mh = 7$, $\ell h = 12$, *and* $\kappa = 5.5$.

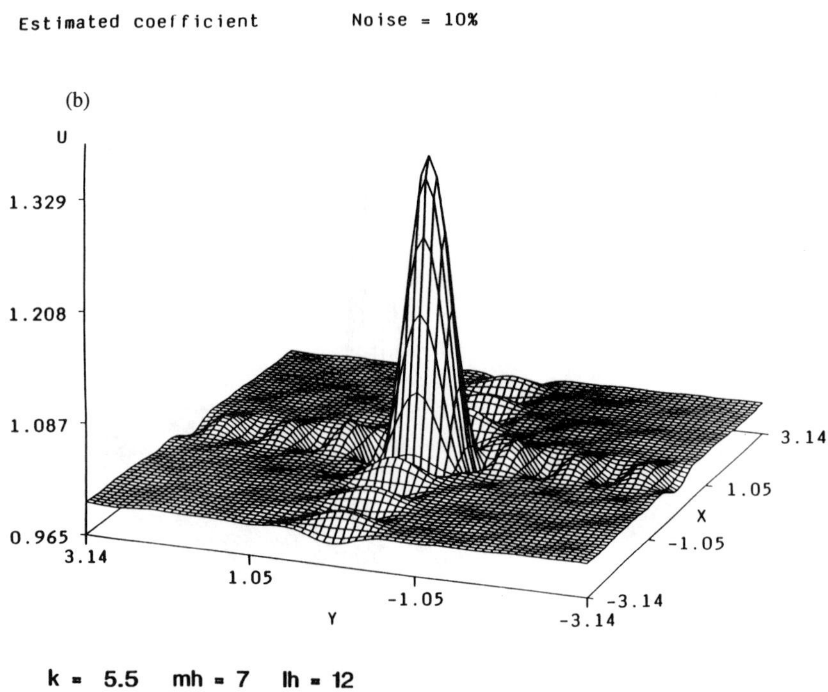


FIGURE 5b. *Test #6.* $\epsilon_{est} + 1$ *obtained after 12 iterations, with 10% of noise. The relative error (4.6) is 5.45%.*

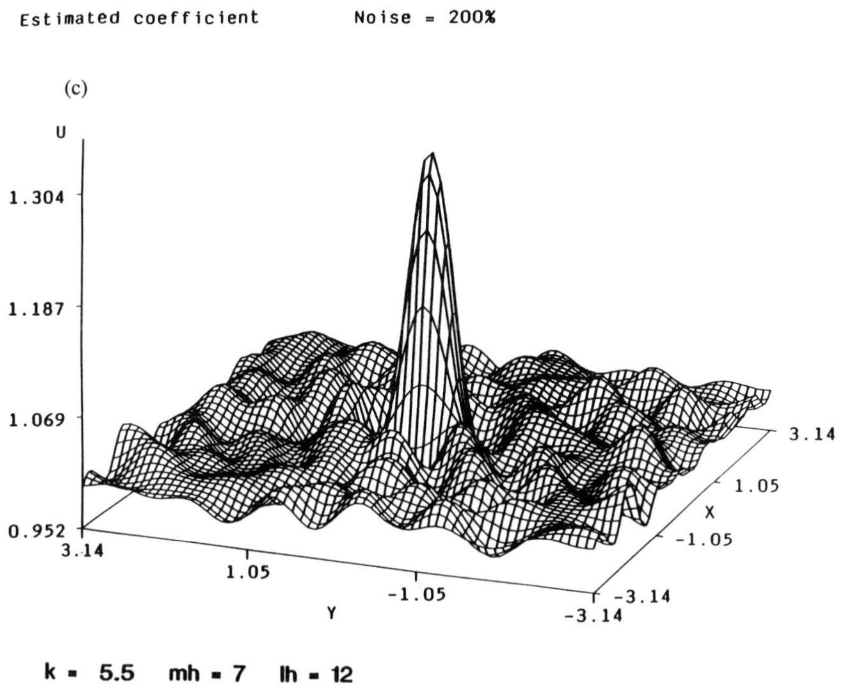


FIGURE 5c. *Test #6. $\epsilon_{est}+1$ obtained after 12 iterations, with 200% of noise. The relative error (4.6) is 68.9%.*

## REFERENCES


[1] S. AGMON, *Spectral properties of Schrodinger operators and scattering theory*, Annali Delta Scuola Normale Superiore di Pisa, Classe di Scienze, Series IV **2** (1975) pp. 151–218.

[2] E. ARNOLD, *Use of advanced biotechnology to visualize the AIDS virus reverse transcriptage enzyme: implication for drug design*, 14th Annual International Conference of the IEEE, Engineering in Medicine and Biology (the final program) 1992, p. 10.

[3] A. BAYLIS, V. LI AND C.S. MORAWETZ, *Scattering by potentials using hyperbolic methods*, Math. Comput. **52** (1989) pp. 321–328.

[4] M. BERGER, *Nonlinearity and functional analysis*, in *Lectures on Nonlinear Problems in Mathematical Analysis*, Academic Press, New York, 1977.

[5] A.L. BUKHGEIM AND M.V. KLIBANOV, *Uniqueness in the large of a class of multidimensional inverse problems*, Sov. Math. Dokl. **24** (1981) pp. 244–247.

[6] T.J. CAVICCHI, S.A. JOHNSON, AND W.D. O'BRIEN, *Application of the sinc basis moment method to the reconstruction of infinite circular cylinders*, IEEE Trans. Ultrasonics Ferroelectrics and Frequency Control **35** (1988) pp. 22–33.

[7] K. CHADAN AND P. SABATIER, *Inverse Problems in Quantum Scattering*, Springer-Verlag, New York, 2nd edition, 1989.

[8] M. CHENEY, *A review of multi-dimensional inverse potential scattering*, in *Inverse Problems in Partial Differential Equations*, SIAM, (edited by D. COLTON, R. EWING, AND W. RUNDELL) 1990.

[9] D. COLTON AND P. MONK, *The inverse scattering problem for acoustic waves in an inhomogeneous medium*, in *Inverse Problems in Partial Differential Equations*, SIAM, (edited by D. COLTON, R. EWING, AND W. RUNDELL) 1990.

[10] D. COLTON AND P. MONK, *The numerical solution of an inverse scattering problem for acoustic waves*, IMA J. Appl. Math. **49** (1990) pp. 163–184.

[11] B. CROSIGNANI AND P. DI PORTO, *Coupled-mode theory approach to depolarization associated with propagation in turbulent medium*, AGARD-CP-419, "Scattering and Propagation in Random Media," Meeting held in Rome, Italy, 18–22 May 1987, AGARD 1988, pp. 1–5.

[12] A. DEVANEY, *Inversion formula for acoustic scattering within the Born approximation*, Opt. Lett. **7** (1982) pp. 11–17.

[13] S.M. FLATTE AND F.D. TAPPERT, *Calculation of the effect of internal waves on oceanic sound transmission*, J. Acoust. Soc. Am. **58** (1975) pp. 1151–1159.

[14] S. GUTMAN AND M.V. KLIBANOV, *Iterative method for multidimensional inverse scattering problems at fixed frequencies*, Inverse Problems (to appear).

[15] S. GUTMAN AND M.V. KLIBANOV, *Regularized Quasi-Newton method for inverse scattering problems*, International J. Math. and Comp. Modelling (to appear).

[16] A. ISHIMARU, *Wave Propagation and Scattering in Random Media, v. 1*, Academic Press, New York, 1978.

[17] R.E. KLEINMAN AND P.M. VAN DEN BERG, *Nonlinearized approach to profile inversion*, Int. J. of Imaging Systems and Technology **2** (1990) pp. 119–126.

[18] M.V. KLIBANOV, *Inverse problems and carleman estimates*, Inverse Problems **8** (1992) pp. 575–596.

[19] M.V. KLIBANOV AND O. IOUSSOUPOVA, *Uniform strict convexity of a cost functional for 3-D inverse scattering problem*, (preprint) 1993.

[20] M.V. KLIBANOV AND J. MALINSKY, *Newton-Kantorovich method for three-dimensional potential inverse scattering problem and stability of the hyperbolic Cauchy problem with time-dependent data*, Inverse Problems **7** (1991) pp. 577–596.

[21] M.V. KLIBANOV, S. GUTMAN AND J. MALINSKY, *Three-dimensional inverse scattering problem in random media*, J. Math. Anal. Applic. (submitted).

[22] M.M. LAVRENT'EV, V.G. ROMANOV AND S.P. SHISHATSKII, *Ill-Posed Problems of Mathematical Physics and Analysis*, American Mathematical Society, Providence, RI, 1986.

[23] R.T. PROSSER, *Formal solutions of inverse scattering problems, IV, Error Estimates*, J. Math. Phys. **23** (1982) pp. 2127–2130.

[24] A. RAMM, *Recovery of the potential from fixed-energy scattering data*, Inverse Problems **4** (1988) pp. 877–886.

[25] S.M. RYTOV, Y.A. KRAVTSOV AND V.I. TATARSKII, *Principles of Statistical Radiophysics*, Vol. 2, (see Chapter 7) Springer-Verlag, New York 1988.

[26] J. SYLVESTER AND G. UHLMANN, *The Dirichlet to Neumann map and its applications*, in *Inverse Problems in Partial Differential Equations*, SIAM (edited by D. COLTON, R. EWING, AND W. RUNDELL) 1990.

[27] W. SYMES, *Layered velocity inversion: a model problem from reflection seismology*, SIAM J. Math. Anal. **7** (1991) pp. 680–716.

[28] F.D. TAPPERT, *Method of parabolic equation*, in *Lecture Notes in Physics*, v. 70, Springer-Verlag, New York, 1977.

[29] A.N. TIKHONOV AND V. YA. ARSENIN, *Solutions of Ill-Posed Problems*, Winston-Wiley, New York, 1977.

# DECAY ESTIMATES FOR THE WAVE EQUATION

VILMOS KOMORNIK*

**Abstract.** Using the multiplier method and a special integral inequality we obtain sharp energy decay rate estimates for the wave equation in the presence of nonlinear distributed or boundary feedbacks. For simplicity we restrict ourselves to the wave equation, but the same approach applies for other systems as well.

**1. Introduction and statement of the main results.** Throughout this paper let $\Omega$ be a bounded open domain in $\mathbb{R}^n$ with a boundary $\Gamma$ of class $C^2$ and let $g : \mathbb{R} \to \mathbb{R}$ be a continuous, non-decreasing function such that $g(0) = 0$.

Let us first consider the wave equation with distributed feedback:

$$u'' - \Delta u + g(u') = 0 \quad \text{in} \quad \Omega \times \mathbb{R}_+, \tag{1.1}$$

$$u = 0 \quad \text{on} \quad \Gamma \times \mathbb{R}_+, \tag{1.2}$$

$$u(0) = u^0 \quad \text{and} \quad u'(0) = u^1 \quad \text{on} \quad \Omega. \tag{1.3}$$

Since $xg(x) \geq 0$ for all $x \in \mathbb{R}$, the system is dissipative in the following sense: the energy $E : \mathbb{R}_+ \to \mathbb{R}_+$ of the solutions, defined by the usual formula

$$E := \tfrac{1}{2} \int_\Omega (u')^2 + |\nabla u|^2 \text{ dx}, \tag{1.4}$$

is non-increasing. Indeed, we have, at least formally,

$$\begin{aligned} E' &= \textstyle\int_\Omega u'u'' + \nabla u \cdot \nabla u' \text{ dx} = \int_\Omega u'\Delta u + \nabla u' \cdot \nabla u - u'g(u') \text{ dx} \\ &= \textstyle\int_\Gamma u'(\partial_\nu u) \text{ d}\Gamma - \int_\Omega u'g(u') \text{ dx} = -\int_\Omega u'g(u') \text{ dx} \leq 0. \end{aligned}$$

Our purpose is to estimate the energy decay rate under suitable growth assumptions on $g$. The following theorem improves several earlier results of Nakao [20], Haraux [8], Haraux and Zuazua [9], Zuazua [23], Conrad, Leblond and Marmorat [6].

THEOREM 1.1. *Let $n \geq 3$, $1 < p < +\infty$ and assume that there are four positive constants $c_1, c_2, c_3, c_4$ such that*

$$c_1|x|^p \leq |g(x)| \leq c_2|x|^{1/p} \quad \textit{if} \quad |x| \leq 1, \tag{1.5}$$

$$c_3|x| \leq |g(x)| \leq c_4|x|^{(n+2)/(n-2)} \quad \textit{if} \quad |x| > 1. \tag{1.6}$$

* Département de Mathématiques, Université Louis Pasteur, 7, rue René Descartes, 67084 Strasbourg Cédex, France. Part of this work was done while the author was visiting the IMA at the University of Minnesota during the fall of 1992. The author thanks the Institute for its hospitality and support.

*Then for every* $(u^0, u^1) \in H_0^1(\Omega) \times L^2(\Omega)$ *the solution of (1.1)–(1.3) satisfies the estimates*

$$E(t) \le Ct^{2/(1-p)}, \quad \forall t > 0 \tag{1.7}$$

*where* $C$ *is a positive constant depending only on* $E(0)$.

*Remark* 1.2.

1) The estimate (1.7) is probably optimal but it has not yet been proved.

2) There is an analogous result for $p = 1$: then (1.7) is replaced by

$$E(t) \le CE(0)e^{-\omega t}, \quad \forall t > 0 \tag{1.8}$$

with some absolute constants $C, \omega > 0$.

3) Analogous results hold for $n = 1, 2$, too: if $n = 2$, then the exponent $(n+2)/(n-2)$ in (1.6) may be replaced by any number $1 \le q < +\infty$; if $n = 1$, then it may be replaced by $q = +\infty$ i.e. we do not need any upper estimate on $g$ near infinity.

4) In Theorem 1.1 the energy decay rate depends only on the growth properties of $g$ near zero.

In many practical applications the feedback function $g$ is bounded; Theorem 1.1 never applies in such cases because of the assumption $c_3 > 0$. F. Conrad asked the author whether there is a variant of Theorem 1.1 which applies for bounded feedback functions as well. The answer is positive:

THEOREM 1.3. *Let* $n \ge 3$, $1 < p' < +\infty$, $0 \le r \le 1$, *and assume that there are four positive constants* $c_1, c_2, c_3, c_4$ *such that*

$$c_1|x|^{p'} \le |g(x)| \le c_2|x| \quad \text{if} \quad |x| \le 1, \tag{1.9}$$

$$c_3|x|^r \le |g(x)| \le c_4|x|^{n/(n-2)} \quad \text{if} \quad |x| > 1. \tag{1.10}$$

*Set*

$$p := \max\{p', (n + (2-n)r)/2, (3n-4)/n\}. \tag{1.11}$$

*Then for every*

$$(u^0, u^1) \in (H^2(\Omega) \cap H_0^1(\Omega)) \times H_0^1(\Omega) \quad \text{such that} \quad g(u^1) \in L^2(\Omega), \tag{1.12}$$

*the solution of (1.1)–(1.3) satisfies the estimates*

$$E(t) \le Ct^{2/(1-p)}, \quad \forall t > 0 \tag{1.13}$$

*where* $C$ *is a positive constant depending on* $\|u^0\|_{H^2(\Omega)}$ *and* $\|u^1\|_{H^1(\Omega)}$.

*Remark* 1.4.

1) The optimality of the estimate (1.13) remains open.

2) Analogous results hold for $n = 1, 2$, too. If $n = 2$, then the exponent $n/(n-2)$ in (1.6) may be replaced by any positive number; if $n = 1$, then we do not need any upper estimate on $g$ near infinity.

3) Observe that the left inequality in (1.10) is always satisfied with $r = 0$. Hence Theorem 1.3 applies for bounded feedback functions, too.

4) The formula (1.11) shows that now the decay rate depends on the growth behavior of $g$ *both* near zero and near infinity.

Now let us consider the wave equation with boundary feedback:

$$u'' - \Delta u = 0 \quad \text{in} \quad \Omega \times \mathbb{R}_+, \tag{1.14}$$

$$u = 0 \quad \text{on} \quad \Gamma_0 \times \mathbb{R}_+, \tag{1.15}$$

$$\partial_\nu u + hg(u') = 0 \quad \text{on} \quad \Gamma_1 \times \mathbb{R}_+, \tag{1.16}$$

$$u(0) = u^0 \quad \text{and} \quad u'(0) = u^1 \quad \text{on} \quad \Omega, \tag{1.17}$$

where $\{\Gamma_0, \Gamma_1\}$ is a partition of $\Gamma$ and $h$ is a given nonnegative function on $\Gamma_1$. The energy $E$, defined by (1.4), is non-increasing again:

$$\begin{aligned} E' &= \textstyle\int_\Omega u'u'' + \nabla u \cdot \nabla u' \ \mathrm{dx} = \int_\Omega u'\Delta u + \nabla u' \cdot \nabla u \ \mathrm{dx} \\ &= \textstyle\int_\Gamma u'(\partial_\nu u) \ \mathrm{d}\Gamma = -\int_{\Gamma_1} hu'g(u') \ \mathrm{d}\Gamma \le 0. \end{aligned}$$

Many works were devoted to the decay properties of this system: cf. e.g. Quinn and Russell [21], Chen [4], Lagnese [16], Triggiani [22], Komornik and Zuazua [15], Bardos, Lebeau and Rauch [1] for the linear case and Zuazua [24], Chen and Wang [5], Lasiecka [17], [18] Conrad and Rao [7], Lasiecka and Tataru [19] for the nonlinear case. Most of the earlier explicit decay *estimates* are contained in Theorem 1.5 below. We assume that there exists a point $x^0 \in \mathbb{R}^n$ such that

$$h(x) = (x - x^0) \cdot \nu(x) \ge 0 \quad \text{on} \quad \Gamma_1, \tag{1.18}$$

$$(x - x^0) \cdot \nu(x) \le 0 \quad \text{on} \quad \Gamma_0 \tag{1.19}$$

and

$$\Gamma_0 \ne \emptyset \quad \text{and} \quad \overline{\Gamma_0} \cap \overline{\Gamma_1} = \emptyset. \tag{1.20}$$

These conditions are satisfied if $\Omega$ is the set-theoretical difference of two star-shaped domains with respect to $x^0$; more general cases are considered in [12] or [14].

We set for brevity

$$H^1_{\Gamma_0}(\Omega) := \{v \in H^1(\Omega) : v = 0 \quad \textit{on} \quad \Gamma_0\}.$$

THEOREM 1.5. *Let $1 < p < +\infty$ and assume that there are four positive constants $c_1, c_2, c_3, c_4$ such that*

$$c_1|x|^p \le |g(x)| \le c_2|x|^{1/p} \quad \textit{if} \quad |x| \le 1, \tag{1.21}$$

$$c_3|x| \le |g(x)| \le c_4|x| \quad \textit{if} \quad |x| > 1. \tag{1.22}$$

*Then for every* $(u^0, u^1) \in H^1_{\Gamma_0}(\Omega) \times L^2(\Omega)$ *the solution of (1.14)–(1.17) satisfies the estimates*

$$E(t) \le Ct^{2/(1-p)}, \quad \forall t > 0 \tag{1.23}$$

*where* $C$ *is a positive constant depending only on* $E(0)$.

*Remark* 1.6.

1) We do not know whether the estimate (1.23) is optimal.

2) There is an analogous result for $p = 1$: then (1.23) is replaced by (1.8).

We have also the following result, analogous to Theorem 1.3 above.

THEOREM 1.7. *Let* $n \ge 3$ *and fix* $k \ge 0$ *such that*

$$(k+1)(k+2) = \frac{2n-2}{n-2}. \tag{1.24}$$

*Assume that* $g$ *is also locally lipschitzian (i.e.* $g \in W^{1,\infty}_{loc}(\mathbb{R})$*). Let* $1 < p' < +\infty$, $0 \le r \le 1$, *and assume that there are four positive constants* $c_1, c_2, c_3, c_4$ *such that*

$$c_1|x|^{p'} \le |g(x)| \le c_2|x| \quad \text{if} \quad |x| \le 1, \tag{1.25}$$

$$c_3|x|^r \le |g(x)| \quad \text{if} \quad |x| > 1, \tag{1.26}$$

$$|g'(x)| \le c_4|x|^k \quad \text{if} \quad |x| > 1. \tag{1.27}$$

*Set*

$$p := \max\{p', 1 + (n-2)(1-r), (k+3)/(k+1)\}. \tag{1.28}$$

*Then for every*

$$(u^0, u^1) \in (H^2(\Omega) \cap H^1_{\Gamma_0}(\Omega)) \times H^1_{\Gamma_0}(\Omega)$$

*such that*

$$\partial_\nu u^0 + hg(u^1) = 0 \quad on \quad \Gamma_1$$

*the solution of* (1.14)–(1.17) *satisfies the estimates*

$$E(t) \le Ct^{2/(1-p)}, \quad \forall t > 0 \tag{1.29}$$

*where* $C$ *is a positive constant depending on* $\|u^0\|_{H^2(\Omega)}$ and $\|u^1\|_{H^1(\Omega)}$.

*Remark* 1.8.

1) Analogous results hold for $n \le 2$, too. If $n = 1$, then we do not need any growth assumption on $g'$; if $n = 2$, then we may replace the exponent $k$ by any positive number in (1.27). In both cases we have (1.29) with $p = p'$.

2) Observe that the inequality in (1.26) is always satisfied with $r = 0$. Hence Theorem 1.7 applies for bounded feedback functions, too.

In what follows we outline the proofs of Theorems 1.5 and 1.7 for $n \ge 3$; for more details and for more general results we refer to [12] and [14]. The proofs of Theorems 1.1 and 1.3 are similar but simpler; see [11].

**2. Well-posedness and regularity.** Let us introduce two separable, real Hilbert spaces $V$ and $H$ by

$$V := H^1_{\Gamma_0}(\Omega) \quad \text{and} \quad H := L^2(\Omega)$$

with

$$\|v\|_V := \Big(\int_\Omega |\nabla v|^2 \,\mathrm{dx}\Big)^{1/2} \quad \text{and} \quad \|v\|_H := \Big(\int_\Omega |v|^2 \,\mathrm{dx}\Big)^{1/2}.$$

As usual, we identify $H$ with $H'$, then we have $V \subset H \subset V'$ with dense and compact imbeddings. We introduce the duality mapping $A : V \to V'$ and for $z, v \in V$ we set

$$\langle Bz, v\rangle_{V',V} := \int_{\Gamma_1} hg(z)v \,\mathrm{d}\Gamma.$$

Since both (1.22) and (1.24), (1.27) imply that

$$|g(x)| \le c|x|^{n/(n-2)} \quad \text{if} \quad |x| \ge 1,$$

using the trace theorem $H^1(\Omega) \hookrightarrow L^{(2n-2)/(n-2)}(\Gamma)$ one can readily verify that $B$ is a well-defined (nonlinear) operator from $V$ into $V'$.

Let us multiply the equation (1.14) formally by $v \in V$. Integrating on $\Omega$ and using the bounday conditions (1.15)–(1.16) we obtain easily that

$$u'' + Au + Bu' = 0 \quad \text{on} \quad \mathbb{R}_+.$$

Putting

$$U = (U_1, U_2) := (u, u') \quad \text{and} \quad \mathcal{A}U := (-U_2, AU_1 + BU_2)$$

we can rewrite this equation in the form

$$U' + \mathcal{A}U = 0 \quad \text{on} \quad \mathbb{R}_+.$$

In view of the definition of the energy it is reasonable to consider $\mathcal{A}$ as an operator acting in the Hilbert space

$$\mathcal{H} := V \times H$$

defined on

$$D(\mathcal{A}) := \{U : \ U_1 \in V, \ U_2 \in V \text{ and } AU_1 + BU_2 \in H\},$$

and to *define* the solution of (1.14)–(1.17) as that of

$$U' + \mathcal{A}U = 0 \quad \text{on} \quad \mathbb{R}_+, \quad U(0) = (u^0, u^1).$$

This is justified by the

LEMMA 2.1. *$\mathcal{A}$ is a maximal monotone operator in $\mathcal{H}$.*

The proof of this lemma is standard. The monotonicity of $\mathcal{A}$ follows easily from the monotonicity of $g$. The proof of the surjectivity of $I + \mathcal{A}$ reduces easily to the surjectivity of $I + A + B : V \to V'$. One can readily verify that $I + A + B$ is monotone and coercive. Also, applying Lebesgue's dominated convergence theorem one can show that $I + A + B$ is hemicontinuous. The surjectivity of $I + A + B$ then follows by applying a well-known theorem of Minty and Browder (cf. e.g. [3]), a nonlinear generalization of the Lax–Milgram theorem.

Applying the theory of maximal monotone operators as e.g. in [2], we obtain the

LEMMA 2.2. *Given $(u^0, u^1) \in H^1_{\Gamma_0}(\Omega) \times L^2(\Omega)$ arbitrarily, the system (1.14)–(1.17) has a unique (weak) solution*

$$u \in C(\mathbb{R}_+; H^1_{\Gamma_0}(\Omega)) \cap C^1(\mathbb{R}_+; L^2(\Omega)).$$

Next we show that if the initial data are more regular, then the solution is more regular, too:

LEMMA 2.3. *Assume that $g$ is locally lipschitzian satisfying*

$$|g'(x)| \le c'|x|^k \quad \textit{if} \quad |x| > 1 \tag{2.1}$$

*for some $c' > 0$ and $0 \le k \le 1/(n-2)$. Then for every*

$$(u^0, u^1) \in (H^2(\Omega) \cap H^1_{\Gamma_0}(\Omega)) \times H^1_{\Gamma_0}(\Omega)$$

*such that*

$$\partial_\nu u^0 + hg(u^1) = 0 \quad \textit{on} \quad \Gamma_1,$$

*the solution of (1.14)–(1.17) satisfies*

$$u \in W^{1,\infty}_{loc}(\mathbb{R}_+; H^1_{\Gamma_0}(\Omega)) \cap W^{2,\infty}_{loc}(\mathbb{R}_+; L^2(\Omega)) \tag{2.2}$$

*and*

$$u \in L^\infty(\mathbb{R}_+; W^{2,2n/(n+(n-2)k)}(\Omega)), \quad u' \in L^\infty(\mathbb{R}_+; H^1_{\Gamma_0}(\Omega)). \tag{2.3}$$

For the proof we need some information on $D(\mathcal{A})$; we write for brevity

$$D_0 := \{U \in V \times V : U_1 \in H^2(\Omega), \partial_\nu U_1 + hg(U_2) = 0 \quad \text{on} \quad \Gamma_1\}$$

and

$$D_k := \{U \in V \times V : U_1 \in W^{2,2n/(n+(n-2)k)}(\Omega), \partial_\nu U_1 + hg(U_2) = 0 \quad \text{on} \quad \Gamma_1\}.$$

LEMMA 2.4. *If $g$ satisfies the hypotheses of Lemma 2.3, then*

$$D_0 \subset D(\mathcal{A}) \subset D_k \tag{2.4}$$

*with continuous imbeddings.*

*Remark* 2.5. Since $D_0$ contains the set

$$D := \{U \in V \times H_0^1(\Omega) : U_1 \in H^2(\Omega) \quad \text{and} \quad \partial_\nu U_1 = 0 \quad \text{on} \quad \Gamma_1\} \tag{2.5}$$

and since the latter set is obviously dense in $V \times H$, it follows from Lemma 2.4 that $D(\mathcal{A})$ is dense in $\mathcal{H}$.

The proof of this lemma is based on the following observation: the equality $AU_1 + BU_2 = f$ means that $U_1$ is the weak solution of the system

$$-\Delta U_1 = f \quad \text{in} \quad \Omega, \tag{2.6}$$

$$U_1 = 0 \quad \text{on} \quad \Gamma_0, \tag{2.7}$$

$$\partial_\nu U_1 = -hg(U_2) \quad \text{on} \quad \Gamma_1. \tag{2.8}$$

The first inclusion in (2.4) is obvious: if $U \in D_0$, then $U_1 \in H^2(\Omega)$ and therefore $f = -\Delta U_1 \in L^2(\Omega) = H$.

To prove the second inclusion in (2.4) it is sufficient to show that $f \in L^{2n/(n+(n-2)k)}(\Omega)$ and $g(U_2) \in W^{1,2n/(n+(n-2)k)}(\Omega)$. Indeed, then we conclude, applying to (2.6)–(2.8) the elliptic regularity theory, that $U_1 \in W^{2,2n/(n+(n-2)k)}(\Omega)$.

The first property is obvious because $f = AU_1 + BU_2 \in L^2(\Omega)$. Also, we have

$$\int_\Omega |g(U_2)|^{2n/(n+(n-2)k)} \,\mathrm{dx} \le \int_\Omega |g(U_2)|^{2n/((n-2)(k+1))} \,\mathrm{dx}$$

$$\le c \int_\Omega 1 + |U_2|^{2n/(n-2)} \,\mathrm{dx} \le c\left(1 + \|U_2\|_{H^1(\Omega)}^{2n/(n-2)}\right) < +\infty$$

since (2.1) implies that

$$|g(x)| \le c\left(1 + |x|^{k+1}\right) \quad \text{if} \quad |x| > 1.$$

Finally, we show that $\nabla g(U_2) \in L^{2n/(n+(n-2)k)}(\Omega)$. Indeed, we have, using the Young inequality,

$$\int_\Omega |\nabla g(U_2)|^{2n/(n+(n-2)k)}\, \mathrm{dx} = \int_\Omega |g'(U_2)\nabla U_2|^{2n/(n+(n-2)k)}\, \mathrm{dx}$$

$$\le c\int_\Omega |g'(U_2)|^{2n/((n-2)k)} + |\nabla U_2|^2\, \mathrm{dx}$$

$$\le c\int_\Omega 1 + |U_2|^{2n/(n-2)} + |\nabla U_2|^2\, \mathrm{dx}$$

$$\le c(1 + \|U_2\|_V^{2n/(n-2)} + \|U_2\|_V^2) < +\infty.$$

Now we are ready to prove Lemma 2.3. We have $(u^0, u^1) \in D_0$ by hypothesis whence $(u^0, u^1) \in D(\mathcal{A})$ by Lemma 2.4. Then it follows from the theory of maximal monotone operators that $(u, u') \in W^{1,\infty}_{loc}(\mathbb{R}_+; \mathcal{H})$ and that the function $\|\mathcal{A}(u, u')\|_{\mathcal{H}}$ is non-increasing on $\mathbb{R}_+$. The first property is equivalent to (2.2) while the second one, together with the relation $D(\mathcal{A}) \subset D_1$, implies (2.3).

**3. Proof of Theorem 1.5.** It is sufficient to prove the estimates for $(u^0, u^1) \in D$ (cf. (34)); the general case then will follow by density.

Also, let us introduce for $\lambda > 0$ the functions $g_\lambda : \mathbb{R} \to \mathbb{R}$ by

$$g_\lambda(x) := \big(x - (id_{\mathbb{R}} + \lambda g)^{-1}(x)\big)/\lambda;$$

then $g_\lambda$ is lipschitzian, non-decreasing, $g_\lambda(0) = 0$, $|g_\lambda| \le |g|$ and

$$g_\lambda(x) \to g(x) \quad \text{as} \quad \lambda \to 0, \quad \forall x \in \mathbb{R}.$$

It follows that for each $\lambda \in [0, 1]$ the function $g_\lambda$ satisfies the same growth conditions (1.21) and (1.22) as $g$, with the same constants $c_1, c_2, c_3, c_4$. Furthermore, denoting by $u_\lambda$ the corresponding solutions of (1.14)–(1.17) with the same $(u^0, u^1) \in D$, the energy function $E_\lambda$ converges to $E$ pointwise on $\mathbb{R}_+$ as $\lambda \to 0$. Indeed, this can be shown by adapting an approximation argument used in [8; pp. 56-62, see especially (II.1.1.23)].

Since $g_\lambda$ is lipschitzian and $(u^0, u^1) \in D \subset D(\mathcal{A}_\lambda)$, the corresponding solution $u_\lambda$ of (1.14)–(1.17) has the regularity properties (2.2) and (2.3). If we show that $E_\lambda$ satisfies the estimates (1.23) with a constant depending (in a bounded way) only on $E_\lambda(0) = E(0)$ and on $c_1, c_2, c_3, c_4$, then the same estimate will hold for $E$, too, by an obvious limit argument.

Hence we may assume in the sequel that $u$ satisfies (2.2) and (2.3).

We need several lemmas.

LEMMA 3.1. *The function $E : \mathbb{R}_+ \to \mathbb{R}_+$ is non-increasing, locally absolutely continuous and*

$$E' = -\int_{\Gamma_1} hu'g(u')\, d\Gamma \quad a.e. \ in\ \mathbb{R}_+. \tag{3.1}$$

*Proof.* Fix $T > 0$ arbitrarily, multiply the equation (1.14) by $u'$ and integrate over $\Omega \times (0, T)$. Integrating by parts and using the boundary conditions (1.15), (1.16) one obtains easily that

$$E(0) - E(T) = \int_0^T \int_{\Gamma_1} hu'g(u') \, \mathrm{d}\Gamma \, \mathrm{d}t, \quad \forall T > 0. \tag{3.2}$$

This implies (3.1) and the absolute continuity of $E$. Since $h \geq 0$ and $xg(x) \geq 0$, (3.2) shows the monotinicity of $E$, too. □

In what follows we shall write for brevity $m(x) := x - x^0$ and

$$\mathrm{d}\Gamma_m := h\mathrm{d}\Gamma = (x - x^0)\nu(x)\mathrm{d}\Gamma.$$

LEMMA 3.2. *Set*

$$Mu := 2(x - x^0) \cdot \nabla u + (n-1)u \tag{3.3}$$

*for brevity. Given $0 \leq S < T < +\infty$ arbitrarily, the following identity holds true:*

$$\begin{aligned} 2\int_S^T E^{(p+1)/2} \, \mathrm{d}t - \int_S^T E^{(p-1)/2} \int_{\Gamma_0} (\partial_\nu u)^2 \, \mathrm{d}\Gamma_m \, \mathrm{d}t \\ = \left[ E^{(p-1)/2} \int_\Omega u' Mu \, \mathrm{d}x \right]_T^S \\ + \tfrac{p-1}{2} \int_S^T E^{(p-3)/2} E' \int_\Omega u' Mu \, \mathrm{d}x \, \mathrm{d}t \\ + \int_S^T E^{(p-1)/2} \int_{\Gamma_1} (u')^2 - |\nabla u|^2 - g(u') Mu \, \mathrm{d}\Gamma \, \mathrm{d}t. \end{aligned} \tag{3.4}$$

*Proof.* We have

$$
\begin{aligned}
0 &= \int_S^T E^{(p-1)/2} \int_\Omega (Mu)(u'' - \Delta u) \, \mathrm{dx} \, \mathrm{dt} \\
&= \left[ E^{(p-1)/2} \int_\Omega u' Mu \, \mathrm{dx} \right]_S^T \\
&\quad - \tfrac{p-1}{2} \int_S^T E^{(p-3)/2} E' \int_\Omega u' Mu \, \mathrm{dx} \, \mathrm{dt} \\
&\quad - \int_S^T E^{(p-1)/2} \int_\Omega u' Mu' + (Mu)(\Delta u) \, \mathrm{dx} \, \mathrm{dt}.
\end{aligned} \tag{3.4'}
$$

Integrating by parts and using the relation div $m = n$ we may transform the internal integral as follows:

$$
\begin{aligned}
&\int_\Omega u' Mu' + (Mu)(\Delta u) \, \mathrm{dx} \\
&= \int_\Omega m \cdot \nabla (u')^2 + (n-1)(u')^2 - \nabla u \cdot \nabla (Mu) \, \mathrm{dx} + \int_\Gamma (Mu) \partial_\nu u \, \mathrm{d}\Gamma \\
&= \int_\Omega m \cdot \nabla (u')^2 + (n-1)(u')^2 - 2|\nabla u|^2 - m \cdot \nabla |\nabla u|^2 \\
&\qquad -(n-1)|\nabla u|^2 \, \mathrm{dx} + \int_\Gamma (Mu) \partial_\nu u d\Gamma \\
&= - \int_\Omega (u')^2 + |\nabla u|^2 \, \mathrm{dx} + \int_\Gamma (m \cdot \nu)((u')^2 - |\nabla u|^2) + (Mu) \partial_\nu u \, \mathrm{d}\Gamma \\
&= - \int_\Omega (u')^2 + |\nabla u|^2 \, \mathrm{dx} + \int_{\Gamma_0} -(m \cdot \nu)|\nabla u|^2 + (2m \cdot \nabla u) \partial_\nu u \, \mathrm{d}\Gamma \\
&\qquad + \int_{\Gamma_1} (u')^2 - |\nabla u|^2 - (Mu) g(u') \, \mathrm{d}\Gamma_m.
\end{aligned}
$$

Since (1.15) implies $\nabla u = \nu \partial_\nu u$ on $\Gamma_0$, hence we deduce, using also the definition of the energy, that

$$\int_\Omega u'Mu' + (Mu)(\Delta u)\ \mathrm{dx} = -2E + \int_{\Gamma_0} (\partial_\nu u)^2\ \mathrm{d}\Gamma_m$$

$$+ \int_{\Gamma_1} (u')^2 - |\nabla u|^2 - (Mu)g(u')\ \mathrm{d}\Gamma_m.$$

Substituting into (3.4') the identity (3.4) follows. □

Our objective is to prove the existence of a positive number $A$, depending only on $c_1, c_2, c_3, c_4$ and $E(0)$ such that

$$\int_t^\infty E^{(p+1)/2}\ \mathrm{dt} \le AE(t) \quad \forall t \in \mathbb{R}_+. \tag{3.5}$$

Indeed, then Theorem 1.5 will follow by applying the following useful

LEMMA 3.3. *Let $E : \mathbb{R}_+ \to \mathbb{R}_+$ be a non-increasing function with $E(0) \neq 0$, and assume that there exist two positive numbers $\alpha$ and $A$ such that*

$$\int_t^\infty E^{\alpha+1} ds \le AE(t) \qquad \text{for all} \qquad t \ge 0. \tag{3.6}$$

*Then, setting*

$$T := AE(0)^{-\alpha} \tag{3.7}$$

*we have*

$$E(t) \le E(0) \left( \frac{T + \alpha T}{T + \alpha t} \right)^{1/\alpha}, \quad \forall t \ge 0. \tag{3.8}$$

SKETCH OF PROOF. (See [11], [14] for more details.) Assume for simplicity that $E(t) > 0$ for all $t \ge 0$. Let us introduce the function

$$F : \mathbb{R}_+ \to \mathbb{R}_+, \qquad F(t) = \int_t^\infty E^{\alpha+1} ds.$$

$F$ is non-increasing, locally absolutely continuous and one deduce from (3.6) easily the inequality

$$(F^{-\alpha})' \ge \alpha A^{-\alpha-1} \quad \text{a.e. in} \quad \mathbb{R}_+.$$

Integrating this inequality we find

$$F(s) \le (F(0)^{-\alpha} + \alpha A^{-\alpha-1} s)^{-1/\alpha} \quad \text{pour tout} \quad s \ge 0. \tag{3.9}$$

Using (3.6) and (3.7) hence we conclude that

$$
\begin{aligned}
(F(0)^{-\alpha}+\alpha A^{-\alpha-1}s)^{-1/\alpha} &\le (A^{-\alpha}E(0)^{-\alpha}+\alpha A^{-\alpha-1}s)^{-1/\alpha}\\
&= (T^{-\alpha}E(0)^{-\alpha^2-\alpha}+\alpha s T^{-\alpha-1}E(0)^{-\alpha^2-\alpha})^{-1/\alpha}\\
&= T^{(\alpha+1)/\alpha}E(0)^{\alpha+1}(T+\alpha s)^{-1/\alpha}.
\end{aligned}
\tag{3.10}
$$

On the other hand, $E$ being non-increasing we have

$$
F(s) \ge (T+\alpha s)E(T+(\alpha+1)s)^{\alpha+1}. \tag{3.11}
$$

Combining (3.9), (3.10) and (3.11) we obtain

$$
(T+\alpha s)E(T+(\alpha+1)s)^{\alpha+1} \le T^{(\alpha+1)/\alpha}E(0)^{\alpha+1}(T+\alpha s)^{-1/\alpha}
$$

whence

$$
E(T+(\alpha+1)s) \le E(0)\Big(1+\frac{\alpha s}{T}\Big)^{-1/\alpha}
$$

for all $s \ge 0$. Putting $t := T+(\alpha+1)s$ hence (3.8) follows. □

The estimate (3.5) will be deduced from the basic identity (3.4). Using the obvious estimate

$$
\Big|\int_\Omega u'Mu \,\mathrm{dx}\Big| = \Big|\int_\Omega u'\{2m\cdot\nabla u+(n-1)u\}\,\mathrm{dx}\Big| \le cE \tag{3.12}
$$

(here and in the sequel $c$ will denote divers constants independent of $(u^0, u^1)$ and of $t \in \mathbb{R}_+$) and the decreasing property of $E$, the first two terms on the right hand side of (3.4) are majorized by $cE^{(p+1)/2}(S)$ and hence also by $cE(S)$. Observing also that the second integral on the left hand side is nonpositive by the assumption (1.19), we conclude from (3.4) that

$$
2\int_S^T E^{(p+1)/2}\,\mathrm{dt} \le cE(S)
$$

$$
+\int_S^T E^{(p-1)/2}\int_{\Gamma_1}(u')^2-|\nabla u|^2-g(u')Mu\,\mathrm{d}\Gamma_m\,\mathrm{dt}.
$$

Since $\mathrm{d}\Gamma_m \ge 0$ on $\Gamma_1$ and

$$
|Mu|^2 \le c|\nabla u|^2+cu^2,
$$

hence we conclude for any fixed $\varepsilon > 0$ the inequality

$$
2\int_S^T E^{(p+1)/2}\,\mathrm{dt} \le cE(S)
$$

$$
+\int_S^T E^{(p-1)/2}\int_{\Gamma_1}(u')^2+\varepsilon u^2+c(\varepsilon)g(u')^2\,\mathrm{d}\Gamma_m\,\mathrm{dt}.
$$

Choosing $\varepsilon$ such that

$$\int_{\Gamma_1} \varepsilon u^2 \, d\Gamma_m \le E$$

($\varepsilon$ may be chosen independently of $u$), hence we obtain that

$$\int_S^T E^{(p+1)/2} \, dt \le cE(S) + c\int_S^T E^{(p-1)/2} \int_{\Gamma_1} (u')^2 + g(u')^2 \, d\Gamma_m \, dt. \tag{3.13}$$

In order to estimate the last integral in (3.13) we shall estimate for each fixed $t \in \mathbb{R}_+$ separately the four integrals

$$\int_{\Gamma_2} (u')^2 \, d\Gamma_m, \quad \int_{\Gamma_3} (u')^2 \, d\Gamma_m, \quad \int_{\Gamma_2} g(u')^2 \, d\Gamma_m, \quad \int_{\Gamma_3} g(u')^2 \, d\Gamma - M$$

where $\{\Gamma_2, \Gamma_3\}$ is a partition of $\Gamma_1$ defined by

$$\Gamma_2 = \{x \in \Gamma_1 : |u'(x)| \le 1\} \quad \text{and} \quad \Gamma_3 = \{x \in \Gamma_1 : |u'(x)| > 1\}. \tag{3.14}$$

Using (1.22) and (3.1) we obtain easily that

$$\begin{aligned}\int_S^T E^{(p-1)/2} \int_{\Gamma_3} (u')^2 + g(u')^2 \, d\Gamma_m \, dt &\le -c\int_S^T E^{(p-1)/2} E' \, dt \\ &\le cE^{(p+1)/2}(S) \le cE(S).\end{aligned} \tag{3.15}$$

Furthermore, using (1.21) we have

$$\begin{aligned}\int_{\Gamma_2} (u')^2 + g(u')^2 \, d\Gamma_m &\le c\int_{\Gamma_2} (u'g(u'))^{2/(p+1)} \, d\Gamma_m \\ &\le c\Big(\int_{\Gamma_2} u'g(u') \, d\Gamma_m\Big)^{2/(p+1)} \le c(-E')^{2/(p+1)}\end{aligned}$$

whence, applying the Young inequality, we obtain for every $\varepsilon > 0$ the estimate

$$\begin{aligned}&\int_S^T E^{(p-1)/2} \int_{\Gamma_2} (u')^2 + g(u')^2 \, d\Gamma_m \, dt \\ &\le c\int_S^T E^{(p-1)/2}(-E')^{2/(p+1)} \, dt \\ &\le \int_S^T \varepsilon E^{(p+1)/2} - c(\varepsilon)E' \, dt \le \varepsilon\int_S^T E^{(p+1)/2} \, dt + c(\varepsilon)E(S).\end{aligned} \tag{3.16}$$

Combining (3.13), (3.15) and (3.16) we find

$$\int_S^T E^{(p+1)/2}\ \mathrm{dt} \le \varepsilon c \int_S^T E^{(p+1)/2}\ \mathrm{dt} + c(\varepsilon)E(S).$$

Chosing $\varepsilon$ such that $\varepsilon c < 1$, hence we deduce that

$$\int_S^T E^{(p+1)/2}\ \mathrm{dt} \le cE(S) \tag{3.17}$$

whence (3.5) follows by letting $T \to +\infty$.

**4. Proof of Theorem 1.7.** Repeating the first part of the proof of Theorem 1.5 (combining (3.13) and (3.16)) we obtain again the estimate

$$\int_S^T E^{\frac{p+1}{2}}\ \mathrm{dt} \le cE(S) + c\int_S^T E^{\frac{p-1}{2}} \int_{\Gamma_3} (u')^2 + |Mu||g(u')|\ \mathrm{d}\Gamma_m\ \mathrm{dt}. \tag{4.1}$$

It remains to show that

$$E^{\frac{p-1}{2}} \int_{\Gamma_3} (u')^2\ \mathrm{d}\Gamma_m \le \varepsilon E^{\frac{p+1}{2}} - c(\varepsilon)E' \tag{4.2}$$

and

$$E^{\frac{p-1}{2}} \int_{\Gamma_3} |g(u')|(|u| + |\nabla u|)\ \mathrm{d}\Gamma_m \le \varepsilon E^{\frac{p+1}{2}} - c(\varepsilon)E' \tag{4.3}$$

for each $\varepsilon > 0$. Indeed, then choosing a sufficiently small $\varepsilon$, (4.1)–(4.3) imply (3.17) and (3.5), and we conclude by applying Lemma 3.3.

We shall write for brevity $\|\ \|_s$ instead of $\|\ \|_{L^s(\Gamma)}$.

Let us first prove (4.2). If $r = 1$, then

$$E^{\frac{p-1}{2}} \int_{\Gamma_3} (u')^2\ \mathrm{d}\Gamma_m \le cE^{\frac{p-1}{2}} \int_{\Gamma_3} u'g(u')\ \mathrm{d}\Gamma_m \le -cE'.$$

If $r < 1$, then put $s := 2/(p+1)$; (1.28) implies that $0 < s < 1$. We prove for each fixed $\varepsilon > 0$ the inequality

$$E^{\frac{p-1}{2}} \int_{\Gamma_3} (u')^2\ \mathrm{d}\Gamma_m \le \varepsilon E^{(p-1)/(2(1-s))} \|u'\|_{(2-(r+1)s)/(1-s)}^{(2-(r+1)s)/(1-s)} - c(\varepsilon)E'. \tag{4.4}$$

Indeed, we have

$$\begin{aligned} E^{\frac{p-1}{2}} \textstyle\int_{\Gamma_3} (u')^2\ \mathrm{d}\Gamma_m &\le cE^{\frac{p-1}{2}} \textstyle\int_{\Gamma_3} |u'|^{2-(r+1)s} |u'g(u')|^s\ \mathrm{d}\Gamma_m \\ &\le cE^{\frac{p-1}{2}} \||u'|^{2-(r+1)s}\|_{1/(1-s)} \||u'g(u')|^s\|_{1/s} \\ &= cE^{\frac{p-1}{2}} \|u'\|_{(2-(r+1)s)/(1-s)}^{2-(r+1)s} \|u'g(u')\|_1^s \\ &= cE^{\frac{p-1}{2}} \|u'\|_{(2-(r+1)s)/(1-s)}^{2-(r+1)s} (-E')^s \\ &\le \varepsilon E^{(p-1)/(2(1-s))} \|u'\|_{(2-(r+1)s)/(1-s)}^{(2-(r+1)s)/(1-s)} - c(\varepsilon)E'. \end{aligned}$$

Using the trace theorem

$$H^1(\Omega) \hookrightarrow L^{2(p-r)/(p-1)}(\Gamma) = L^{(2-(r+1)s)/(1-s)}(\Gamma)$$

(which follows from our choice (1.28) of $p$) and (2.3) we deduce from (4.4) that

$$E^{\frac{p-1}{2}} \int_{\Gamma_3} (u')^2 \, \mathrm{d}\Gamma_m \leq c\varepsilon E^{\frac{p+1}{2}} - c(\varepsilon)E'$$

yielding (4.2) (with another $\varepsilon$).

Next we prove (4.2). Putting for brevity $q := k+1$ we have

$$\begin{aligned} E^{\frac{p-1}{2}} & \int_{\Gamma_3} |g(u')|(|u| + |\nabla u|) \, \mathrm{d}\Gamma_m \\ & \leq cE^{\frac{p-1}{2}} \|g(u')\|_{1+q^{-1}} \left( \|u\|_{1+q} + \|\nabla u\|_{1+q} \right) \\ & \leq cE^{\frac{p-1}{2}} \left( \|u\|_{1+q} + \|\nabla u\|_{1+q} \right) \|(u'g(u'))^{1/(1+q^{-1})}\|_{1+q^{-1}} \\ & \leq cE^{\frac{p-1}{2}} \left( \|u\|_{1+q} + \|\nabla u\|_{1+q} \right) (-E')^{1/(1+q^{-1})} \\ & \leq \varepsilon E^{(p-1)(1+q)/2} \left( \|u\|_{1+q} + \|\nabla u\|_{1+q} \right)^{1+q} - c(\varepsilon)E'. \end{aligned}$$

It remains to show that

$$E^{\frac{(p-1)(1+q)}{2}} \left( \|u\|_{1+q} + \|\nabla u\|_{1+q} \right)^{1+q} \leq cE^{\frac{p+1}{2}}. \tag{4.5}$$

Using (2.3) and applying the trace theorem

$$W^{1,2n/(n+(n-2)k)}(\Omega) \hookrightarrow L^{k+2}(\Gamma)$$

(this follows from our assumption (1.24)) we obtain that

$$\|u\|_{1+q} + \|\nabla u\|_{1+q} \leq c\|u\|_{W^{2,2n/(n+(n-2)k)}(\Omega)} < c;$$

therefore (4.5) reduces to

$$E^{\frac{(p-1)(1+q)}{2}} \leq cE^{\frac{p+1}{2}}.$$

And this is true because $(p-1)(1+q) \geq p+1$ by the choice (1.28) of $p$.

## REFERENCES


[1] C. Bardos, G. Lebeau and J. Rauch, *Contrôle et stabilisation dans les problèmes hyperboliques*, dans J.-L. Lions, *Contrôlabilité exacte, stabilisation et perturbation des systèmes distribués, volume I*, Masson, Paris, 1988, pp. 492–537.

[2] H. Brézis, *Opérateurs maximaux monotones et semi-groupes de contractions dans les espaces de Hilbert*, North Holland, Amsterdam, 1973.

[3] F.E. Browder, *Problèmes non-linéaires*, Les Presses de l'Université de Montréal, 1966.

[4] G. Chen, *Energy decay estimates and exact boundary value controllability for the wave equation in a bounded domain*, J. Math. Pures Appl. 58(1979), pp. 249–274.

[5] G. Chen and H.K. Wang, *Asymptotic behavior of solutions of the one-dimensional wave equation with a nonlinear elastic dissipative boundary stabilizer*, SIAM J. Control Opt. 22(1989), pp. 758–775.

[6] F. Conrad, J. Leblond and J.P. Marmorat, *Stabilization of second order evolution equations by unbounded nonlinear feedback* in *Proc. of the Fifth IFAC Symposium on Control of Distributed Parameter Systems* (eds., A. El Jai and M. Amouroux) Perpignan, June 1989, pp. 101–116.

[7] F. Conrad and B. Rao, *Decay of solutions of wave equation in a star-shaped domain with nonlinear boundary feedback*, Asymptotic Analysis 7 (1993), pp. 159–177.

[8] A. Haraux, *Semilinear hyperbolic problems in bounded domains, Mathematical reports*, (J. Dieudonné, editor) Harwood Academic Publishers, Gordon and Breach, New York, 1987.

[9] A. Haraux and E. Zuazua, *Decay estimates for some semilinear damped hyperbolic problems*, Arch. Rat. Mech. Anal.100(2) (1988), pp. 191–206.

[10] V. Komornik, *Rapid boundary stabilization of the wave equation*, SIAM J. Control Opt. 29(1991), pp. 197–208.

[11] V. Komornik, *Decay estimates for the wave equation with internal damping* in *Proceedings of the Conference on Control Theory*, Vorau, Austria 1993, Birkhäuser, International Series Num. Analysis (to appear).

[12] V. Komornik, *On the nonlinear boundary stabilization of the wave equation*, Chinese Annals of Math. 14B(2) (1993), pp. 153–164.

[13] V. Komornik, *On the nonlinear boundary stabilization of Kirchoff plates*, Nonlinear Diff. Equations and Appl. (to appear).

[14] V. Komornik, *Exact controllability and stabilization* in *The Multiplier Method*, Masson-John Wiley, Paris 1994.

[15] V. Komornik and E. Zuazua, *A direct method for the boundary stabilization of the wave equation*, J. Math. Pures Appl. 69(1990), pp. 33–54.

[16] J. Lagnese, *Decay of solutions of wave equations in a bounded region with boundary dissipation*, J. Diff. Equations 50(1983), pp. 163–182.

[17] I. Lasiecka, *Stabilization of wave equations with nonlinear dissipative damping on the boundary*, J. Diff. Equations 79(1989), pp. 340–381.

[18] I. Lasiecka, *Global uniform decay rates for the solutions to wave equation with nonlinear boundary conditions* (to appear).

[19] I. Lasiecka and R. Tataru, *Uniform boundary stabilization of semilinear wave equation with nonlinear boundary conditions, Differential and Integral Equations*, (to appear).

[20] M. Nakao, *On the decay of solutions of some nonlinear dissipative wave equations in higher dimensions*, Math. Z. 193(1986), pp. 227–234.

[21] J.P. Quinn and D.L. Russell, *Asymptotic stability and energy decay rates for solutions of hyperbolic equations with boundary damping*, Proc. Roy. Soc. Edinburgh Sect. A 77(1977), pp. 97–127.

[22] R. Triggiani, *Wave equation on a bounded domain with boundary dissipation: an operator approach*, J. Math. Anal. Appl. 137(1989), pp. 438–461.

[23] E. ZUAZUA, *Stability and decay for a class of nonlinear hyperbolic problems*, Asymptotic Anal. 1,2(1988), pp. 161–185.

[24] E. ZUAZUA, *Uniform stabilization of the wave equation by nonlinear boundary feedback*, SIAM J. Control Opt. 28(1990), pp. 466–477.

# ASYMPTOTIC BEHAVIOR AND ATTRACTORS FOR NONLINEAR VON KÁRMÁN PLATE EQUATIONS WITH BOUNDARY DISSIPATION

IRENA LASIECKA*

**Abstract.** We consider the problem of asymptotic behavior of the solutions to a fully nonlinear von Kármán plate with boundary damping. The results presented depend on whether a forcing term (representing a load or force attached to the plate) appears in the equation. In the absence of this term, the uniform decay rates for the solutions are established. In the second case, when a constant load is attached to the plate, the existence of finite dimensional attractor is demonstrated. The role of geometric conditions and its relation to the presence of "light" interior damping is also discussed.

**Key words.** von Kármán plate, uniform decay rates, finite dimensional attractors, nonlinear boundary damping

**AMS(MOS) subject classifications.** 35K35

**1. Introduction.** Let $\Omega$ be a bounded open domain in $R^2$ with smooth boundary $\Gamma$. Consider the von Kármán system (large "displacement" model for thin plates) in the variables $w(t,x)$ and $\chi(t,x)$):

$$\left.\begin{array}{ll}(1.1a)\; w_{tt}-\gamma^2\Delta w_{tt}+\Delta^2 w+Dw_t=[w,\chi(w)]+F & \text{in } Q\\ \quad w(0,\cdot)=w_0;\, w_t(0,\cdot)=w_1 & \text{in } \Omega\\ (1.1b)\; \Delta w+(1-\mu)B_1 w=-f_1\left(\frac{\partial}{\partial\nu}w_t\right) & \text{on } \Sigma\\ \frac{\partial}{\partial\nu}\Delta w+(1-\mu)B_2 w-\gamma^2\frac{\partial}{\partial\nu}w_{tt}=g(w_t)-\frac{\partial}{\partial\tau}f_2\left(\frac{\partial}{\partial\tau}w_t\right)+\beta w & \text{on } \Sigma\end{array}\right\},$$

where $Q\equiv\Omega\times(0,\infty)$ and $\Sigma\equiv\Gamma\times(0,\infty)$. Here, $0<\mu<\frac{1}{2}$ is Poisson's ratio, $\beta>0$, and the operators $B_1$ and $B_2$ are given by

$$\left.\begin{array}{rcl}B_1 w &=& 2n_1n_2w_{xy}-N_1^2w_{yy}-n_2^2w_{xx}\\ B_2 w &=& \frac{\partial}{\partial r}\left[\left(n_1^2-n_2^2\right)w_{xy}+n_1n_2(w_{yy}-w_{xx})\right]\end{array}\right\}.$$

The operator $D: L_2(\Omega)\to L_2(\Omega)$ is self-adjoint, positive, and bounded: $H^1(\Omega)\to H^{-1}(\Omega)$. The function $F\in L_2(\Omega)$ represents the action of a load/force applied to the plate, and functions $f_i, g$ are monotone increasing and continuous. $\chi(\omega)$ is the Airy's stress function which satisfies the system

* Department of Mathematics, Thornton Hall, University of Virginia, Charlottesville, VA 22903. Partially funded by NSF Grant DMS-8902811.

of equations

$$\left.\begin{aligned} \Delta^2\chi &= -[w,w]; \quad \text{in } \Omega \\ \chi &= \frac{\partial}{\partial\nu}\chi = 0 \quad \text{on } \Sigma \end{aligned}\right\}. \tag{1.2}$$

where

$$[\phi,\psi] = \frac{\partial^2\phi}{\partial x^2}\frac{\partial^2\psi}{\partial y^2} + \frac{\partial^2\phi}{\partial x^2}\frac{\partial^2 y}{\partial x^2} - 2\frac{\partial^2\phi}{\partial x\partial y}\frac{\partial^2\psi}{\partial x\partial y}.$$

This system of equations describes the transversal displacement $w$ and the Airy-stress function $\lambda$ of a vibrating plate (for a derivation of the model see [Lag-1] and references therein). The second term in the equation (1.1a) represents rotational inertia. This term is proportional to $\gamma^2$ which, in turn, is proportional to the square of the thickness of the plate. Since the parameter $\gamma$ is "small," it may be in some cases neglected. The right side of boundary conditions in (1.1b) represents the nonlinear feedback controls, whose purpose is to induce a certain desired asymptotic behavior of the plate (such as stability, convergence to an equilibrium, etc.) With dynamics described by (1.1) we associate the following energy function

$$\begin{aligned} (1.3)\ E(t) &= \frac{1}{2}\int_\Omega \{|w_t|^2 + \gamma^2|\nabla w_t|^2\}\,d\Omega + \frac{\beta}{2}\int_\Gamma w^2 d\Gamma + \frac{1}{2}a(w,w) \\ &+ \frac{1}{2}\int_\Omega |\Delta\chi(w)^2 d\Omega, \end{aligned}$$

where $a(w,w) \equiv \int_\Omega |\Delta w|^2 + (1-\mu)[2w_{xy}^2 - w_{xx}^2 - w_{yy}^2]d\Omega.$

We also define the Hilbert space $H \equiv H^2(\Omega)\times H^1(\Omega)$ endowed with a norm

$$|(w_0,w_1)|_H^2 \equiv a(w_0,w_0) + \int_\Gamma w_0^2 d\Gamma + \int_\Gamma (w_1^2 + \gamma^2|\nabla w_1|^2)d\Omega.$$

The following well-posedness result is available.

PROPOSITION 1.1. (See [F-L] and [Las.1].) *Given initial data* $(w_0,w_1) \in H^2(\Omega)\times H^1(\Omega) \equiv H$, *there exists a unique solution* $(w,w_t)$ *to* (1.1) *with* $w \in C([0,\infty);H^2(\Omega))$ *and* $w_t \in C([0,\infty);H^1(\Omega))$.

The main topic of this paper is asymptotic behavior of the solutions when $t\to\infty$. Of particular interest are questions pertinent to uniform decay of the solutions and to an existence of compact attractors (see [H.1]).

Problems related to the asymptotic behavior of solutions to both, linear and nonlinear plate equations, have attracted considerable attention in recent years (see [Lag. 1] and references therein). In what follows we shall focus on von Kármán plate model as described by (1.1), (1.2).

In the case when the boundary conditions are *homogeneous*, asymptotic behavior of the solutions to von Kármán plate model with "strong" interval damping was considered by Chueshov in [Ch. 1], [Ch. 2]. Indeed, models treated by Chueshov are von Kármán plate equations with the homogeneous clamped boundary conditions

$$w = \frac{\partial w}{\partial \nu} = 0 \quad \text{on} \quad \Sigma,$$

and with the dissipative mechanism introduced via "strong" internal damping. This is to say that the operator $D$ in (1.1a) takes the form

$$Dw_t = \alpha^2 \Delta w_t \quad \text{when} \quad \gamma \neq 0 \quad \text{(see [Ch.2])}; \tag{1.4}$$

$$Dw_t = \alpha^2 w_t \quad \text{when} \quad \gamma = 0 \quad \text{(see [Ch. 1])}. \tag{1.5}$$

In both cases, Chueshov was able to prove convergence of the solutions to an appropriate finite-dimensional attractor. For the problems when rotational inertia are accounted for (i.e., $\gamma \neq 0$), this convergence result applies to all "weak" (i.e., of finite energy) solutions (see [Ch. 2]). If $\gamma = 0$ in the model (1.1), then this result (see [Ch. 1]) applies only to "strong" solutions (i.e., $w \in C(0,T); H^4(\Omega))$, $w_t \in C[(0,T); H^2(\Omega)))$. Global existence of such strong solutions, has been demonstrated in [Ch. 1] (see also [S.1] and [K-S]).

A problem which has attracted a considerable attention in recent years is that of boundary stabilization, i.e., when dissipation mechanism acts on the boundary of $\Omega$ as in (1.1). The present paper is devoted to a discussion of related results. We shall first consider the case when no load is applied to the plate (i.e., $F = 0$ in Eqn. (1.1)). Then, the uniform decay rates for the solutions can be established. In Section 2, which is mostly a review section, we collect and summarize some of these uniform stabilization results. Sections 3 and 4 deal with a more general case when $F \neq 0$. The main result in this direction, existence of finite-dimensional attractors is stated at the beginning of Section 3. The proofs are given in Sections 3 and 4.

**2. Uniform decay rates for the von Kármán model (1.1) with $F = 0$.** The problem of boundary stabilization for the von Kármán system was treated, for the first time, by Lagnese in [Lag.1]. Indeed, in Chapter 5 of [Lag.1] it is shown that the energy of the solutions to (1.1), (1.2) with $\gamma = 0$, $D = 0$, $F = 0$, and the *linear* boundary feedbacks decays to zero exponentially. Precise statement of this result is given below. We recall the following "star-shaped" condition imposed on $\Omega$.

(H) There exists $x_0 \in R^n$ such that $(x - x^0) \cdot \nu > 0$ on $\Gamma$.

THEOREM 2.1. [[Lag.1]. p. 118] *Assume the geometric hypothesis* (H). *Let* $w$ *be a solution to* (1.1a), (1.2) *with* $\gamma = 0$, $D = 0$, $F = 0$, *and the following boundary conditions on* $\Sigma$,

$$\Delta w + (1-y)\beta_1 w = -\alpha \frac{\partial}{\partial \nu}\, w_t \tag{2.1}$$

$$\frac{\partial}{\partial \nu}\, \Delta w + (1-y)\beta_2 w = \lambda_0 (m\cdot\nu) w_t + \beta (m\cdot\nu) w - \alpha \frac{\partial}{\partial \tau}\left((m\cdot\gamma)\frac{\partial w}{\partial \tau}\right), \tag{2.2}$$

*where the constants* $\alpha$, $\beta$, $\lambda_0$ *are positive.*

*Assume, moreover, that the solution* $w$ *satisfies the following regularity assumptions*

$$\begin{aligned} &\text{(i)} && w \in L^\infty((0,T); H^4(\Omega)),\\ &\text{(ii)} && w_t \in L^\infty((0,T); H^2(\Omega)),\\ &\text{(iii)} && w_{tt} \in L^\infty((0,T); L_2(\Omega)), \end{aligned} \tag{2.3}$$

*There is a number* $a_0 > 0$, *and, for each* $\alpha$, *a number* $\beta_0(\alpha) > 0$ *such that if* $0 < \alpha \le \alpha_0$, $0 < \beta \le \beta_0(a)$, *and* $\lambda_0 > 0$, *then*

$$E(t) \le Ce^{-wt} E(0); \qquad w > 0. \tag{2.4}$$

□

*Remark 2.1.* By keeping track of various constants appearing in the proof of Theorem 2.1 (see [Lag.1], p. 91–96) one can obtain explicit estimates for the constants $C$; $w$ in inequality (2.3).

*Remark 2.2.* A careful analysis of the proof of Theorem 2.1 reveals that the estimate (2.3) remains true for the solutions which are less regular than (2.3)(i). Indeed, it is enough to have $w \in L^\infty(0,T); H^3(\Omega)) \cap W^{1,\infty}((0,T); H^2(\Omega)) \cap W^{2,\infty}((0,T); L_2(\Omega))$.

*Remark 2.3.* The existence of weak solutions (i.e., of finite energy) for the problem considered in Theorem 2.1 can be proved by using Galerkin's method (see Theorem 2.1 in [Lag.1]). However, as noted in Remark 2.2 of [Lag.1], such solutions are not, in general, unique. A consequence of this is that the estimate (2.3) of Theorem 2.1 cannot be extended to hold for all weak solutions (finite energy solutions do not depend continuously on the initial data). Another delicate issue is that of *regularity* of the solutions. More precisely, what are the conditions on the initial data $\{w_0, w_1\}$ that guarantee that every solution to the problem in Theorem 2.1 possesses regularity properties listed in (2.2). Again, as remarked in [Lag.1], p. 113 "... regularity of solutions is (apparently) and open question .... " We shall cast more light on this later (see Theorem 2.4).

The proof of Theorem 2.1 in [Lag.2] is based on Liapunov function method combined with the appropriately modified multiplier's technique, used earlier in the context of linear theory (see [Lag.1], Chapters 3,4). In fact, the same, essentially, technique has proved successful in [P-T] for establishing the exponential decay rates valid for a more complex von Kármán system accounting for in plane accelerations (see also [Lag.2] where a static version of this problem is treated). These results are proved for "star-shaped domains" (i.e., under hypothesis (H)), linear boundary feedbacks, and they are valid for *regular* solutions only (i.e., at least $w \in L^\infty((0,T); H^3(\Omega))$. On the other hand, the existence of such *regular* solutions has been apparently an open problem. In the special case of annular domains $\Omega$, [P-T] was able to prove the required regularity of the solutions. This was accomplished by adopting the one-dimensional arguments of L-L].

Needless to say that the requirement of existence of smooth solutions is restrictive in nonlinear problems. Indeed, even if the initial data are regular, the corresponding solutions may not be smooth at all. This is certainly the case, when one considers nonlinear boundary feedbacks. The above considerations gave rise to the following questions. What one can say about the asymptotic behavior of the solutions to nonlinear model (1.1) in the absence of information on the additional smoothness of the solutions? Is it possible to derive the uniform decay rates which are valid not *only* for *regular* solutions but for weak solutions as well (typically, the finite energy solutions)? Are the uniform decay estimates still valid for the models accounting for rotational inertia (i.e., $\gamma > 0$ in (1.1)) and the nonlinear boundary feedbacks?

The results collected below attempt to answer these questions. In addition, issues related to lack of geometric conditions (i.e., hypothesis (H)) will be dealt with. As it can be seen, coping with these kind of questions requires different techniques than previously employed. The power of microlocal analysis together with a development of a very general approach for proving decay rates in nonlinear p.d.e problems (used earlier in [Las-Tat] in the context of nonlinear wave equation) are critical to the solution of the problem.

In what follows, we consider the nonlinear model (1.1), (1.2) accounting for rotational inertia forces (i.e., $\gamma > 0$) and with boundary feedbacks $f_1, f_2, g$ subject to the following assumptions.

The functions $g$ and $f_i$ are continuous, monotone and are subject to the following conditions:

$$\text{(A)} \qquad \left.\begin{array}{lll} g(s)s > 0, & \text{for} & s \neq 0 \\ f_i(s)s > 0, & \text{for} & s \neq 0 \\ m|s| \leq |f_i(s)| \leq M|s|, & \text{for} & |s| > R, i = 1,2 \\ m|s|^2 \leq g(s)s \leq M|s|^{r+1}, & \text{for} & |s| > R \end{array}\right\}$$

where $r$ and $R$ are arbitrary positive constant.

We shall assume that the "damping" operator $D : H^1(\Omega) \to H^{-1}(\Omega)$ is injective, i.e.,

$$Dw = 0 \implies w = 0. \tag{2.5}$$

*Remark 2.4.* $(Dw)(x) \equiv b(x)w$ where $b \in L^\infty(\Omega)$ and $b(x) > 0$ a.e. in $\Omega$ complies, in particular, with (2.4).

To state our results, we introduce the following notation. Define function $h(x)$ by:

$$h(x) \equiv h_0(x) + h_1(x) + h_2(x),$$

where $h_i(x)$ are concave, strictly increasing functions with $h_i(0) = 0$ such that

$$\begin{aligned} &h_0(sg(s)) \geq s^2 + g^2(s), \quad |s| \leq 1; \\ &h_i(sf_i(s)) \geq s^2 + f_i^2(s), \quad |s| \leq 1, i = 1, 2. \end{aligned}$$

Such functions can be easily constructed (see [Las-Tat]). Then $h(x)$ enjoys the same properties, i.e., it is concave, strictly increasing, and $h(0) = 0$. Define next

$$\widetilde{h}(x) \equiv h\left(\frac{x}{\text{meas}\Sigma_T}\right), \quad \text{where} \quad \Sigma_T \equiv \Gamma \times (0, T). \tag{2.6}$$

Since $\widetilde{h}$ is monotone increasing, for every $c \geq 0$, $cI + \widetilde{h}$ is invertible. Setting

$$p(x) \equiv (cI + \widetilde{h})^{-1}(Kx), \tag{2.7}$$

where $K$ is a positive constant, we see that $p$ is a positive, continuous, strictly increasing function with $p(0) = 0$.

We are now in a position to state the following result proved in [H-L] (see also [H-L.1]).

THEOREM 2.2. [H-L] *Consider* (1.1), (1.2) *with* $F = 0$, $\gamma > 0$, *nonlinear functions* $f_i, g$ *subject to* (A) *and the damping operator* $D$ *subject to* (2.5). *Then, the energy functional* $E(t)$ *(see* (1.3)*) associated with any* weak *solution guaranteed by Proposition* 1.1 *satisfies the following estimate: there exists* $T_0 > 0$ *such that*

$$E(t) \leq S\left(\frac{t}{T_0} - 1\right) \quad \textit{for} \quad t > T_0, \tag{2.8}$$

*where* $S(t) \to 0$ *as* $t \to \infty$, *and it is the solution (contraction semigroup) of the differential equation*

$$\begin{cases} \dfrac{d}{dt} S(t) + q(S(t)) = 0, \\ S(0) = E(0), \end{cases} \tag{2.9}$$

*with $q(x)$ given by*

$$q(x) \equiv x - (I + p)^{-1}(x) \quad \text{for} \quad x > 0.$$

*In this case, the constant $K$ will generally depend on $E(0)$ and the constant $c = \dfrac{1}{meas\, \Sigma_T}(m^{-1} + M)$. Moreover, all the constants involved are* <u>*independent*</u> *on the value of $\gamma > 0$.* □

*Remark 2.5.* Notice that the result of Theorem 2.2 does not require any geometric restrictions imposed on $\Omega$, that the nonlinear functions $f_i, g$ are not subject to any growth condition at the origin (a typical assumption used in the context of stability) and that the estimate (2.7) is valid for *all weak* solutions.

*Remark 2.6.* One could also consider feedback controls acting on a portion of the boundary only. In this case, the appropriate geometric condition imposed (see [B-L.2]) on the *uncontrolled* part of the boundary are needed.

*Remark 2.7.* In the special case, when the growth at the origin of the nonlinear boundary feedbacks is specified, one can compute explicitly the governing decay rates for the energy function $E(t)$. Indeed, this can be easily accomplished by constructing the appropriate function $q$ and solving the nonlinear ODE problem (2.8). For instance, if the nonlinear functions $f_i$ and $g$ have a linear growth at the origin, then (2.7) specializes to

$$E(t) \leq C(E(0))e^{-wt} \quad \text{for some } w > 0.$$

If, instead, these nonlinearities are of the polynomial growth at the origin (say $\sim x^p, p > 1$, then $E(t) \leq C(E(0))t^{\frac{2}{1-p}}$.

In the case of *linear* boundary feedbacks, the result of Theorem 2.2 [which amounts, now, to exponential decay rates] was proved earlier in [B-L.2]. In this particular situation, one can show (see [F-L]) that for the initial data $\{w_0, w_1\} \in H^4(\Omega) \times H^3(\Omega)$ and subject to the appropriate compatibility conditions on the boundary, the corresponding solutions are classical, i.e., $w \in C((0,\infty); H^4(\Omega))$, $w_{tt} \in C((0,\infty); H^2(\Omega)$. This regularity was critical in [B-L] for justification of p.d.e. estimates. However, in the general nonlinear case, as treated by Theorem 2.2, weak solutions obtained from the existence theorem (see Proposition 1.1) do not display sufficient regularity which would permit to justify the necessary p.d.e. arguments. To cope with this difficulty, [H-L] resorts to a very special limiting approximation argument (used earlier in [Las-Tat] in the context of the wave equation), which allows to bypass the necessity of having regular solutions.

It is interesting to note the role played by the "light" internal damping modeled by the operator $D$. The presence of this light damping is physically motivated, since most vibrating materials possess some degree of interior damping. On the other hand, it is obvious that this *mild* damping along (i.e., without the dissipation on the boundary would not suffice to stabilize uniformly the plate ("strong" internal damping which would cause the energy to decay uniformly would have to be coercive). However, the presence of this mild damping appears essential in proving the effectiveness of the boundary dissipation. This is at the level of applying certain unique continuation type of argument. Indeed, in order to dispense all together with "light" damping (i.e., taking $D = 0$) one would need the following "Unique Continuation property" (UCP) for the von Kármán system.

(UCP) PROPERTY. *We say that the problem* (1.1), (1.2) *satisfies (UCP) property if any solution* $w \in \mathcal{D}'(Q_T)$ *which satisfies* (1.1a) *with* $F = 0$, $D = 0$, (1.2) *and the following (over-determined) boundary conditions on* $\Sigma_T$,

$$\begin{aligned}&w_t = \frac{\partial}{\partial\nu} w_t = 0;\\ &\Delta w + (1-\mu)B_1 w = 0;\\ &\frac{\partial}{\partial\nu}\Delta w + (1-\mu)B_2 w - \beta w = 0,\end{aligned}$$

*satisfies also* $w \equiv 0$ *in* $Q$.

By using computations similar to these in [Lag.1] on p. 110–118, one can show that under the geometric hypothesis (H) the (UCP) property is satisfied. In this case, the result of Theorem 2.2 holds true *without* any interior damping (i.e., one can take $D = 0$). Whether the (UCP) property remains true *without* any geometric conditions is, as far as we know, an open question.

Notice that the estimate (2.8) of Theorem 2.2 is *independent* on the value of the parameter $\gamma > 0$. This fact has far-reaching consequences. First of all, it shows that the effectiveness of the stabilizing boundary feedback does not deteriorate for small values of $\gamma > 0$ (this is a "robust" aspect of stabilization). Similar property was demonstrated in [Lag.1], p. 88 for the linear Kirchhoff plate model with the linear feedback depending on $\gamma > 0$). Second, it allows to obtain uniform decay rates for the limit problem, i.e., when $\gamma = 0$. More specifically, we obtain

COROLLARY 2.1. *For each value of the parameter* $\gamma > 0$ *and linear functions* $f_i, g$, *consider the problem in Theorem 2.2. Then, the solution* $w_\gamma$ *(resp.* $w_{\gamma t}$*) converges weak* in* $L^\infty((0,T); H^2(\Omega))$ *(resp.* $L^\infty((0,T); L^2(\Omega))$*) to function* $w(t)$ *(resp.* $w_t$*), which is a weak solution to the limit problem (1.1) with* $\gamma = 0$. *Moreover, there exist constants* $C$, $w > 0$ *such that*

$$|w(t)|_{H^2(\Omega)} + |w_t(t)|_{L_2\Omega)} \le Ce^{-wt}\big[|w_0|_{H^2(\Omega)} + |w_1|_{L_2(\Omega)}\big]. \qquad \square$$

Corollary 2.1 follows easily from the results of Theorem 2.2 combined with the limiting argument of [H-L]. It is precisely at the level of the passage to the limit, when the linearity of the boundary feedbacks is exploited. This raises the following questions: What can be said about the limit problem (i.e., $\gamma = 0$) with *nonlinear* boundary feedbacks? In this case, as we recall, weak solutions may not be unique. However, in the class of the solutions which are only "slightly" more regular than the finite energy solutions, the uniqueness property holds. Precise statement of this result is given below.

LEMMA 2.1. (Uniqueness) [H-L-2] *Let* $\{w, w_t\}$ *be any weak solution corresponding to the model* (1.1), (1.2) *with* $\gamma = 0$ *and such that for some* $\varepsilon > 0$,

$$w \in C((0,T); H^{2+\varepsilon}(\Omega)); w_t \in C((0,T); L_2(\Omega)). \tag{2.10}$$

*Then, such solution is unique within this class.* □

The next result shows that for such solutions the decay estimates of Theorem 2.2 are valid. Indeed, by arguments identical to these in [H-L.1] we obtain.

THEOREM 2.3. *Consider model* (1.1), (1.2) *with* $F = 0$, $\gamma = 0$ *and the damping operator* $D$ *subject to the assumption (2.4). The nonlinear functions* $f_1, g$ *satisfy hypothesis* (A) *and* $f_2 = 0$. *Then, any solution* $w$ *such that (2.10) holds, satisfies the uniform decay estimate of Theorem 2.2.*

□

*Remark 2.8.* If the system considered in Theorem 2.3 satisfies the (UCP) property (with $\gamma = 0$), then the statement of Theorem 2.3 holds true with $D = 0$. This is a case, in particular, when the geometric condition (H) is satisfied.

We conclude this section by providing sufficient conditions for the initial data $\{w_0, w_1\}$, which guarantee the additional regularity of the solutions to the limit equation (i.e., $\gamma = 0$).

THEOREM 2.4. [H-L.2] *With reference to the problem considered in Theorem 2.3, and assuming additionally that functions* $f_1, g$ *are continuously differentiable we have: for any initial data such that*

$$\begin{cases} w_0 \in H^4(\Omega); \ w_1 \in H^2(\Omega) \ \textit{and} \\ \Delta w_0 + (1-\mu)B_1 w_0 = -f_1\left(\dfrac{\partial}{\partial \nu} w_t\right) \quad \textit{on} \quad \Gamma, \end{cases} \tag{2.11}$$

*there exists a unique, global solution* $w$ *with the following regularity properties*

$$w \in C((0,T); H^3(\Omega)); \ w_t \in C((0,T); H^2(\Omega)). \tag{2.12}$$

□

*Remark 2.9.* In the case of *homogeneous* boundary conditions, the existence of classical solutions ($w \in C((0,T); H^4(\Omega \cap C^1((0,T); H^2(\Omega))$ was established in [K-S] and [Ch.1].

As a corollary of Theorem 2.4, we immediately obtain

COROLLARY 2.2. *Assume that the initial data $w_0, w_1$ satisfy (2.9) and that the nonlinear functions $f_i, g$, in addition to hypothesis (A), are continuously differentiable. Then, the statement of Theorem 2.3 holds true.*

## 3. Asymptotic behavior of solutions to (1.1), (1.2) with a load applied to the plate (i.e., $F \neq 0$)

**3.1. Statement of the result.** Before we state our results we recall a few definitions from [H.1] and [H-R].

Let $\{T(t) : t \geq 0\}$ be a $C_0$ semigroup on a Hilbert space $H$. The positive orbit of a bounded set $B \subset H$ is $\gamma^+(B) \equiv \bigcup_{t\geq 0} T(t)B$.

The $w$ limit set of a bounded set $B \subset H$ is

$$w(B) \equiv \bigcap_{\tau \geq 0} C\ell\gamma^+(T(\tau)B).$$

A set $\mathcal{A}$ is said to be a global attractor for $T(t)$ if

(i) $\mathcal{A}$ is invariant under $T(t)$ (i.e., $T(t)\mathcal{A} = \mathcal{A}$).

(ii) $\mathcal{A}$ attracts all bounded sets of $H$, i.e., for any bounded set $B \in H$ and any $\varepsilon > 0$ there is $t_0(B, \mathcal{A}, \varepsilon)$ such that $T(t)B \subset \mathsf{U}(\mathcal{A}, \varepsilon)$ for $t \geq t_0$ where $\mathsf{U}(\mathcal{A}, \varepsilon)$ denotes $\varepsilon$-neighborhood of $\mathcal{A}$ in $H$.

If the set $\mathcal{A}$ is minimal with respect to these properties it is called minimal global attractor.

Our main result is the following

THEOREM 3.1. *With reference to problem* (1.1), (1.2), *where $f_i$ and $g$ are linear and $\gamma > 0$.*

*(i) There exists a locally compact minimal global attractor $\mathcal{A}$ (for the semigroup $T(t)$ generated by* (1.1)) *in $H^2(\Omega) \times H^1(\Omega)$ given by*

$$\mathcal{A} = \cup\{w(B) : B \in H^2(\Omega) \times H^1(\Omega),\ B \ \textit{bounded}\}. \tag{3.1}$$

*(ii) For each bounded set $B \subset H^2(\Omega) \times H^1(\Omega)$, $w(B) \in B(0, R)$ is of finite Hausdorff dimension.* □

*Remark 3.1.* The statement in part (i) of Theorem 3.1 is valid also for the nonlinear feedback $f_i$ and $g$ (see [Las.2]). □

The remainder of this section is devoted to the proof of Theorem 3.1. Here we shall outline only the main steps. The first step in the proof employs an idea due to Hale (see [H.1], [H-R]) which is to write the semigroup $T(t)$ as a sum $S(t) + K(t)$ where $S(t)$ is a uniformly stable semigroup and $K(t)$ is compact. This allows us to apply Theorem 1.1 of [H-R] (see also Theorem 3.4.6 in [H.1]) which asserts the existence of a locally compact minimal global attractor. The proof of (local) finite dimensionality of the attractor including the estimates for its dimension is more involved. It employs the result due to Ladyzhenskaya [Lad.1] which, in turn, permits to evaluate explicitly the dimension of the attractor provided that one first shows that the projection of the flow on a certain subspace of finite codimension has the Lipschitz norm less than one.

**3.2. Proof of part (i) of Theorem 3.1.** *Step 1: Exponential decay rates for the Kirchhoff model.* Consider the following equation of the Kirchhoff plate with nonlinear functions $f_i, g$ subject to the hypothesis (A):

$$(3.2)\quad \begin{cases} w_{tt} - \gamma^2 \Delta w_{tt} + \Delta^2 w = 0 & \text{in } Q; \\ w(0,\cdot) = w_0;\, w_t(0,\cdot) = w_1 & \text{in } \Omega; \\ \Delta w + (1-\mu)B_1 w = -f_1\left(\dfrac{\partial}{\partial\nu} w_t\right) & \text{on } \Sigma; \\ \dfrac{\partial}{\partial\nu}\Delta w + (1-\mu)B_2 w - \gamma^2 \dfrac{\partial}{\partial\nu} w_{tt} = g(w_t) - \dfrac{\partial}{\partial\tau} w_t + \beta w & \text{on } \Sigma. \end{cases}$$

THEOREM 3.2. *For any $w_0, w_1 \in H^2 \times H^1(\Omega) \equiv H$ there exists a unique solution to* (3.2), $(w, w_t) \in \mathbf{C}([0,\infty); H^2(\Omega) \times H^1(\Omega))$ *such that the estimate* (2.8) *(with appropriately modified energy function) holds. In particular case, when the feedbacks $f_i$ and $g$ are linear, then there exists constants $C$, $w > 0$ such that*

$$\|w(t)\|_{H^2(\Omega)} + \|w_t(t)\|_{H^1(\Omega)} \le Ce^{-wt}\left[\|w_0\|_{H^2(\Omega)} + \|w_1\|_{H^1(\Omega)}\right], \quad t \ge 0. \ \Box$$

For star-shaped domains (i.e., subject to the geometric conditions on the boundary) and for the nonlinear feedbacks subject to the polynomial growth conditions at the origin (in addition to hypothesis (A)), the statement of Theorem 3.2 has been proved in [Lag.1]. The present version of this theorem can be obtained from the arguments of [H-L] supported by microlocal analysis estimates for the traces of the solutions to the linear Kirchoff plate (see [L-T]).

*Step 2: Existence of the global attractor.* Let $\widehat{w}(t) \equiv (w(t), w_t(t))$ denotes the solution to (1.1) due to initial data $\widehat{w}_0 \equiv (w_0, w_1)$ (the existence of $\widehat{w}(t)$ is guaranteed by Proposition 1.1). Let $T(t)$ denote the nonlinear semigroup associated with (1.1), i.e.:

$$T(t)\widehat{w}_0 \equiv \widehat{w}(t) \quad \text{for} \quad \widehat{w}_0 \in H \equiv H^2(\Omega) \times H^1(\Omega).$$

We first rewrite the problem (3.2) in a semigroup form. To accomplish this we introduce the following operators

$A_N : L_2(\Omega) \to L_2(\Omega)$ given by

$$(3.3)\qquad A_N u = -\Delta u;\ \mathcal{D}(A_N) \equiv \left\{u \in H^2(\Omega);\ \frac{du}{d\nu}|\Gamma = 0\right\}.$$

$\mathcal{A} : L_2(\Omega) \to L_2(\Omega)$, a self-adjoint positive operator defined by

$$(3.4)\qquad \mathcal{A} \equiv \Delta^2 u;\ \mathcal{D}(\mathcal{A}) = \{u \in H^4(\Omega);\ \Delta u + (1-\mu)B_1 u = 0;$$

$$\frac{\partial}{\partial\nu}\Delta u + (1-\mu)B_2 u = u \text{ on } \Gamma\}.$$

$G_i : L_2(\Gamma) \to L_2(\Omega)$; $i = 1, 2$, where $G_i$-Green's maps are defined as below.

$G_1 g = v$ iff $\Delta^2 v = 0$ in $\Omega$ and

$$\Delta v + (1-\mu)B_1 v = 0 \text{ on } \Gamma; \frac{\partial}{\partial\nu}\Delta v + (1-\mu)B_2 v - v = g; \quad \text{on } \Gamma,$$

and

$$G_2 g = v \text{ iff } \Delta^2 v = 0 \text{ in } \Omega;\ \Delta v + (1-\mu)B_1 v = g \text{ on } \Gamma;$$
$$\frac{\partial}{\partial\nu}\Delta v + (1-\mu)B_2 v - v = 0 \text{ on } \Gamma.$$

It is well known from elliptic theory [L-M] that the maps $G_i$ are well defined and bounded.

With the above definition, one easily shows (see [B.1]) that the equation (3.2) can be written as a system of equations for $\widehat{w} = (w, w_t)$,

$$\begin{cases} \frac{d}{dt}\widehat{w} + A\widehat{w} = 0; \\ \widehat{w}(0) = (w_0, w_1), \end{cases} \tag{3.5}$$

where the generator $A : H \to H$ is defined by

$$A\begin{pmatrix} u \\ v \end{pmatrix} \equiv \begin{bmatrix} -v \\ (1+\gamma^2 A_N)^{-1}\mathcal{A}[u + G_1 G_1^* \mathcal{A}v - G_1 \frac{\partial^2}{\partial\tau^2} G_1^* \mathcal{A}v + G_2 G_2^* \mathcal{A}v] \end{bmatrix};$$

$$\mathcal{D}(A) = \left\{(u,v) \in [H^2(\Omega)]^2; u + G_1 G_1^* \mathcal{A}v - G_1 \frac{\partial^2}{\partial\tau^2} G_1^* \mathcal{A}v + G_2 G_2^* \mathcal{A}v \in \mathcal{D}\left(\mathcal{A}^{\frac{3}{4}}\right)\right\}.$$

Let $e^{-At}$ be a $C_0$-semigroup generated by $A$. Now we write

$$T(t) = S(t) + K(t), \tag{3.6}$$

where $S(t) \equiv e^{-At}$.

Straightforward computations (see [B-L.2]) show that $K(t)$ is given by the following variation of parameter formula

$$K(t)(\widehat{w}_0) \equiv \int_0^t e^{-A(t-s)} \begin{bmatrix} 0 \\ (I + \gamma^2 A_N)^{-1}[w(s), \chi w(s)] \end{bmatrix} ds, \tag{3.7}$$

$$+ \int_0^t e^{-A(t-s)} \begin{bmatrix} 0 \\ (I + \gamma^2 A_N)^{-1} F \end{bmatrix} ds,$$

where $(w(t), w_t(t)) \equiv T(t)\widehat{w}_0$.

In order to establish an existence of a locally compact global attractor (as claimed in part (i) of Theorem 3.1), we shall use Theorem 1.1 of [H-R]. Indeed, according to Theorems 1.1 and 1.3 of [H-R], the conclusion of part (i) in Theorem 1 follows provided:

$$\gamma^+(B) \text{ are bounded for every } B \text{ bounded in } H. \tag{3.8}$$

$$S(t) \text{ is uniformly stable, i.e., } |S(t)|_{\mathcal{L}(H)} \leq Ce^{-wt} \text{ for some constants } C, w < 0. \tag{3.9}$$

$$K(t) : H \to H \text{ is compact.} \tag{3.10}$$

In view of Theorem 3.2 it suffices to show (3.8) and (3.10).

*Proof of (3.8).* From the regularity theory in [F-L] we know that with initial data $\{w_0, w_1\} \in H^4(\Omega) \times H^3(\Omega)$ and subject to the appropriate compatibility conditions, the corresponding solution satisfies

$$w(t) \in H^4(\Omega);\ w_t(t) \in H^3(\Omega);\ w_{tt} \in H^2(\Omega). \tag{3.11}$$

Now let $w(t)$ be such a smooth (classical) solution to (1.1). We multiply (1.1) by $w_t(t)$ and integrate from 0 to "$t$". The application of Green's formula, together with rather straightforward computations gives

$$E(t) + \int_0^t \int_\Gamma |w_t|^2 + |\nabla w_t|^2 d\Sigma = E(0). \tag{3.12}$$

Let $\widehat{w} = (w_0, w_1) \in H$. Then for $0 < \varepsilon < \frac{1}{2}$,

$$[w_0, w_1] \in L_1(\Omega) \subset H^{-1-\varepsilon}(\Omega) \equiv (H^{1+\varepsilon}(\Omega))' = \mathcal{D}\left(A_N^{\frac{1}{2}+\frac{\varepsilon}{2})}\right)', \tag{3.13}$$

where duality is considered with respect to $L_2(\Omega)$ inner product. By elliptic theory and (3.13) we infer

$$|\chi(w_0)|_{H^{3-\varepsilon}(\Omega)} \leq C|w_0|^2_{H^2(\Omega)},$$

hence

$$|\Delta\chi(w_0)|_{L_2(\Omega)} \leq C|w_0|^2_{H^2(\Omega)}. \tag{3.14}$$

This proves that

$$E(0) \leq C\left[|w_0|^4_{H^2(\Omega)} + |w_0|^2_{H^2(\Omega)} + |w_1|^2_{H^1(\Omega)}\right]. \tag{3.15}$$

Inequality (3.15), together with the regularity result of Proposition 1.1, allows us to extend (3.12) to the initial data in $H$. From (3.12) and (3.15) we infer that the orbits of bounded sets in $H$ are bounded $H$, i.e., (3.8) holds true.

*Proof of (3.10).* Let $|\widehat{w}|_H = |(w_0, w_1)|_H \leq R$. Then, by Proposition 1.1 and (see (3.14)) $w(t)$ the solution to (1.1) satisfies

$$|\chi w(t)|_{C([0,t];H^{3-\epsilon}(\Omega))} \leq C|w(t)|^2_{C([0,t];H^2(\Omega)]} \leq CR^2,$$

and (see [B-L]; [Lag.1]),

$$(3.16) \qquad |[w(t), \chi(w(t))]|_{C([0,t]H^{-\epsilon}(\Omega))} \leq CR^3; \qquad \forall\, 0 < \varepsilon < 1.$$

By elliptic regularity $(I + \gamma^2 A_N)^{-1} \in \mathcal{L}(H^{-\varepsilon}(\Omega) \to H^{2-\varepsilon}(\Omega))$. Hence

$$(3.17) \qquad |(I + \gamma^2 A_N)^{-1}[w(t), \lambda(w(t))]|_{C([0,t];H^{2-\epsilon}(\Omega)]} \leq CR^3.$$

Taking $\varepsilon < 1$ recalling compactness of the imbedding $H^{2-\varepsilon}(\Omega) \subset H^1(\Omega)$¡ and Mazur's Theorem, we conclude that $K(t)$ given by (3.7) is compact for each $t > 0$.

The proof of part (i) of Theorem 3.1 is thus completed. □

**4. Finite dimensionality of $w(B)$: Proof of part (ii) in Theorem 3.1.** Here we shall use the following result due to Ladyzhenskaya [Lad.1].

THEOREM 4.1. *(See Theorem* 1.1 *in* [Lad.1].*) Let $M$ be a compact set in a Hilbert space $\mathcal{H}$ and assume that on it there is defined a transformation $V : M \to V(M) \subset \mathcal{H}$ such that $M \subseteq V(M)$ and for, $v, \widetilde{v} \in M$*

$$(4.1) \qquad |V(v) - V(\widetilde{v}|_{\mathcal{H}} \;\leq\; \ell|v - \widetilde{v}|_{\mathcal{H}},$$

$$(4.2) \qquad |Q_n V(v) - Q_n V(\widetilde{v})|_{\mathcal{H}} \;\leq\; \delta|v - \widetilde{v}|_{\mathcal{H}}; \qquad \delta < 1,$$

*where $Q_n$ is the orthoprojection in $\mathcal{H}$ onto the subspace $\mathcal{H}_n^I$ of codimension $n$. Then, the set $M$ has a zero $\alpha$-measure for any $\alpha > \alpha_0 \equiv n\ell n\left(\dfrac{2c\ell^2}{1-\delta^2}\right)$ $/\ell n\dfrac{2}{1+\delta^2}$, and, consequently, its Hausdorff measure does not exceed $\alpha_0$.*

□

*Remark 4.1.* To prove finite dimensionality of the attractor, one could use a general result of [M.1]. However, by using the theorem of [M.1] (see also Thm. 2.8.1 in [H.1]), it is difficult to obtain the explicit estimates for the dimension of the set.

Let $V = T(t_0)$ for $t_0$ sufficiently large and let $M = w(B)$ where $B$ is a bounded set in $H$. Set $\mathcal{H} \equiv \mathcal{D}\left(\mathcal{A}^{\frac{1}{2}}\right) \times \mathcal{D}\left(\mathcal{A}^{\frac{1}{4}}\right)$. Topologically and algebraically $\mathcal{H}$ is equivalent to $H$ (see [G.1]). By the result of Theorem 3.1, part (i), we know that the set $M$ is compact and $M = V(M)$.

Define $Q_n = I - P_n$ where

$$P_n \equiv \begin{bmatrix} \mathcal{P}_n & 0 \\ 0 & \mathcal{P}_n \end{bmatrix}$$

and

$$\mathcal{P}_n x = \sum_{i=1}^{n} (x, \phi_n)_{L_2(\Omega)} \phi_n,$$

where $\phi_n$ are eigenfunctions of the generator $\mathcal{A}$ defined by (3.4).

Since the resolvent of $\mathcal{A}$ is compact and $\mathcal{A}$ is self-adjoint on $L_2(\Omega)$, $Q_n$ is an orthonormal projection in $\mathcal{H}$ onto the span $\{\phi_{n+1}, \phi_{n+2}, \ldots\}$.

To apply the result of Theorem 4.1, we need to verify conditions (4.1) and (4.2). This will be done with the help of the following supporting lemmas.

LEMMA 4.1. *For any $Q > 0$,*

$$|Q_n A^{-Q}|_{\mathcal{L}(\mathcal{H})} \leq C \lambda_n^{-\alpha Q},$$

*where $\lambda_n$ are the eigenvalues corresponding to $\mathcal{A}$ indexed in increasing order and with multiplicity taken into account, and $\alpha \equiv \frac{1}{4} - \varepsilon$; $\varepsilon$ is arbitrarily small.*

LEMMA 4.2. *For any $\varepsilon_0 < \frac{1}{2}$ the following inequality holds*

$$\left| A^{\varepsilon_0} \begin{bmatrix} 0 \\ (1+\gamma A_N)^{-1}[\mathcal{F}(w) - \mathcal{F}(\widetilde{w})] \end{bmatrix} \right|_{\mathcal{H}} \leq C[|w|^3_{H^2(\Omega)} + |\widetilde{w}|^3_{H^2(\Omega)}] |w - \widetilde{w}|_{\mathcal{D}(\mathcal{A}^{\frac{1}{2}})},$$

*where $\mathcal{F}(w) \equiv [w, \chi(w)]$.*

The proofs of Lemma 4.1 and Lemma 4.2 are relegated to the end of this section. Assuming, for a moment, validity of these lemmas we shall show that (4.1) and (4.2) hold.

*Proof of (4.1).* Let $v, \widetilde{v} \in M = w(B) \in B(0, R)$ where $B(0, R)$ denotes a ball with a radius $R$ in $\mathcal{H}$. From (3.6) and (3.6),

(4.3) $T(t)v - T(t)\widetilde{v}$

$$= e^{At}(v - \widetilde{v}) + \int_0^t e^{-A(t-s)} \begin{bmatrix} 0 \\ (1+\gamma^2 A_N)^{-1}[\mathcal{F}w(s) - \mathcal{F}\widetilde{w}(s)] \end{bmatrix} ds,$$

where

$$(w(t), w_t(t)) \equiv T(t)v; (\widetilde{w}(t), \widetilde{w}_t(t)) \equiv T(t)\widetilde{v}.$$

By using regularity results in (3.13), (3.14), it is straightforward to show (see [B-L] or [Lag.1]) that the nonlinear operator $\mathcal{F}$ is locally Lipschitz. More precisely, for $0 < \rho \leq 1$,

$$(4.4)\quad |\mathcal{F}(w) - \mathcal{F}(\widetilde{w})|_{H^{-\rho}(\Omega)} \leq L\left(|w|^3_{H^2(\Omega)} + |\widetilde{w}|^3_{H^2(\Omega)}\right)|w - \widetilde{w}|_{H^2(\Omega)},$$

where, we reall, $H^{-\rho}(\Omega) \equiv (H^\rho(\Omega))'$ and duality is understood with respect to $L_2(\Omega)$ inner product. This, combined with (3.8) and equivalence of the topologies in $H$ and $\mathcal{H}$ yields, in particular,

$$(4.5)\quad |\mathcal{F}(w(t)) - \mathcal{F}(\widetilde{w}(t))|_{H^{-1}(\Omega)} \leq \left(|v|^3_{\mathcal{H}} + |\widetilde{v}|^3_{\mathcal{H}}\right)|w(t) - \widetilde{w}(t)|_{\mathcal{D}(\mathcal{A}^{\frac{1}{2}})}$$

for a suitable constant $L$ (not necessarily the same as in (4.4)).

Combining (4.3), (4.5) with the result of Theorem 3.2 yields

$$\begin{aligned}(4.6)\quad |T(t)v - T(t)\widetilde{v}|_{\mathcal{H}} &\leq C|T(t)v - T(t)\widetilde{v}|_H \\ &\leq C\Big\{|v - \widetilde{v}|_{\mathcal{H}}e^{-wt} + L(|v|^3_{\mathcal{H}} + |\widetilde{v}|^3_{\mathcal{H}}) \\ &\qquad \int_0^t e^{-w(t-s)}|w(s) - \widetilde{w}(s)|_{\mathcal{D}(\mathcal{A}^{\frac{1}{2}})}ds\Big\},\end{aligned}$$

where we have used the fact that $e^{-At}$ is a contraction on $H$ and that

$$(4.7)\qquad (1+\gamma^2 A_N)^{-1} \in \mathcal{L}(H^{-1}(\Omega)'; H^1(\Omega)).$$

Inequality (4.6), in particular, implies

$$e^{wt}|w(t) - \widetilde{w}(t)|_{\mathcal{D}(\mathcal{A}^{\frac{1}{2}})} \leq C\left[|v - \widetilde{v}|_{\mathcal{H}} + LR^3\int_0^t e^{ws}|w(s) - \widetilde{w}(s)|_{\mathcal{D}(\mathcal{A}^{\frac{1}{2}})}ds\right].$$

By Gronwell's inequality applied to (4.7),

$$(4.8)\qquad |w(t) - \widetilde{w}(t)|_{\mathcal{D}(\mathcal{A}^{\frac{1}{2}})} \leq Ce^{(LR^3 - w)t}|v - \widetilde{v}|_{\mathcal{H}}$$

Combining now the results of (4.6) and (4.8) yields

$$(4.9)\qquad |T(t_0)v - T(t_0)\widetilde{v}|_{\mathcal{H}} \leq C[e^{-wt_0} + e^{(LR^3 - w)t_0}|v - \widetilde{v}|_{\mathcal{H}},$$

which proves (4.1) with a constant $\ell \equiv C[e^{-wt_0} + e^{(LR^3-w)t_0}]$.

*Proof of (4.2).* From (3.6), (3.7),

$$|Q_n[V(v) - V(\widetilde{v})]|_{\mathcal{H}} \leq |Q_n e^{-At_0}(v - \widetilde{v})|_{\mathcal{H}}$$

$$+\int_0^{t_0}\left|Q_n e^{-A(t_0-s)}\begin{bmatrix}0\\(1+\gamma^2A_N)^{-1}[\mathcal{F}(w(s))-\mathcal{F}(\widetilde{w}(s))]\end{bmatrix}\right|_{\mathcal{H}} ds$$

$$\leq C|e^{-At_0}(v-\widetilde{v})|_H+|Q_nA^{-\varepsilon_0}|_{\mathcal{L}(\mathcal{H})}$$

$$\cdot\int_0^{t_0}\left|e^{-A(t_0-s)}A^{\varepsilon_0}\begin{bmatrix}0\\(1+\gamma^2A_N)^{-1}[\mathcal{F}(w(s))-\mathcal{F}(\widetilde{w}(s))]\end{bmatrix}\right|_{\mathcal{H}} ds$$

(by Lemma 4.2 and Theorem 3.2,)

$$\leq C\left[|e^{-wt_0}(v-\widetilde{v})|_H+|Q_nA^{-\varepsilon_0}|_{\mathcal{L}(\mathcal{H})}LR^3\int_0^{t_0}e^{-w(t_0-t)}|w(t)-\widetilde{w}(t)|_{H_2(\Omega)}dt\right]$$

(by (4.8),)

$$\leq C\left[|e^{-wt_0}(v-\widetilde{v})|_H+|Q_nA^{-\varepsilon_0}|_{\mathcal{L}(\mathcal{H})}e^{(-w+LR^3)t_0}|v-\widetilde{v}|_{\mathcal{H}}\right], \tag{4.10}$$

Selecting in (4.10) $t_0$ and $n$ such that $Ce^{-wt_0}=\frac{1}{2}$; $C|Q_nA^{\varepsilon_0}|_{\mathcal{L}(\mathcal{H})}e^{(LR^3-w)t_0}$ $=\frac{1}{2}$; yields $\delta=\frac{3}{4}$, $t_0=\frac{\ell n2C}{w}$ and $\frac{1}{\lambda_n^\alpha}\leq\frac{1}{2}e^{\frac{LR^3}{w}}$. Taking into account the asymptotic behavior of $\lambda_n\approx n$, we obtain $n\sim 2^{\frac{1}{\alpha}}e^{\frac{LR^3}{w}}$. This proves Theorem 3.1. □

It remains to prove Lemmas 4.1 and 4.2.

*Proof of Lemma 4.1. Step 1.* We shall prove that

$$|Q_nA^{-1}|_{\mathcal{L}(\mathcal{H})}\leq f(n) \tag{4.11}$$

where $f(n)=\lambda_n^{-\frac{1}{4}+\varepsilon}$.

Direct computations involving the definition of the operator $A$ given in Section 3 give

$$Q_nA^{-1}=\begin{pmatrix}-(I-\mathcal{P}_n)\mathcal{A}^{-1}\mathcal{B}\mathcal{B}^*; & (I-\mathcal{P}_n)\mathcal{A}^{-1}(1+\gamma^2A_N)\\ I-\mathcal{P}_n; & 0\end{pmatrix}, \tag{4.12}$$

where $\mathcal{B}:[L_2(\Gamma)]^3\to\mathcal{D}(\mathcal{A})'$ is defined by $\mathcal{B}g\equiv\mathcal{A}G_1g_1+\mathcal{A}G_2g_2+\mathcal{A}G_1\frac{\partial}{\partial\tau}g_3$ and for $u\in\mathcal{D}(\mathcal{A}^{\frac{1}{2}})$ we have

$$\mathcal{A}^{-1}\mathcal{B}\mathcal{B}^*u\equiv G_1G_1^*\mathcal{A}u-G_1\frac{\partial^2}{\partial\tau^2}G_1^*\mathcal{A}u+G_2G_2^*\mathcal{A}u.$$

Hence

$$(4.13)\quad |Q_n A^{-1}\begin{pmatrix} u \\ v \end{pmatrix}|_{\mathcal{H}} \leq |(I-\mathcal{P}_n)\mathcal{A}^{-1}\mathcal{B}\mathcal{B}^* u|_{\mathcal{D}(\mathcal{A}^{\frac{1}{2}})} + |(I-\mathcal{P}_n)\mathcal{A}^{-1}(1+\gamma^2 A_N)v|_{L(\mathcal{A}^{\frac{1}{2}})} + |(I-\mathcal{P}_n)u|_{\mathcal{D}(\mathcal{A}^{\frac{1}{4}})}.$$

We shall estimate each term on RHS of (4.13).

$$(4.14)\quad |(I-\mathcal{P}_n)u|_{\mathcal{D}(\mathcal{A}^{\frac{1}{4}})} = |\mathcal{A}^{\frac{1}{4}}(I-\mathcal{P}_n)u|_{L_2(\Omega)} = |(I-\mathcal{P}_n)\mathcal{A}^{\frac{1}{4}}u|_{L_2(\Omega)} = |(I-\mathcal{P}_n)\mathcal{A}^{-\frac{1}{4}}\mathcal{A}^{\frac{1}{2}}u|_{L_2(\Omega)}.$$

Elementary computations give

$$(4.15)\quad |(I-\mathcal{P}_n)\mathcal{A}^{-\alpha}|_{\mathcal{L}(L_2(\Omega))} \leq \frac{1}{\lambda_n^{\alpha}}\,; \qquad \alpha \geq 0,$$

where $\lambda_n$ is an eigenvalue corresponding to $\phi_n$. From (4.14) and (4.15),

$$(4.16)\quad |(I-\mathcal{P}_n u|_{\mathcal{D}(\mathcal{A}^{\frac{1}{4}})} \leq \frac{1}{\lambda_n^{\frac{1}{4}}}|\mathcal{A}^{\frac{1}{2}}u|_{L_2(\Omega)}.$$

$$(4.17)\quad |(I-\mathcal{P}_n)\mathcal{A}^{-1}(1+\gamma^2 A_N)v|_{\mathcal{D}(\mathcal{A}^{\frac{1}{2}})}$$
$$= |(I-\mathcal{P}_n)\mathcal{A}^{-\frac{1}{2}}(1+\gamma^2 A_N)v|_{L_2(\Omega)}$$
$$\leq |(I-\mathcal{P}_n)\mathcal{A}^{-\frac{1}{4}}|_{\mathcal{L}(L_2(\Omega))}|\mathcal{A}^{-\frac{1}{4}}(1+\gamma^2 A_N)v|_{L_2(\Omega)}$$
$$\leq |(I-\mathcal{P}_n)\mathcal{A}^{-\frac{1}{4}}|_{\mathcal{L}(L_2(\Omega))}|\mathcal{A}^{-\frac{1}{4}}(1+\gamma^2 A_N)\mathcal{A}^{-\frac{1}{4}}|_{\mathcal{L}(L_2(\Omega))}|\mathcal{A}^{\frac{1}{4}}v|_{L_2(\Omega)}.$$

From elliptic theory and the equivalence of $\mathcal{D}(\mathcal{A}^{\frac{1}{4}} \sim H^1(\Omega)$ (see [G.1]),

$$(4.18)\quad \mathcal{A}^{-\frac{1}{4}}(I+\gamma^2 A_N)\mathcal{A}^{-\frac{1}{4}} \in \mathcal{L}(L_2(\Omega)).$$

(recall $I+\gamma^2 A_N : H^1(\Omega) \to (H^1(\Omega))' \equiv H^{-1}(\Omega)$ is bounded).

Inequalities (4.16), (4.17), and (4.18) imply

$$(4.19)\quad |(I-\mathcal{P}_n\mathcal{A}^{-1}(1+\gamma^2 A_N)v|_{\mathcal{D}(\mathcal{A}^{\frac{1}{2}})} \leq \frac{C}{\lambda_n^{\frac{1}{4}}}|\mathcal{A}^{\frac{1}{4}}v|_{L_2(\Omega)}.$$

$$(4.20)\quad |(I-\mathcal{P}_n\mathcal{A}^{-1}\mathcal{B}\mathcal{B}^* u|_{\mathcal{D}(\mathcal{A}^{\frac{1}{2}})}$$
$$\leq |(I-\mathcal{P}_n)\mathcal{A}^{\frac{1}{2}}\left\{G_1\left[u|_\Gamma - \frac{\partial^2}{\partial\tau^2}u|_\Gamma\right] + G_2\frac{\partial}{\partial\nu}u|_\Gamma\right\}|_{L_2(\Omega)},$$

where we have used (see [B.1]) that for $u \in H^2(\Omega)$,

$$
\begin{aligned}
G_1^* \mathcal{A} u &= -u|_\Gamma; \\
G_2^* \mathcal{A} u &= \frac{\partial}{\partial \nu} u|_\gamma.
\end{aligned} \tag{4.21}
$$

Elliptic theory (see [L-M] and [G.1]) provides

$$G_2 \in \mathcal{L}(H^{\frac{1}{2}}(\Gamma) \to H^3(\Omega) \subset \mathcal{D}(\mathcal{A}^{\frac{3}{4}-\varepsilon})); \tag{4.22a}$$

$$G_1 \in \mathcal{L}(H^{-\frac{1}{2}}(\Gamma) \to H^3(\Omega) \in \mathcal{D}(\mathcal{A}^{\frac{3}{4}-\varepsilon})), \tag{4.22b}$$

where $\varepsilon > 0$ can be taken arbitrarily small.

Hence, by the Trace Theorem,

$$
\begin{aligned}
\left| G_1 \frac{\partial^2}{\partial \tau^2} w \right|_{H^3(\Omega)} &\le C \left| \frac{\partial^2}{\partial \tau^2} u \right|_{H^{-\frac{1}{2}}(\Gamma)} \le C |u|_\Gamma|_{H^{\frac{3}{2}}(\Gamma)} \\
&\le C |u|_{H^2(\Omega)} \le C |u|_{\mathcal{D}(\mathcal{A}^{\frac{1}{2}})};
\end{aligned} \tag{4.23}
$$

$$\left| G_2 \frac{\partial}{\partial \nu} w \right|_{H^3(\Omega)} \le C \left| \frac{\partial}{\partial \nu} u \right|_{H^{\frac{1}{2}}(\Gamma)} \le C |u|_{H^2(\Omega)} \le C |u|_{\mathcal{D}(\mathcal{A}^{\frac{1}{2}})} \tag{4.24}$$

By (4.15), (4.23), (4.22), and (4.24),

$$
\begin{aligned}
&\left| (I - \mathcal{P}_n) \mathcal{A}^{\frac{1}{2}} G_1 \left[ u|_\Gamma - \frac{\partial^2}{\partial \tau^2} u|_\Gamma \right] \right|_{L_2(\Omega)} \\
&\le |(I - \mathcal{P}_n) \mathcal{A}^{-\frac{1}{4}+\varepsilon}|_{\mathcal{L}(L_2(\Omega))} \\
&\cdot \left[ |\mathcal{A}^{\frac{3}{4}-\varepsilon} G_1 \frac{\partial^2}{\partial \tau^2} u|_{L_2(\Omega)} + |\mathcal{A}^{\frac{3}{4}-\varepsilon} G_1 u|_\Gamma|_{L_2(\Omega)} \right] \\
&\le C \frac{1}{\lambda_n^{\frac{1}{4}-\varepsilon}} |u|_{\mathcal{D}(\mathcal{A}^{\frac{1}{2}})}.
\end{aligned} \tag{4.25}
$$

Similarly, recalling (4.24),

$$
\begin{aligned}
\left| (I-\mathcal{P}_n) \mathcal{A}^{\frac{1}{2}} G_2 \frac{\partial}{\partial \nu} u|_\Gamma \right|_{L_2(\Omega)} &\le |(I-\mathcal{P}_n) \mathcal{A}^{\frac{1}{4}-\varepsilon}|_{\mathcal{L}(L_2(\Omega))} |\mathcal{A}^{\frac{3}{4}-\varepsilon} G_2 \frac{\partial}{\partial \nu} u|_{L_2(\Omega)} \\
&\le \frac{C}{\lambda_n^{\frac{1}{4}-\varepsilon}} |u|_{\mathcal{D}(\mathcal{A}^{\frac{1}{2}})}.
\end{aligned} \tag{4.26}
$$

Collecting (4.20), (4.25), and (4.26) yields

$$|(I - \mathcal{P}_n \mathcal{A}^{-1} \mathcal{B} \mathcal{B}^* u|_{\mathcal{D}(\mathcal{A}^{\frac{1}{2}})}. \le \frac{C}{\lambda_n^{\frac{1}{4}-\varepsilon}} |u|_{\mathcal{D}(\mathcal{A}^{\frac{1}{2}})}. \tag{4.27}$$

After combining (4.13) with (4.16), (4.19), and (4.27), inequality (4.11) follows with $f(n) = \dfrac{C}{\lambda_n^{\frac{1}{4}-\varepsilon}} \to 0$ as $n \to \infty$.

*Step 2*. We shall use the Generalized Heinz Inequality from [K.1]. Indeed, using the notation of Theorem 1 in [K.1], we set: $\delta = \delta' \equiv H$; $B \equiv A^{*-1}$; $T \equiv Q_n^*$; $A \equiv I$. Since $A^*$ is a dissipative generator on $H$, $B$ is $m$-accreative. $Q_n^*$ (the adjoint of $Q_n$ with respect to the $H$-topology) satisfies $T \equiv Q_n^* : H \to H$ is bounded. (This follows from the fact that $Q_n$ is an orthonormal projection on $\mathcal{H}$ and the topologies in $\mathcal{H}$ and $H$ are equivalent). Moreover, the result of Step 1 yields

$$\begin{aligned} (4.28) \quad |BT|_{\mathcal{L}(H)} &= |A^{*-1}Q_n^*|_{\mathcal{L}(H)} = |Q_nA^{-1}|_{\mathcal{L}(H)} \\ &\leq C|Q_nA^{-1}|_{\mathcal{L}(\mathcal{H})} \leq Cf(n) \to 0 \text{ as } m \to \infty. \end{aligned}$$

Applying the result of Theorem 1 in [K.1],

$$\begin{aligned} (4.29) \qquad |B^QTu|_H &\leq C_1|f(n)|^Q|u|_H\,, \quad \text{or} \\ |A^{*-Q}Q_n^*|_{\mathcal{L}(H)} &\leq C_1|f(n)|^Q \to 0 \text{ as } n \to \infty, \end{aligned}$$

or equivalently,

$$(4.30) \qquad |Q_nA^{-Q}|_{\mathcal{L}(H)} \leq C_1|f(n)|^Q.$$

Inequality (4.30) and the equivalence of the topologies on $\mathcal{H}$ and $H$ yields the result in Lemma 4.1. □

*Proof of Lemma 4.2. Step 1.* We shall prove that for $0 \leq Q < \dfrac{1}{2}$ and $0 \leq \widetilde{Q} < Q$ we have

$$(4.31) \qquad H^{2+Q}(\Omega) \times H^{1+Q}(\Omega) \subset \mathcal{D}(A^{\widetilde{Q}}).$$

Recall

$$(4.32)\ \mathcal{D}(A) = \{(u,v) \in \mathcal{D}(\mathcal{A}^{\frac{1}{2}}) = H^2(\Omega); u + \mathcal{A}^{-1}\mathcal{B}\mathcal{B}^*u \in \mathcal{D}(\mathcal{A}^{\frac{3}{4}})\},$$

where $\mathcal{A}^{-1}\mathcal{B}\mathcal{B}^*$ is given below (4.12).

Let us introduce the space

$$(4.33) \qquad X \equiv \mathcal{D}(\mathcal{A}^{\frac{3}{4}}) \times H_0$$

where $H_0 \equiv \{v \in H^2(\Omega);\ \nabla v|_\Gamma = 0\}$ and $H_0$ is equipped with $H^2(\Omega)$ norm.

Since for $v \in H_0$,

$$(4.34) \qquad \mathcal{A}^{-1}\mathcal{B}\mathcal{B}^*v = -G_1v|_\Gamma \in H^5(\Omega),$$

and (see [G.1]),

$$\mathcal{D}(\mathcal{A}^{\frac{3}{4}}) = \{u \in H^3(\Omega);\ \Delta u + (1-\mu)B_1 u = 0 \text{ on } \Gamma\},$$

we obtain (recalling the definition of $G_1$),

$$\mathcal{A}^{-1}\mathcal{B}\mathcal{B}^* v \in \mathcal{D}(\mathcal{A}^{\frac{3}{4}}) \ \text{ for } \ v \in H_0. \tag{4.35}$$

This, together with (4.32) and (4.33), implies

$$X \subset \mathcal{D}(A). \tag{4.36}$$

Therefore, for $Q < \dfrac{1}{2}$,

$$[H, X]_Q \subset [H, \mathcal{D}(A)]_Q. \tag{4.37}$$

But

$$\begin{aligned} [H, X]_Q &= [\mathcal{D}(\mathcal{A}^{\frac{1}{2}}) \times H^1(\Omega, \mathcal{D}(\mathcal{A}^{\frac{3}{4}}) \times H^0]_Q \\ &= [\mathcal{D}(\mathcal{A}^{\frac{1}{2}}), \mathcal{D}(\mathcal{A}^{\frac{3}{4}})]_Q \times [H^1(\Omega), H^0]_Q \\ &= \mathcal{D}(\mathcal{A})^{\frac{1}{2}+\frac{1}{4}Q} \times H^{1+Q}(\Omega), \end{aligned} \tag{4.38}$$

where the last inequality holds only for $Q < \dfrac{1}{2}$. On the other hand, by using Theorem on p. 25 in [Tr.1], we obtain that for $\widetilde{Q} < Q$.

$$[H, \mathcal{D}(A)]_Q \quad \subset \quad (H, \mathcal{D}(A))_{Q,2} \tag{4.39}$$

$$\subset \quad (H; \mathcal{D}(A))_{Q,\infty} \subset (H, \mathcal{D}(A))_{\widetilde{q},1} \tag{4.40}$$

p. 101, [Tr.1]

$$\subset \mathcal{D}(A^{\widetilde{Q}}).$$

Combining (4.37)–(4.40) and recalling (see [G.1]) $\mathcal{D}(A^{\frac{1}{2}+\frac{1}{4}Q}) \simeq H^{2+Q}(\Omega)$ yields

$$H^{2+Q}(\Omega) \subset \mathcal{D}(A^{\widetilde{Q}}) \tag{4.41}$$

for any $0 \le Q < \dfrac{1}{2}$ and $\widetilde{Q} < Q$.

The proof of (4.31) is thus completed.

*Step 2*. From (4.4) and elliptic regularity

$$(1 + \gamma^2(A_N)^{-1} \in \mathcal{L}(H^{-\rho}(\Omega) \to H^{2-\rho}(\Omega)); \qquad 0 \le \rho \le 1,$$

we obtain

$$
\begin{aligned}
(4.42) \qquad & |(1+\gamma^2 A_N)^{-1}[\mathcal{F}(w)-\mathcal{F}(\widetilde{w})]|_{H^{2-\rho}(\Omega)} \\
& \leq C\Big(|w|_{H^2(\Omega)}||\widetilde{w}|_{H^2(\Omega)}\Big)|w-\widetilde{w}|_{H^2(\Omega)}; \qquad 0<\rho\leq 1.
\end{aligned}
$$

Taking $2-\rho = 1+Q$ (i.e., $\rho = 1-Q$) with $Q < \frac{1}{2}$ and applying inequality (4.31) of Step 1 yields the result of Lemma 4.2 with $\varepsilon_0 = \widetilde{Q} < Q < \frac{1}{2}$. □

## REFERENCES


[B.1] M.E. Bradley, *Uniform stabilization at a nonlinearly perturbed Kirchhoff plate*, Appl. Math. Opt. 26 (1982), pp. 1–19.

[B-L.1] M.E. Bradley and I. Lasiecka, *Local exponential stabilization of a nonlinearly perturbed von Kármán plate* , J. Nonlinear Anal.: Tech., Meth. and Appl. 18 (4) (1992), pp. 333–343.

[B-L.2] M.E. Bradley and I. Lasiecka, *Global decay rates for the solutions to a von Kármán plate without geometric conditions*, Journal of Mathematical Analysis and Applications, to appear.

[Ch.1] I.D. Chueshov, *Strong solutions and the attractor of the von Kármán equations*, Math USSR Sbornik 69 (1991), No. 1, pp. 25–26.

[Ch.2] I.D. Chueshov, *Finite dimensionality of an attractor in some problems of nonlinear shell theory*, Math USSR Sbornik 61 (1988), pp. 419–428.

[F-L] A. Favini and I. Lasiecka, *Second order abstract evolution equations*, in the Proceedings of 2nd International Conference on Differential Equations and Banach Spaces, University of Bologne, July 1981, LNPAM, vol 148, Marcel Dekker, 1993, pp 65–85.

[H-L-2] M.A. Horn and I. Lasiecka, *Global existence and uniqueness of regular solutions to the von Kármán system with nonlinear boundary dissipation.* Proceedings of the IFIP Conference on Control of Deterministic and Stochastic PDE's, Trento, 1993, LNAPAM, vol 165, Marcel Dekker 1994, pp. 99–121.

[G.1] P. Grisvard, *Elliptic Problems in Nonsmooth Domains*, Pitman, London, 1985.

[H.1] J.K. Hale, *Asymptotic Behavior of Dissipative Sets*, AMS, Providence, 1988.

[H-R] J.K. Hale and G. Raquel, *Attractors for dissipative evolutionary equations*, Preprint CDSNS91-72. Center for Dynamical Systems and Nonlinear Studies, Georgia Institute of Technology, 1992.

[H-L] M.A. Horn and I. Lasiecka, *Global stabilization of the dynamic von Kármán plate with nonlinear boundary feedback*, Applied Mathematics and Optimization, vol 31, 1995, pp. 57–84

[H-L.1] M.A. Horn and I. Lasiecka, *Nonlinear boundary stabilization of a von Kármán plate equations.* LNPAM, vol 152, Marcel Dekker, 1994, pp 581-605.

[K.1] T. Kato, *A Generalization of the Heinz Inequality*, Proc. Japan Acad. 37 (1961), pp. 305–308.

[K-S] H. Koch and A. Stahel, *Global existence of classical solutions to the dynamical von Kármán equation*, Preprint Nr. 650, Universität Heidelberg, 1991.

[Lad.1] O.A. Ladyzhenskaya, *Finite-dimensionality of bounded invariant sets for Navier-Stokes systems and other dissipative sets*, J. Soviet Math. 28 (1985), pp. 714–726.

[Lag.1] J. Lagnese, *Boundary Stabilization of Thin Plates*, Society for Industrial and Applied Mathematics, Philadelphia, 1989.

[Lag.2] J. Lagnese, *Uniform asymptotic energy estimates for the solutions of the equation of dynamical plane elasticity with nonlinear dissipation of the boundary*, Nonlinear Analysis, 16(1) (1991), pp. 35–54.

[Las-Tat] I. Lasiecka and D. Tataru, *Uniform boundary stabilization of semilinear wave equations with nonlinear boundary damping*, Journal of Differential and Integral Equations, vol 6, 1993, pp. 507–533

[Las.1] I. Lasiecka, *Existence and uniqueness of the solutions to second order abstract equations with nonlinear and nonmonotone boundary conditions*, Journal of Nonlinear Analysis, Theory, Methods and Applications, to appear.

[Las.2] I. Lasiecka, *Finite-dimensionality of attractors associated with von Kármán plate equations and boundary damping*, Journal of Differential Equations, to appear.

[Lio.1] J.L. Lions, *Controllabilite Exacte, Perturbations et Stabilization des Systems Distribueès*, Masson, Paris, 1989.

[L-M] J.L. Lions and E. Magenes, *Non-Homogeneous Boundary Value Problems and Applications*, vol. 1, Springer-Verlag, New York, 1972.

[L-T] I. Lasiecka and R. Triggiani, *Sharp trace estimates of solutions to Kirchoff and Euler-Bernoulli equations*, Applied Mathematics and Optimization, 28 (1993), pp. 277–306.

[M.1] R. Mañe, *On the dimension of the compact invariant sets of certain nonlinear maps*, Lecture Notes in Math., vol. 838, Springer-Verlag (1981), pp. 230–242.

[P.1] A. Pazy, *Semigroups of Linear Operators and Applications to Partial Differential Equations*, Springer-Verlag, New York, 1983.

[P-T] J. Puel and M. Tucsnak, *Boundary stabilization for the von Kármán equations*, C.R. Acad. Sc. Paris, t. 314, Serie I (1992), pp. 609–612.

[T.1] H. Triebel, *Interpolation Theory, Function Spaces, Differential Operators*, North Holland, 1978.

# STABILIZATION OF THE KORTEWEG-DE VRIES EQUATION ON A PERIODIC DOMAIN

DAVID L. RUSSELL* AND BING-YU ZHANG†

**Abstract.** We study solutions of the Korteweg-de Vries (KdV) equations

$$u_t + uu_x + u_{xxx} = f$$

and

$$u_t + uu_x + u_{xxx} = 0$$

for $t \geq 0$ and $0 \leq x \leq 1$ where the subscripts denote partial derivatives. In the first case, periodic boundary conditions are imposed at 0 and 1, and the distributed control $f$ is assumed to be generated by a linear feedback control law conserving the "volume" or "mass" $\int_0^1 u(x,t)dx$ which monotonically reduces the "energy" $\int_0^1 u(x,t)^2 dx$. For the second equation a feedback boundary control is applied having the same properties. In both cases we obtain uniform exponential decay of the solutions to a constant state.

## 1. Introduction.

The Korteweg-de Vries equation

$$u_t + uu_x + u_{xxx} = 0$$

was first derived by Korteweg and de Vries [12] in 1895 as a model to describe long water waves propagating in a channel of shallow cross section. The equation has been extensively studied since the 1960's. It turns out the equation is not only a good model for some water waves but also a very useful model for many other other waves. In general the equation can be used as an approximation model in many studies when one wishes to include and balance a weak nonlinearity and a weak dispersive effect [15], [21]. In this paper we study some control problems of the KdV equation. We will consider both distributed and boundary control of the equation.

The *distributed control case* consists in a study of the forced KdV equation

$$u_t + uu_x + u_{xxx} = v \tag{1.1}$$

in which, the forcing function $v = v(x,t)$, with possible restrictions on the function which we will introduced as we proceed, appears as the control function by which solutions are to be influenced. We study the system on the interval $0 \leq x \leq 1$, $t \geq 0$, with periodic boundary conditions

$$\partial_x^j u(1,t) = \partial_x^j(0,t), \qquad j = 0,\, 1,\, 2 \tag{1.2}$$

* Department of Mathematics, Virginia Polytechnic Institute and State University, Blacksburg, Virginia 24061-0123.

† Department of Mathematical Sciences, University of Cincinnati, Cincinnati, Ohio 45221.

so that $x = 1$ is effectively identified with $x = 0$, yielding a spatial domain equivalent to a circle.

To explain the control problem to be studied, a certain amount of background is required. For appropriately smooth solutions of the unforced equation $v \equiv 0$) on $-\infty < x < +\infty$, or on the indicated domain with the periodic boundary conditions (1.2), and extensions of those boundary conditions as required to express further smoothness, it is known [16] that there is an infinite set of conserved integral quantities of which only first three interest us here

$$\int_0^1 u(x,t)dx, \tag{1.3}$$

$$\int_0^1 u(x,t)^2 dx, \tag{1.4}$$

$$\int_0^1 \left( (u_x(x,t))^2 - \frac{1}{3} u(x,t)^3 \right) dx. \tag{1.5}$$

From the historical origins [12] of the KdV equation, involving the behavior of water waves in a shallow channel, it is natural to think of (1.3) as expressing conservation of volume or mass. That context suggests that the control problem might also be appropriately studied in a volume conserving context. For smooth solutions of (1.1) satisfying boundary conditions (1.2) we have

$$\begin{aligned} \frac{d}{dt} \int_0^1 u(x,t)dx &= \int_0^1 \left(-u(x,t)u_x(x,t) - u_{xxx}(x,t) + v(x,t)\right) dx \\ &= \int_0^1 \left( \left[-u(x,t)^2 - u_{xx}(x,t)\right]_x + v(x,t) \right) dx \\ &= \int_0^1 v(x,t)dx, \end{aligned}$$

from which we conclude that volume is conserved in the controlled system (1.1)–(1.2) just in case, as we assume henceforth,

$$\int_0^1 v(x,t)dx = 0. \tag{1.6}$$

For the control system just introduced the *open loop control problem* consists in use of the indicated control function to transfer the system, during $[0, T]$, between a given initial state $u(x,T) = \psi(x)$, in an appropriate function space of system states, necessarily, in view of the volume conserving control actions under consideration, such that

$$\int_0^1 \phi(x)dx = \int_0^1 \psi(x)dx. \tag{1.7}$$

It is familiar that each control problem has associated with it a corresponding *stabilization problem* in which the control is required, for each $t$, to be a functional of fixed form (i.e., independent of $t$) of the state $u(.,t)$ and the objective of the control action is to cause the system to approach, as $t \to \infty$, a particular set of equilibrium states. The stabilization problem in our current context has as its goal that the controlled solution $u(x,t)$ should tend, as $t \to \infty$, to the constant state

$$\bar{u}(t) \equiv [\phi], \quad [\phi] \equiv \int_0^1 \phi(x)dx.$$

For both control problems posed we are interested in a particular type of *feedback* control action. The prototype of the control action considered in the distributed control case may be described as follows. Let $I = [a,b]$ be a subinterval of $[0,1]$. We require

$$v(x,t) \equiv 0, \quad x \notin I,$$

and determine $v$ on $I$, for $K > 0$, by

$$v(x,t) = -K(u(x,t) - [u]_I(t)), \tag{1.8}$$

where

$$[u]_I(t) \equiv \frac{1}{b-a}\int_a^b u(x,t)dx.$$

If $u(x,t)$ is thought of as describing the depth profile of a body of fluid, this mode of control may be thought of as a pumping action confined to the interval $[a,b]$; the average depth over $I$ is determined, i.e., $[u]_I(t)$, and fluid is extracted or injected at points $x \in I$, depending on whether $u(x,t)$ is $> [u]_I(t)$ or $< [u]_I(t)$, respectively, in the proportional manner indicated by (1.8). It is easy to verify that (1.6) is satisfied and therefore the volume is conserved when this control action is used.

We will review in Section 2 our work [18], and recent joint work with V. Komornik [11] on the distributed control system described above. In [18], the associated linear system (dropping the nonlinear term $uu_x$) has been considered. There the open loop system is shown *exactly controllable* and the closed loop system is proved to be asymptotically stabled with exponential decay rate. The nonlinear system was treated in [11] where we showed that the closed loop system is stabilized with exponential decay rate if the full domain control action is used (i.e., $I = [0,1]$). The problem whether the closed loop system is stabilized remains open if the control acts only on a subinterval $[a,b]$ of $[0,1]$.

It is also possible to pose a volume-conserving open loop boundary control problem for the system described by the KdV equation. Let us take $v(x,t) \equiv 0$ in (1.1) and apply the boundary condition (1.2) for $j = 0$ and $j = 2$ but not for $j = 1$. Then we have

$$u_t + uu_x + u_{xxx} = 0$$

and it is easy to see that $[u(.,t)] := \int_0^1 u(x,t)dx$ is a constant and

$$\frac{d}{dt}\left\{\frac{1}{2}\int_0^1 (u(x,t) - [u(.,t)])^2\, dx\right\}$$

$$= \frac{1}{2}(u_x(1,t) + u_x(0,t))(u_x(1,t) - u_x(0,t)). \tag{1.9}$$

If we suppose that the boundary control mechanism takes the form

$$u_x(1,t) - u_x(0,t) = h(t),$$

which, since 1 and 0 are identified in this context, is the same as

$$u_x(0^-,t) - u_x(0^+,t) = h(t),$$

then (1.9) becomes

$$\frac{d}{dt}\left\{\int_0^1 (u(x,t) - [u(.,t)])^2\, dx\right\} = (u_x(1,t) + u_x(0,t))h(t).$$

An appropriate feedback mechanism is then

$$v(t) = -\Re(u_x(1,t) + u_x(0,t)), \qquad \Re > 0, \tag{1.10}$$

which is clearly the same as imposing the *closed loop boundary condition*

$$u_x(1,t) = \alpha u_x(0,t), \qquad \alpha = \frac{1-\Re}{1+\Re} \tag{1.11}$$

($0 < |\alpha| < 1$ since $\Re > 0$), resulting in

$$\begin{aligned} \frac{d}{dt}\left(\int_0^1 u^2(x,t)dx\right) &= -\Re(u_x(1,t) + u_x(0,t))^2 \\ &= (\alpha^2 - 1)u_x^2(0,t) \le 0. \end{aligned}$$

This suggests the feedback control law (1.10) introduces a dissipation mechanism via the boundary into the system and the resulting closed loop system

$$\begin{cases} u_t + uu_x + u_{xxx} = 0, \quad x \in (0,1),\ t \ge 0 \\ u(x,0) = \phi(x) \\ u(1,t) = u(0,t), \quad u_x(1,t) = \alpha u_x(0,t), \quad u_{xx}(1,t) = u_{xx}(0,t) \end{cases} \tag{1.12}$$

may be stabilized. Indeed, it has been shown in [18] that the associated linear system of (1.10) is stabilized with uniform exponential decay rate and the small amplitude solutions of the nonlinear system (1.12) also decay exponentially to a constant state [11].

It is worth pointing out that the system (1.12) is not only interesting to study from the point of view of boundary stabilization, but also from the point of view of the smoothing effects of dispersive wave systems. We recall that for the KdV equation posed on a whole real line, it is well known in this now "standard" theory [2], [3], [6] that evolution of solutions smoothes compactly supported initial data. This occurs through the dispersive effects associated with the third order operator, resulting in solution components of increasingly high "spatial" frequency being more rapidly dispersed to infinity as the solution evolves. However, there is no counterpart of this smoothing action on strictly periodic domains where boundary conditions

$$\partial_x^j u(1,t) = \partial_x^j u(0,t),$$

minimally applying for $j = 0, 1, 2$, are imposed on the process. Intuitively one could say that in this compact domain situation there is nowhere for the high frequency solution components to "escape to". More rigorously one would observe that since the process is invariant under simultaneous time ($t$) and space ($x$) reversal, each instance of a non-smooth initial state evolving into a later smooth state is matched by a corresponding smooth initial state evolving into a non-smooth state (cf. [20]). The only way to restore the smoothing property is to build into the system a selective dissipative mechanism which increasingly attenuates higher frequency solution components. The system (1.12) with boundary dissipation will do the job to restore smoothing properties.

In Section 3, we will consider a system similar to (1.12) but described by the generalized KdV equation as follows:

$$(1.13)\quad \begin{cases} u_t + u^k u_x + u_{xxx} = 0, \quad x \in (0,1),\ t \geq 0 \\ u(x,0) = \phi(x) \\ u(1,t) = u(0,t), \quad u_x(1,t) = \alpha u_x(0,t), \quad u_{xx}(1,t) = u_{xx}(0,t) \end{cases}$$

where $k \geq 1$ is an integer. The governing equation of the system (1.11) when $k = 2$ is the so-called modified KdV equation. The method developed in [11] to deal with the system (1.12) is easily adapted to study the system (1.13) with few changes. The well-posedness of the system (1.13) will be established in some suitable Banach spaces and small amplitude solutions will be proved to decay exponentially to a constant state, the mean value of the initial state.

Finally some interesting open questions are given in Section 4 to conclude our discussion.

**2. Distributed control.** Consider first the following linear system associated with to the system (1.1)–(1.2)

$$
(2.1) \quad \begin{cases} u_t + u_{xxx} = v \\ u(x,0) = \phi(x) \\ u(1,t) = u(0,t), \quad u_x(1,t) = u_x(0,t), \quad u_{xx}(1,t) = u_{xx}(0,t) \end{cases}
$$

with $v = v(x,t)$ takes the following restricted form

$$
(2.2) \qquad v(x,t) = Gf := -g(x)\left\{ f(x,t) - \int_0^1 g(y) f(y,t)dy \right\},
$$

where $f = f(x,t)$ is considered as a control function, and $g$ is a given piecewise continuous non-negative function defined for $x$ in $[0,1]$ with

$$
(2.3) \qquad \int_0^1 g(x)dx = 1
$$

and

$$
(2.4) \qquad \text{supp } g = [a,b] \subset [0,1].
$$

It is easy to check that

$$
\int_0^1 (Gf)(x)dx = 0, \qquad \text{for any } f \in L^2(0,1)
$$

and hence the volume of solutions of (2.1) is conserved. We have the following exact controllability result for this open loop system:

THEOREM 2.1. *Let $T > 0$ be given and let it be assumed that the function $g$ associated with $G$ is of class $C^0$ on $[0,1]$ ( is of class $C_p^3$ ($g$, $g'$ and $g''$ are all periodic on $[0,1]$)). Given any initial state $u_T \in L^2(0,1)$ ($u_T \in H_p^3(0,1)$) with $[u_T] = [\phi]$, there exists a control $f \in L^2((0,1); L^2(0,1))$ ($f \in L^2((0,T); H_p^3)$ ) such that the solution $u$ of (2.1) satisfies the terminal condition*

$$
u(.,T) = u_T \qquad \text{in } L^2(0,1) \ ( \ H_p^3(0,1) \ ).
$$

The proof of Theorem 2.1 proceeds by using the bases

$$
\phi_k(x) = e^{2\pi i k x}, \qquad -\infty < k < \infty
$$

of the space $L^2(0,1)$ to change the exact controllability problem to an equivalent moment problem. Details can be found in [18].

Taking $f = u$ in (2.1) we obtain a closed loop system

$$(2.5)\quad \begin{cases} u_t + u_{xxx} = Gu \\ u(x,0) = \phi(x) \\ u(1,t) = u(0,t), \quad u_x(1,t) = u_x(0,t), \quad u_{xx}(1,t) = u_{xx}(0,t) \end{cases}$$

This system is asymptotically stable as we have indicated.

THEOREM 2.2. *There exist $\alpha$, $\beta > 0$ such that, for any $\phi \in L^2(0,1)$ ( $\phi \in H_p^3(0,1)$ ), the unique solution $u(.,t)$ of (2.5) satisfies*

$$\|u(.,t) - [\phi]\|_{L^2(0,1)} \le \beta e^{-\alpha t}\|\phi - [\phi]\|_{L^2(0,1)}, \quad t \ge 0.$$

$$\Big( \quad \|u(.,t) - [\phi]\|_{H_p^3(0,1)} \le \beta e^{-\alpha t}\|\phi - [\phi]\|_{H_p^3(0,1)}, \quad t \ge 0.\Big)$$

*Proof.* see [18].

Now we turn to consideration of distributed control for the nonlinear system

$$(2.6)\quad \begin{cases} u_t + uu_x + u_{xxx} = -\Re(v - [v]_I) \\ u(x,0) = \phi(x) \\ u(1,t) = u(0,t), \quad u_x(1,t) = u_x(0,t), \quad u_{xx}(1,t) = u_{xx}(0,t) \end{cases}$$

where $v = v(x,t)$ is considered as a control function. We assume here that the $I = [0,1]$ and thus the control acts on the whole interval $[0,1]$. □

THEOREM 2.3. *There is a positive number $T$ such that, for each pair of initial and terminal states, $\phi$ and $\psi$, in $H_p^3$ with $[\phi] = [\psi]$, there is a control $v$ in $C([0,T], H_p^2)$ such that the unique solution $u$ of (2.6) achieves at time $T$ the terminal state $\psi$.*

The proof of Theorem 2.3 is based on the exact controllability of the corresponding linear system. See [11] for the details of the proof.

Again, choosing $v = u$ in (2.6), we obtain a closed loop system

$$(2.7)\quad \begin{cases} u_t + uu_x + u_{xxx} = -K(u - [u]_I) \\ u(x,0) = \phi(x) \\ u(1,t) = u(0,t), \quad u_x(1,t) = u_x(0,t), \quad u_{xx}(1,t) = u_{xx}(0,t) \end{cases}$$

THEOREM 2.4. *Let $m > 0$ be a given integer. For every $\phi \in H_p^m$ the solution $u(.,t)$ of (2.7) converges to $[\phi]$ in $H_p^m$ and $u_t(.,t)$ converges to 0 in*

$H_p^{m-3}$ *as* $t \to \infty$. *More precisely, for every positive number* $K' < K$ *there exists a positive constant* $C$ *such that*

$$\|u(.,t) - [\phi]\|_{H_p^m \times H_p^{m-3}} \leq Ce^{-K't}, \quad t \geq 0.$$

The proof of Theorem 2.4 is based the following important lemma which is proved by making use of of an infinite sequence of polynomial conservation laws for the KdV equation obtained in [14], [16]. We refer to [11] for details of the proof.

LEMMA 2.5. *Let* $\phi \in H_p^m$ *and consider the solution* $u(.,t)$ *of (2.7). Then*

1. *we have in* $(0,\infty)$ *the identity*

$$\frac{d}{dt}\int_0^1 (u-[u])^2 dx = -2K\int_0^1 (u-[u])^2 dx;$$

2. *for every* $n = 1, 2, ..., m$ *there exist two polynomials*

$$Q_{n-1} = Q_{n-1}(v_0, v_1, ..., v_{n-1})$$

*and*

$$R_{n-1} = R_{n-1}(v_0, v_1, ..., v_{n-1})$$

*in* $n$ *variables such the following identity holds in* $(0,\infty)$*:*

$$\begin{aligned} &\frac{d}{dt}\int_0^1 \left(u_{,n}^2 + Q_{n-1}(u-[u], u_{,1}, ..., u_{,n-1})\right) dx \\ = \; & -2K\int_0^1 \left(u_{,n}^2 + R_{n-1}(u-[u], u_{,1}, ..., u_{,n-1})\right) dx \end{aligned}$$

*where*

$$u_{,j} := \frac{\partial^j u}{\partial x^j};$$

3. *each term of* $Q_{n-1}(v_0, v_1, ..., v_{n-1})$ *and likewise each term of* $R_{n-1}(v_0, v_1, ..., v_{n-1})$ *being always of degree less than four.*

**3. Boundary control.** In this section we consider the following initial boundary value problem for the generalized KdV equation

$$\begin{cases} u_t + u^k u_x + u_{xxx} = 0, & x \in (0,1), \quad t \geq 0 \\ u(x,0) = \phi(x) & \\ u(1,t) = u(0,t), \quad u_x(1,t) = \alpha u_x(0,t), & u_{xx}(1,t) = u_{xx}(0,t) \end{cases} \tag{3.1}$$

where $|\alpha| < 1$ and the integer $k \geq 1$.

The study of the nonlinear system (3.1) is based on understanding of the following associated linear system:

$$\begin{cases} \partial_t w + \partial_x^3 w = 0, \quad 0 < x < 1,\ t \geq 0 \\ w(x,0) = \phi(x) \\ w(1,t) = w(0,t), \quad w_x(1,t) = \alpha w_x(0,t), \quad w_{xx}(1,t) = w_{xx}(0,t) \end{cases} \tag{3.2}$$

whose solution is given by the semigroup $S_\alpha(t)$ in $L^2(0,1)$, where $S_\alpha(t)$ is generated by the operator $A_\alpha$ defined by

$$(A_\alpha u)(x) = -u'''(x), \tag{3.3}$$

with

$$\begin{aligned} \mathcal{D}(A_\alpha) &\equiv \{w \in H^3(0,1) \mid w(1) = w(0), \quad w'(1) = \alpha w'(0), \\ &\qquad w''(1) = w''(0)\} \\ &\equiv H_\alpha^3 \end{aligned} \tag{3.4}$$

PROPOSITION 3.1. *The operator $A_\alpha$ generates a strongly differentiable semigroup $S_\alpha(t)$ in $L^2(0,1)$. Moreover, there exists $\beta > 0$, and $M > 0$ such that for any $\phi \in L^2(0,1)$,*

$$||S_\alpha(t)\phi - [\phi]||_{L^2} \leq Me^{-\beta t}||\phi - [\phi]||_{L^2}, \qquad t \geq 0. \tag{3.5}$$

*Proof.* see [18].

Thus any solution $w(x,t) = S_\alpha(t)\phi$ of (3.2) with $\phi \in L^2(0,1)$ is infinitely smooth in both the time and the space variables for $t > 0$, which is one of the smoothing properties we wished to restore by introducing dissipation mechanism through the boundary conditions. □

The following proposition (see [19] for the proof) gives detailed information about the eigenvalues of $A_\alpha$ and its ajoint $A_\alpha^*$.

PROPOSITION 3.2. *The operators $A_\alpha$, $A_\alpha^*$ have compact resolvents and, for $\alpha \neq -1/2$ and $|\alpha| < 1$, complete sets of eigenvectors, respectively,*

$$\{\phi_k \mid -\infty < k < +\infty\}, \qquad \{\psi_k \mid -\infty < k < +\infty\}$$

*which, normalized so that (with $\delta_{k,j}$ the Kronecker delta )*

$$\psi_j^* \phi_k = \delta_{k,j}$$

*form dual Riesz bases for $L^2[0,1]$. The corresponding eigenvalues of $A_\alpha$ have the asymptotic form*

$$\lambda_k = \left(8\pi^3 k^3 + O(k^2)\right) i - 12\pi^2 k^2 r + O(k), \quad k \to \infty \tag{3.6}$$

*where*

$$r = -\log\left|\frac{1+2\alpha}{2+\alpha}\right| > 0.$$

DEFINITION 3.3. We denote by $H^n_\alpha$ the Hilbert space consisting of functions $w$ in the Sobolev space $H^n[0,1]$ which obey boundary conditions of the form

$$w^{(3j)}(1) = w^{(3j)}(0), \quad w^{(3j+2)}(1) = w^{(3j+2)}(0), \quad w^{(3j+1)}(1) = \alpha w^{(3j+1)}(0), \tag{3.7}$$

as long as the indicated derivatives are of order $\le n-1$. The norm and inner product in $H^n_\alpha$ are $\| \ \|_n$, $(\,.\,)_n$, inherited from $H^n[0,1]$.

PROPOSITION 3.4. *For $\alpha \neq -1/2$, a function $w \in L^2$ also lies in $H^n_\alpha$ if and only if, represented in the form*

$$w = \sum_{k=-\infty}^{\infty} c_k \phi_k$$

*we have*

$$\sum_{k=-\infty}^{\infty} |k^n c_k|^2 < \infty. \tag{3.8}$$

*Moreover, $\|w\|^2_{H^n_\alpha}$ is a equivalent to the sum*

$$\sum_{k=-\infty}^{\infty} \left[|c_k|^2 + |k^n c_k|^2\right],$$

*i.e., there are positive constants $e_n$, $E_n$ (these will, in general, depend on $\alpha$) such that, uniformly for $w \in H^n_\alpha$,*

$$e_n^2 \sum_{k=-\infty}^{\infty} \left[|c_k|^2 + |k^n c_k|^2\right] \le \|w\|^2_{H^2_\alpha} \le E_n^2 \sum_{k=-\infty}^{\infty} \left(|c_k|^2 + |kc_k|^2\right). \tag{3.9}$$

*Proof.* see [19].

Let $\{\phi_k(x)\}$ be the Riesz bases of $L^2(0,1)$ defined in Proposition 3.2. For any $s \ge 0$ and $p \ge 1$, define

$$H^{s,p}_\alpha = \left\{ w = \sum_{k=-\infty}^{\infty} c_k\phi_k \,\middle|\, \sum_{k=-\infty}^{\infty} (1+|k|^{ps})\,|c_k|^p < \infty \right\}$$

with the norm

$$\|w\|^p_{H^{s,p}_\alpha} \equiv \sum_{k=-\infty}^{\infty} |c_k|^p (1+|k|^{ps}).$$

$H_\alpha^{s,p}$ is continuously imbedded into the space $H_\alpha^{s',p}$ if $s' < s$. Obviously, if $p \geq 2$, then

$$H_\alpha^s \subset H_\alpha^{s,p}, \qquad \text{for any } s \geq 0. \tag{3.10}$$

In the case that $p = 2$, we denote $H_\alpha^{s,p}$ by $H_\alpha^s$. Clearly, if $s$ is an integer, then the space $H_\alpha^s$ agrees with the space $H_\alpha^n$ defined by Definition 3 with the equivalent norm. □

For the semigroup $S_\alpha(t)$ generated by $A_\alpha$ in the space $L^2(0,1)$, we have the following properties.

PROPOSITION 3.5. *Assume $\alpha \neq 1/2$ and let $s \geq 0$ and $T > 0$ be given. Then there exists a $B_s > 0$ such that*

$$||S_\alpha(t)w_0||_s \leq ||w_0||_s, \tag{3.11}$$

$$\int_0^T ||S_\alpha(t)w_0||_{s+1}^2 dt \leq B_s^2 ||w_0||_s^2 \tag{3.12}$$

*for any $w_0 \in H_\alpha^s$,*

$$\sup_{0\leq t\leq T} ||\int_0^t S_\alpha(t-\tau)f(.,\tau)d\tau||_{s+1} \leq B_s \left[\int_0^T ||f(.,\tau)||_s^2 d\tau\right]^{\frac{1}{2}} \tag{3.13}$$

*uniformly for $f \in L^2(0,T;H_\alpha^s)$,*

$$\sup_{0\leq t<\infty} ||\int_0^t S_\alpha(t-\tau)f(.,\tau)d\tau||_{s+1} \leq B_s \sup_{0\leq t<+\infty} ||f(.,t)||_s \tag{3.14}$$

*for any $f \in L^\infty(0,\infty;H_\alpha^s)$ with*

$$\int_0^1 f(x,t)dx \equiv 0.$$

*Proof.* see [19].

*Remark* 3.6. (3.12) and (3.13) are analogous to the Kato type smoothing effects (but in global). We refer to [6] and [8] for the local Kato type smoothing property for solutions of the KdV equation posed on the whole real line $R$. The result (3.12), together with the results of [18] to the effect that $S_\alpha$ is a $C^\infty$ semigroup (easily re-proved with the framework used here except for the singular case $\alpha = -1/2$), expresses the fundamental smoothing property of the linear system (3.2). □

Now we turn to consideration of the nonlinear system (3.1). First we show that the initial boundary value problem (3.1) is well-posed in certain function spaces. From now on we assume that $\alpha \neq -1/2$.

THEOREM 3.7. *For any $\phi \in H^1_\alpha$ there exists a $T = T(||\phi||_1) > 0$ such that*

*(i). the IVP (3.1) has a unique solution*

$$u \in X_T := C(0,T; H^1_\alpha) \cap L^\infty(0,T; H^1_\alpha)$$

*where $T \to \infty$ as $||\phi||_1 \to 0$;*

*(ii). for any $T' < T$, there exists a neighborhood $U$ of $\phi$ in $H^1_\alpha$ such that the map*

$$K: \quad w_0 \to u(.,t)$$

*from $U$ to $X_{T'}$ is Lipschitz continuous.*

*Remark* 3.8. In fact we are able to show that the map $K$ is analytic from $U$ to $X_{T'}$ (cf. [23]).

*Proof of Theorem 3.7:* Since the proof is very similar to the one given in [19], only a sketch is given here.

In the equation (3.1), if we move the nonlinear term $uu_x$ to the right side of the equation and treat it as an inhomogeneous term, then the *variation of parameter* formula yields for $u$ the equivalent integral equation

$$u(.,t) = S_\alpha(t)\phi - \int_0^t S_\alpha(t-\tau)(u^k u_x)(.,\tau)d\tau. \tag{3.15}$$

We study this equation by the familiar method of looking for a fixed point of the corresponding map

$$\begin{aligned} u(.,t) &= (Fv)(.,t) \\ &:= S_\alpha(t)\phi - \int_0^t S_\alpha(t-\tau)\left(v^k v_x\right)(.,\tau)d\tau \end{aligned} \tag{3.16}$$

in which $\phi$ appears as a parameter and is considered fixed.

Let

$$S_{T,b} = \left\{ v \in X_T | \quad \sup_{0\le t\le T} ||v(.,t)||_1 \le b \right\} \tag{3.17}$$

for some $b > 0$, $T > 0$ to be determined. We shall show that for appropriately chosen $b$ and $T$, the map defined by (3.16) is a contraction from $S_{T,b}$ to $S_{T,b}$ and as a consequence, its fixed point is the desired solution of the IVP (3.1).

Applying (3.11) and (3.13) to (3.16) we have

$$\begin{aligned} \sup_{0\le t\le T} ||Fv||_1 &\le c||\phi||_1 + c\left(\int_0^T ||v^k v_x||^2 dt\right)^{1/2} \\ &\le c||\phi||_1 + cT^{/12}\left(\sup_{0\le t\le T} ||v(t)||_1\right)^{k+1} \end{aligned} \tag{3.18}$$

for some $c > 0$ independent of $T$ and $v$.

If one chooses

$$b = 2c\|\phi\|_1 \tag{3.19}$$

and $T > 0$ such that

$$cT^{1/2}b^k < 1/2 \tag{3.20}$$

then it follows from (3.18) that

$$\sup_{0\le t\le T} \|Fv\|_1 \le b$$

and therefore $F$ is a map from $S_{T,b}$ into $S_{T,b}$. The contraction property of the map $F$ follows similarly. As a consequence, its fixed point is the unique solution of (3.1). See [19] for more details of the proof. □

The next theorem gives a regularity result for solutions of the system (3.1).

THEOREM 3.9. *Let a series of differential operators $\{P_k\}_0^n$ be defined as follows:*

$$\begin{cases} P_0(\psi) = \psi \\ P_1(\psi) = \frac{1}{k}\partial_x(\phi^k) + \partial_x^3\phi \\ \cdots\cdots\cdots \\ P_n(\psi) = \frac{1}{k}\partial_x\left(\sum_{j=0}^{n-1}\begin{pmatrix} j \\ n-1\end{pmatrix} P_j(\psi)P_{n-j}(\psi)\right) + \partial_x^3 P_{n-1}(\psi) \end{cases}$$

*for any $\psi \in H^{3n}$.*

*Let $n \ge 1$ be given. For any $\phi \in H_\alpha^3 \cap H^{3n+1}$ satisfying*

$$P_j(\phi) \in H_\alpha^3 \cap H^{3(n-j)+1}, \quad j = 0, 1, 2, ..., n-1$$

*and*

$$P_n(\phi) \in H_\alpha^1,$$

*there exists a $T = T(\|\phi\|_{3n+1}) > 0$ such that (3.1) has a unique solution*

$$u \in C(0, T; H_\alpha^3 \cap H^{3n+1}).$$

*Moreover,*

$$\partial_t^j u \in C(0, T; H_\alpha^3 \cap H^{3(n-j)+1}), \quad j = 1, 2, ..., n-1$$

*and*

$$\partial_t^n u \in C(0,T; H^1_\alpha).$$

The proof of the above theorem is same as the proof of [19, Theorems 4.3 and 4.4] with few changes and therefore is omitted here.

The above theorem is a local result since the time interval $(0,T)$ on which the solution exists depends on the size of its initial value. The following theorem shows that solutions exist globally if the norm of the initial state is small.

THEOREM 3.10. *For any $n \geq 0$, there exists a $\gamma = \gamma(n) > 0$ such that if $\phi \in H^3_\alpha \cap H^{3n+1}$ with $||\phi||_{3n+1} \leq \gamma(n)$ and satisfies*

$$P_j(\phi) \in H^3_\alpha \cap H^{3(n-j)+1}, \quad j = 0,1,2,...,n-1,$$

*and*

$$P_n(\phi) \in H^1_\alpha,$$

*then (3.1) has a unique solution*

$$u \in C(0,\infty; H^3_\alpha \cap H^{3n+1}) \cap L^\infty(0,\infty; H^{3n+1}),$$

*and satisfies*

$$\sup_{0\leq t<\infty} ||u(.,t)||_{3n+1} \leq \theta_n(||\phi||_{3n+1}) \tag{3.21}$$

*where $\theta_n(.)$ is continuous function from $R^+$ to $R^+$ and $\theta_n(0) = 0$.*

*In addition,*

$$\partial_t^k u \in C(0,\infty; H^3_\alpha \cap H^{3(n-k)+1}), \quad k = 1,2,...,n-1$$

*and*

$$\partial_t^n u \in C(0,\infty; H^1_\alpha) \cap C(0,\infty; H^1_\alpha).$$

*Proof.* see [19].

In the remainder of this section we consider asymptotic property of small amplitude solutions of (3.1) as $t \to \infty$. We have the following □

THEOREM 3.11. *There exists a $\delta > 0$ such that for any $\phi \in H^1_\alpha$ with $||\phi||_1 < \delta$, the corresponding unique solution of (3.1) satisfies*

$$||u(.,t) - [\phi]||_{L^2} \leq ce^{-\rho t}||\phi - [\phi]||_{L^2}, \quad \text{for any } t \geq 0 \tag{3.22}$$

*where $c > 0$ and $\rho > 0$ are independent of $\phi$.*

*Proof.* A similar result for the KdV equation ($k = 1$ in (3.1) is established in [19] by using an infinite dimensional version of the second method of Lyapounov. However, the method developed in [19] can be used here with few modifications. Therefore we have omitted the proof. □

**4. Concluding remarks.** In summary, in this paper, we have discussed the stabilization problem of the KdV equation on a periodic domain. Both distributed control and boundary control actions are considered. The corresponding linear systems have been shown in both cases to be well stabilized with uniform exponential decay rates. For the nonlinear system, if the control acts on the whole interval $(0,1)$, the system is globally stabilized with exponential decay rates. But for the boundary control case, only small amplitude solutions have been shown to be stabilized with exponential decay. There are still some questions remaining open which warrant further pursuit. We list two of them here to end our discussion.

**Problem 1 (local control action):** Consider the distribute control system (1.1) with the periodic boundary condition (1.2) but we assume control $v = v(x,t)$ acts only on parts of the interval $[0,1]$, say $[a,b] \subset [0,1]$ and takes the following restricted form:

$$v = Gu := -Kg(x)\left\{u(x,t) - \int_0^1 g(s)u(s,t)ds\right\},$$

where $K > 0$ is a given constant and $g(x)$ is a nonnegative smooth function with *supp* $g = [a,b]$ and

$$\int_0^1 g(x)dx = 1.$$

More precisely, we consider the following closed loop system

$$(4.1)\quad \begin{cases} u_t + uu_x + u_{xxx} = G(u), & x \in (0,1), \quad t \geq 0 \\ u(x,0) = \phi(x) \\ u(1,t) = u(0,t), \quad u_x(1,t) = u_x(0,t), \quad u_{xx}(1,t) = u_{xx}(0,t) \end{cases}$$

The system (4.1) is known [11] to be globally well posed in the space $H_p^n$ for any $n \geq 1$. In Section 2, we have already seen that the corresponding linear system is asymptotically stable with exponential decay rate. For the nonlinear system, if $[a,b] = [0,1]$, then (4.1) is also stabilized with exponential decay rate. These lead us to conjecture that the system (4.1) with local control action is also stabilized, i.e., any solution $u$ of (4.1) satisfies

$$(4.2)\qquad \lim_{t\to\infty} \|u(.,t) - [\phi]\|_{L^2} = 0.$$

In [11] we showed, using the LaSalle Invariance Principle [13], [22] and a unique continuation result for the KdV equation [24], that (4.2) would be true if $H^2$ norm of the solution $u$ is uniformly bounded for $t \in (0,\infty)$.

The other problem concerns the following closed loop boundary control system

$$(4.3) \quad \begin{cases} u_t + u^k u_x + u_{xxx} = 0, & x \in (0,1), \quad t \geq 0 \\ u(x,0) = \phi(x) \\ u(1,t) = u(0,t), \quad u_x(1,t) = \alpha u_x(0,t), & u_{xx}(1,t) = u_{xx}(0,t) \end{cases}$$

where $|\alpha| < 1$, and integer $k \geq 1$.

Theorem 3.9 in Section 3 states that (4.3) is locally well-posed in the space $H^1_\alpha$ if $\alpha \neq -1/2$. Only small amplitude solutions of (4.3) are known to exist for all $t \geq 0$ and to have the property that they decay exponentially to the mean value of their initial states.

**Problem 2:** Do large amplitude solutions of (4.3) exist globally in time or blow up in finite time? What is the asymptotic behavior as $t \to \infty$ of a solution $u$ of (4.3) which exists for all $t \geq 0$?

Usually, global existence results follow from local existence results and global a priori estimates. However for the system (4.3), the only known a priori estimate is

$$\int_0^1 u^2(x,t)dx + \int_0^t (1-\alpha^2)u_x^2(0,s)ds = \int_0^1 \phi^2(x)dx, \quad t \geq 0,$$

which is not enough to obtain global existence of solutions of (4.3) in the space $H^1_\alpha$.

## REFERENCES


[1] J.L. Bona and R. Smith, *The initial value problem for the Korteweg-de Vries equation*, Roy. Soc. London, Ser. A, **278** (1978), pp. 555–601.

[2] A. Cohen, *Solutions of the Korteweg-de Vries equation from irregular data*, Duke Math. J., **45** (1978), pp. 149–81.

[3] P. Constantin and J.C. Saut, *Local smoothing properties of dispersive equations*, J. Amer. Math. Soc., **1**(1988), pp. 413–446.

[4] W. Craig, T. Kappeler and W.A. Strauss, *Gain of regularity for equations of KdV type*, Analyse Non Linéaire, Ann. Inst. Henri Poincaré, **9**(1992), pp. 147–186.

[5] L.F. Ho and D.L. Russell, *Admissible input elements for systems in Hilbert space and a Carlson measure criterion*, SIMA J. Contrl. & Opt., **21**(1981), pp. 614–640.

[6] T. Kato, *On the Cauchy problem for the (generalized) Korteweg-de Vries equations*, Advances in Mathematics Supplementary Studies, Studies in Applied Math., **8**(1983), pp. 93–128.

[7] C.E. Kenig, G. Ponce and L. Vega, *Oscillatory integrals and regularity of dispersive equations*, Indiana University Math. J., **40**(1991), pp. 37–69.

[8] C.E. Kenig, G. Ponce and L. Vega, *Well-posedness of the initial value problem for the Korteweg-de Vries*, J. Amer. Math. Soc., **4**(1991), pp. 323–347.

[9] C.E. Kenig, G. Ponce and L. Vega, *Well-posedness and scattering results for the generalized Korteweg-de Vries equation via the contraction principle*, Comm. Pure Appl. Math., **46** (1993), pp. 527–620.

[10] V. Komornik, D.L. Russell and B.-Y. Zhang, *Stabilisation de l' equation de Korteweg-de Vries*, C. R. Acad. Sci. Paris, t. **312**(1991), pp. 841–843.

[11] V. Komornik, D.L. Russell and B.-Y. Zhang, *Control and stabilization of the Korteweg-de Vries equation on a periodic domain*, (to appear in) Journal of Differential Equations.

[12] D.J. Korteweg and G. de Vries, *On the change of form of long waves advancing in a rectangular canal, and on a new type of long stationary waves*, Philos. Mag., 5, **39**(1895), pp. 422–423.

[13] J.P. LaSalle and S. Lefschetz, *Stability by Liapunov's direct method with applications*, Academic Press, New York, 1961.

[14] P.D. Lax, *Integrals of nonlinear equations of evolution and solitary waves*, Comm. Pure and Appl. Math., **21**(1968), pp. 467–490.

[15] R.M. Miura, *The Korteweg-de Vries equation: a survey of results*, SIAM Review, **18**(1976), pp. 412–459.

[16] R.M. Miura, S.C. Gardner and M.D. Kruskal, *Korteweg-de Vries equation and generalizations, II. Existence of conservation laws and constant of motion*, J. Math. Phys., **9**(1968), pp. 1204–1209.

[17] D.L. Russell, *Mathematics of Finite Dimensional Control Systems; Theory and Design*, Marcel Dekker, Inc., New York, 1979.

[18] D.L. Russell and B.-Y. Zhang, *Controllability and stabilizability of the third order linear dispersion equation on a periodic domain*, SIAM J. Contrl. & Opt., **31** (1993), pp. 659–676.

[19] D.L. Russell and B.-Y. Zhang, *Smoothing and decay properties of solutions of the Korteweg-de Vries equations on a periodic domain with point dissipation*, IMA Preprint Series # 1083, December, 1992 (to appear in) J. Math. Anal. Appl.

[20] J.C. Saut and R. Temam, *Remarks on the Korteweg-de Vries equation*, Israel J. Math., **24**(1976), pp. 78–87.

[21] A.C. Scott, F.Y. Chu and D.W. McLaughlin, *The soliton: a new concept in applied sciences*, Proc. IEEE, **61**(1973), pp. 1443–1483.

[22] M. Slemrod, *A note on complete controllability and stabilizability for linear control systems in Hilbert space*, SIAM J. Contrl. & Opt., **12**(1974), pp. 500–508.

[23] B.-Y. Zhang, *Taylor series expansion for solutions of the Korteweg-de Vries equation with respect to their initial values*, IMA preprint, series # 1015, August 1992 (to appear in) J. Funct. Anal.

[24] B.-Y. Zhang, *Unique continuation for the Korteweg-de Vries equation*, SIAM J. Math. Anal. **23**(1992).

# A NOTE CONCERNING BOUNDARY EFFECTS AND LONG TIME VIBRATIONS OF LAYERED MEDIA

FADIL SANTOSA* AND MICHAEL VOGELIUS†

**Abstract.** In this work, we study the vibration of, and wave propagation in, a periodically laminated composite slab. Special emphasis is given to the long time behavior of the vibration. It is found that a phase shift phenomenon occurs as a result of the interaction of the disturbance with the microstructure and the boundaries of the slab. We provide an asymptotic analysis to explain this phenomenon which we also illustrate by numerical calculations.

**1. Introduction.** The purpose of this note is to demonstrate and explain a phase shift phenomenon that is connected with low frequency vibration of (periodically) layered, composite media. The media we study are governed by a scalar wave equation, but similar behaviour can be found in media governed by the full system of linear elastodynamics.

As an example let us consider a layered composite slab which occupies $\Omega = \{x = (x', z) : x' \in \mathbb{R}^{n-1}, 0 \leq z \leq 1\}$, and let us assume that the layers are parallel to the boundary: in other words, the two material parameters $a > 0$ (the "stiffness") and $\rho > 0$ (the density) are assumed to be periodic functions of the variable $z$ only. The microstructure is of size $\epsilon$, so that the composite has "stiffness" and density given by

$$a\left(\frac{z}{\epsilon}\right) \quad \text{and} \quad \rho\left(\frac{z}{\epsilon}\right)$$

respectively. For simplicity we take the function to be periodic with period 1, *i.e.*,

$$a(y+1) = a(y), \quad \rho(y+1) = \rho(y).$$

The displacement $u(x,t)$ is the solution to

$$\rho\left(\frac{z}{\epsilon}\right)\frac{\partial^2 u}{\partial t^2} = \nabla \cdot a\left(\frac{z}{\epsilon}\right)\nabla u, \quad x \in \Omega, \tag{1.1}$$

with boundary condition

$$u(x,t)|_{x\in\partial\Omega} = 0.$$

For the purpose of clarity of presentation, we choose the simplest initial conditions

$$u(x,0) = f(z), \quad \frac{\partial u}{\partial t}(x,0) = g(z).$$

* Department of Mathematical Sciences, University of Delaware, Newark, DE 19716. Partially supported by National Science Foundation Grants DMS-90-11076, DMS-92-10489 and AFOSR Contract 91-NM-875.

† Department of Mathematics, Rutgers University, New Brunswick, NJ 08903; and LMC, Université Joseph Fourier, 38041 Grenoble, France. Partially supported by National Science Foundation Grant DMS-92-02042 and AFOSR Contract 89-NM-605.

The problem is to find $u(x,t) \equiv u(z,t)$ for $t > 0$. Note that with this simplified initial data the solution only depends on one spatial variable. Problems involving more general initial conditions (functions that also depend on $x'$) will have solutions that depend on all spatial variables. Since the coefficients $a$ and $\rho$ do not depend on $x'$, we can use Fourier transformation to reduce such a multidimensional problem into a family of one dimensional problems. Such a problem may therefore still be treated by an approach very similar to that which we will use for the one dimensional problem.

A most direct representation for $u(z,t)$ is in terms of an eigenfunction expansion. Consider the spectral problem

$$(1.2) \qquad (a\left(\frac{z}{\epsilon}\right) v_z)_z = -\lambda\rho\left(\frac{z}{\epsilon}\right) v \;\text{ in } (0,1) \;\text{ with } v(0) = v(1) = 0.$$

The (generalized) eigenpairs for this boundary value problem are denoted by $(v_m^\epsilon, \lambda_m^\epsilon)$, $m = 1, 2, \ldots$. The eigenvalues $\lambda_m^\epsilon$ are listed in increasing order, $\lambda_1^\epsilon < \lambda_2^\epsilon < \ldots$. The $\rho(\cdot/\epsilon)$-orthogonal set of eigenfunctions have been normalized by

$$\int_0^1 [v_m^\epsilon]^2 dz = 1.$$

The displacement, $u$, has the representation

$$(1.3) \qquad u(z,t) = \sum_{m=1}^{\infty} b_m^{\pm} v_m^\epsilon(z) \exp(\pm i\sqrt{\lambda_m^\epsilon}\, t).$$

Note that we have explicitly indicated the dependence of the eigenpairs on $\epsilon$.

We are interested in *low frequency vibration* of our system, hence we will assume that the initial data $f(z)$ and $g(z)$ are such that

$$\sum_{m>M} |b_m^{\pm}|^2 \le C\epsilon^2.$$

This implies that we get an $O(\epsilon)$ approximation, uniformly in $t$, by terminating the sum in (1.3) at $M$.

The basic idea behind homogenization is that if the microstructure is small (that is $\epsilon \ll 1$) then one should be able to obtain a relatively good asymptotic approximation to $u(z,t)$. Associated to (1.2) is the homogenized spectral problem

$$Av_{zz}^{(0)} = -\lambda^{(0)}\overline{\rho}v^{(0)} \;\text{ in } (0,1) \;\text{ with } v^{(0)}(0) = v^{(0)}(1) = 0.$$

The constants $A$ and $\overline{\rho}$ are given by the formulae

$$A = \int_0^1 a(y)(1 - \chi'(y))\, dy \;\text{ and }\; \overline{\rho} = \int_0^1 \rho(y)\, dy.$$

It is well known that (after appropriate change of $v_m^\epsilon$ by multiplication by $\pm 1$)

$$\lambda_m^\epsilon \to \lambda_m^{(0)} \text{ and } v_m^\epsilon \to v_m^{(0)} \quad \text{as } \epsilon \to 0;$$

indeed one has the estimates

$$|\lambda_m^\epsilon - \lambda_m^{(0)}| \leq C_m \epsilon \text{ and } ||v_m^\epsilon - v_m^{(0)}||_{L^2(\Omega)} \leq C'_m \epsilon,$$

provided $a$ and $\rho$ are sufficiently regular (*cf.* [1]).

With the low-frequency assumption on $f(\cdot)$ and $g(\cdot)$ we now obtain

$$u(z,t) = \sum_{m=1}^{M} b_m^{\pm} v_m^{(0)}(z) \exp(\pm i \sqrt{\lambda_m^\epsilon} t) + O(\epsilon), \tag{1.4}$$

where $||O(\epsilon)||_{L^2(\Omega)} \leq C\epsilon$, with $C$ independent of $\epsilon$ and $t$. Thus the sum in (1.4) represents an $O(\epsilon)$ approximation to the solution of (1.1), valid uniformly in $t$. For $t \ll 1/\epsilon$ we may substitute $\lambda_m^{(0)}$ for $\lambda_m^\epsilon$ in this sum, and still obtain an approximation which is $o(1)$. For $t \approx 1/\epsilon$ such substitution no longer provides an acceptable approximation — one may encounter an "erroneous" phase shift.

The purpose of this note is to describe the limiting behaviour of the ratio

$$\frac{\lambda_m^\epsilon - \lambda_m^{(0)}}{\epsilon} \tag{1.5}$$

for any $m$. We shall refer to the limits of this ratio (or $\epsilon\times$ the limits) as first order corrections. In a case when the first order correction is not zero, substitution of $\lambda_m^{(0)}$ for $\lambda_m^\epsilon$ in (1.4) will lead to an erroneous phase shift at time $t \approx 1/\epsilon$. Explicit knowledge of the possible first order corrections will allow us to obtain acceptable approximations to $u$ for $t \approx 1/\epsilon$, eliminating the aforementioned phase shift. Such approximations will not require knowledge of the exact eigenvalues $\lambda_m^\epsilon$ and they will presumably be valid for $t \ll 1/\epsilon^2$.

For a layered composite, with layers parallel to the slab boundaries we shall demonstrate that the ratio (1.5) has a continuum of limit points and we shall, by means of a numerical example, illustrate how these limit points may be used to obtain good long time approximations for the vibration problem. Exactly which limits are attained (as $\epsilon \to 0$ along some sequence) depends on the the way in which the boundary $z = 1$ asymptotically falls within a period cell. For a layered composite occupying a bounded domain, part of whose boundary is flat and parallel to the layers one will also find a continuum of limits. The same phenomenon (a continuum of limits) is found for a general (non-layered) periodic composite occyping a bounded domain $\Omega$, provided the boundary of $\Omega$ has a flat part with a rational normal. Rationality is measured relative to a coordinate system in which the composite is periodic (with period 1, say) in the axis-directions.

It is important to realize that for a bounded domain whose boundary has no flat parts (or at least no flat parts with rational normals) it is most likely true that the ratio (1.5) has only a single limit, namely zero. If the composite is layered it is very likely true that it is necessary to have a flat part of the boundary which is parallel to the layers in order to obtain a continuum of limits (as opposed to the single limit zero). We refer to the discussion towards the end of the next section for more details.

The present analysis is carried out for a Dirichlet boundary condition, and we have not at this point made a careful study of any other type of boundary condition. It is probable that other boundary conditions will lead to a similar phenomenon (the possibility of a continuum of first order corrections as well as a single correction, depending on the structure of $\partial\Omega$) but the actual values will of course depend on the boundary conditions. In particular, even if there is only a single correction (say, because the boundary has no flat parts) this correction may well be nonzero for other types of boundary conditions.

**2. Asymptotics.** We now give a brief sketch of the asymptotic multiscale expansion which leads to formulae for the possible first order corrections to the eigenvalues. We consider eigenvalues corresponding to a fixed integer $m$, for simplicity we drop the subscript $m$ from the eigenvalues. As an ansatz we take $v^\epsilon(z)$ to be of the form

$$V_\epsilon(z,y)|_{y=z/\epsilon},$$

where $V_\epsilon(z,y)$ is a periodic function of $y$ with period 1. Next, we assume that $V_\epsilon(z,y)$ and $\lambda^\epsilon$ have power series expansions in $\epsilon$

$$\begin{aligned} V_\epsilon(z,y) &= v^{(0)}(z,y)+\epsilon v^{(1)}(z,y)+\epsilon^2 v^{(2)}(z,y)+\epsilon^3 v^{(3)}(z,y)+\cdots, \\ \lambda^\epsilon &= \lambda^{(0)}+\epsilon\lambda^{(1)}+\epsilon^2\lambda^{(2)}+\cdots. \end{aligned}$$

The chain rule for differentiation gives

$$\frac{d}{dz}v^\epsilon = \frac{d}{dz}V_\epsilon(z,y)|_{y=z/\epsilon}+\frac{1}{\epsilon}\frac{d}{dy}V_\epsilon(z,y)|_{y=z/\epsilon}\,.$$

Upon substitution of these formulas into (1.2) and collection of terms of order $\epsilon^{-2}$ and $\epsilon^{-1}$ we obtain the equations

$$\begin{aligned} O(\epsilon^{-2}) &: (a(y)v_y^{(0)})_y = 0 \\ O(\epsilon^{-1}) &: (a(y)v_y^{(1)})_y+(a(y)v_y^{(0)})_z+(a(y)v_z^{(0)})_y = 0\,. \end{aligned} \tag{2.1}$$

These equations require that $v^{(0)}$ be independent of $y$ and that $v^{(1)}$ have the representation

$$v^{(1)}(z,y) = -\chi(y)\frac{dv^{(0)}}{dz}(z)+w^{(1)}(z), \tag{2.2}$$

where the auxiliary function $\chi(y)$ is the 1-periodic solution to

$$(a(y)\chi_y)_y = (a(y))_y \ .$$

Collecting term of order $\epsilon^0$ and $\epsilon^1$ we obtain

$$\begin{aligned} O(\epsilon^0) : (a(y)v_y^{(2)})_y &+ (a(y)v_y^{(1)})_z + (a(y)v_z^{(1)})_y \\ &+ (a(y)v_z^{(0)})_z = -\lambda^{(0)}\rho(y)v^{(0)}, \\ O(\epsilon^1) : (a(y)v_y^{(3)})_y &+ (a(y)v_y^{(2)})_z + (a(y)v_z^{(2)})_y \\ &+ (a(y)v_z^{(1)})_z = -\lambda^{(0)}\rho(y)v^{(1)} - \lambda^{(1)}\rho(y)v^{(0)}. \end{aligned} \tag{2.3}$$

For any 1-periodic function $g(y)$ let $\overline{g}$ denote the "cell" average $\int_0^1 g(y)dy$. A necessary and sufficient condition for the solvability of the first equation of (2.3) (in terms of a 1-periodic function $v^{(2)}$) is that the left and the right side have the same "cell" averages. After insertion of the formula (2.2) for $v^{(1)}$ this yields

$$Av_{zz}^{(0)} = \overline{a(1-\chi')}v_{zz}^{(0)} = -\lambda^{(0)}\overline{\rho}v^{(0)} \quad \text{in } (0,1).$$

The boundary conditions for $v^{(0)}$ are

$$v^{(0)}(0) = v^{(0)}(1) = 0.$$

By solution of the first equation of (2.3) we now get that $v^{(2)}$ has the form

$$v^{(2)}(z,y) = \eta(y)v_{zz}^{(0)}(z) - \chi(y)w_z^{(1)}(z) + w^{(2)}(z), \tag{2.4}$$

where $\eta(y)$ is the 1-periodic solution to

$$(a(y)\eta_y)_y = a(y)\chi'(y) + (a\chi)'(y) - a(y) + \frac{\rho(y)}{\overline{\rho}}A.$$

A necessary and sufficient condition for the solvability of the second equation of (2.3) (in terms of a 1-periodic function $v^{(3)}$) is that the left and the right side have the same "cell" averages. After insertion of the formulae (2.2) and (2.4) for $v^{(1)}$ and $v^{(2)}$ this yields

$$Aw_{zz}^{(1)} + \lambda^{(0)}\overline{\rho}w^{(1)} = (-\overline{a\eta'} + \overline{a\chi})v_{zzz}^{(0)} + \lambda^{(0)}\overline{\rho\chi}v_z^{(0)} - \lambda^{(1)}\overline{\rho}v^{(0)}. \tag{2.5}$$

A simple calculation gives that

$$\overline{a\eta'} = \overline{a\chi} - A\frac{\overline{\rho\chi}}{\overline{\rho}}.$$

Insertion of this identity into (2.5) results in the following equation for $w^{(1)}$

$$\begin{aligned} Aw_{zz}^{(1)} + \lambda^{(0)}\overline{\rho}w^{(1)} &= A\tfrac{\overline{\rho\chi}}{\overline{\rho}}v_{zzz}^{(0)} + \lambda^{(0)}\overline{\rho\chi}v_z^{(0)} - \lambda^{(1)}\overline{\rho}v^{(0)} \\ &= -\lambda^{(1)}\overline{\rho}v^{(0)}. \end{aligned} \tag{2.6}$$

The boundary conditions for $w^{(1)}$ are

$$w^{(1)}(0) = \chi(0)v_z^{(0)}(0), \quad w^{(1)}(1) = \chi(\frac{1}{\epsilon})v_z^{(0)}(1). \tag{2.7}$$

so that $v^{(1)}(z, z/\epsilon) = 0$ at $z = 0, 1$. Because of (2.6) and (2.7) the function $w^{(1)}$ has the form $w^{(1)} = W + \tilde{\phi}_\epsilon$, where

$$\tilde{\phi}_\epsilon = z\chi(\frac{1}{\epsilon})v_z^{(0)}(1) + (1-z)\chi(0)v_z^{(0)}(0),$$

and $W$ is a solution to

$$\begin{aligned} AW_{zz} + \lambda^{(0)}\overline{\rho}W &= -\lambda^{(1)}\overline{\rho}v^{(0)} - \lambda^{(0)}\overline{\rho}\tilde{\phi}_\epsilon, \\ W(0) &= W(1) = 0. \end{aligned} \tag{2.8}$$

A necessary and sufficient condition for the solvability of (2.8) is that the right hand side of the differential equation be orthogonal to the eigenfunction $v^{(0)}$, *i.e.*,

$$\lambda^{(1)} = \lambda_\epsilon^{(1)} = -\lambda^{(0)} \int_0^1 \tilde{\phi}_\epsilon v^{(0)}\, dz$$

Here we use the fact that $v^{(0)}$ is normalized so that $\int_0^1 [v^{(0)}]^2\, dz = 1$. If $\lambda_*^{(1)}$ denotes any of the possible limit points of the ratio $(\lambda^\epsilon - \lambda^{(0)})/\epsilon$, we thus have

$$\lambda_*^{(1)} = -\lambda^{(0)} \int_0^1 \phi_* v^{(0)}\, dz, \tag{2.9}$$

where $\phi_*$ is a limit point for the family of functions $\tilde{\phi}_\epsilon$. Based on the formula for $\tilde{\phi}_\epsilon$, and the continuity of $\chi$, it follows immediately that the set of possible $\phi_*$'s have the form

$$\phi_* = z\chi(\delta_0)v_z^{(0)}(1) + (1-z)\chi(0)v_z^{(0)}(0), \tag{2.10}$$

where $\delta_0$ ranges over the interval $[0, 1]$. If $\phi_\epsilon$ denotes the solution to

$$\frac{d}{dz}(a(\frac{z}{\epsilon})\frac{d}{dz}\phi_\epsilon) = 0 \text{ in } (0,1),$$

with boundary conditions

$$\phi_\epsilon(0) = \chi(0)v_z^{(0)}(0) \quad \phi_\epsilon(1) = \chi(\frac{1}{\epsilon})v_z^{(0)}(1),$$

then it is not difficult to see that $\phi_\epsilon - \tilde{\phi}_\epsilon \to 0$ as $\chi(1/\epsilon) \to \chi(\delta_0)$. The functions $\phi_*$ may therefore equally well be characterized as the limit points of the family $\phi_\epsilon$. In giving a rigorous proof of the representation (2.9),

(2.10) for the possible first order corrections it is actually the functions $\phi_\epsilon$ that enter as opposed to $\tilde{\phi}_\epsilon$ from the above "formal" asymptotic expansion, [3]. From a rigorous proof one initially obtains that all possible limits $\lambda_*^{(1)} = \lim_{\epsilon_k \to 0}(\lambda^{\epsilon_k} - \lambda^{(0)})/\epsilon_k$ are of the form

$$\lambda_*^{(1)} = -\lambda^{(0)} \lim_{\epsilon_k \to 0} \frac{\int_0^1 \rho(\frac{z}{\epsilon_k})\phi_{\epsilon_k} v^{(0)}\, dz}{\int_0^1 \rho(\frac{z}{\epsilon_k})[v^{(0)}]^2\, dz}, \tag{2.11}$$

where the right hand side denotes any possible limit along a sequence $\epsilon_k \to 0$. Since $\phi_{\epsilon_k}$ (or at least a subsequence thereof) converges strongly in $L^2$ to some $\phi_*$ and since $\rho(\frac{z}{\epsilon_k})$ converges weakly in $L^2$ to $\overline{\rho}$ it follows that all possible first order corrections are of the form

$$\lambda_*^{(1)} = -\lambda^{(0)} \frac{\int_0^1 \overline{\rho}\phi_* v^{(0)}\, dz}{\int_0^1 \overline{\rho}[v^{(0)}]^2\, dz} = -\lambda^{(0)} \int_0^1 \phi_* v^{(0)}\, dz,$$

as predicted by the "formal" expansion.

For an arbitrary periodic composite (not necessarily layered) occupying a bounded domain $\Omega \subset \mathbb{R}^n$, it is still possible to show that the first order corrections, $\lambda_*^{(1)}$ (at least corresponding to any simple eigenvalue) have a representation similar to (2.11), where $\phi_\epsilon$ are the solutions to

$$\begin{aligned} &\nabla \cdot (a(\tfrac{x}{\epsilon})\nabla\phi_\epsilon) = 0 \quad \text{in } \Omega, \\ &\phi_\epsilon = \chi_j(\tfrac{x}{\epsilon})\frac{\partial v^{(0)}}{\partial x_j} \quad \text{on } \partial\Omega, \end{aligned} \tag{2.12}$$

and $\chi_j$, $1 \le j \le n$, are the periodic cell functions satisfying $\nabla_y \cdot (a(y)\nabla_y \chi_j) = \nabla_y \cdot (a(y)\nabla_y y_j)$. If the boundary of $\Omega$ has no flat parts (or at least no flat parts with rational normals) then it is well known that $\chi_j(\frac{x}{\epsilon})|_{\partial\Omega}$ converge weakly in $L^2(\partial\Omega)$ towards the full spatial average $\overline{\chi_j}$, *cf.* [1] page 88. It is therefore quite natural to conjecture that the solutions to (2.12), $\phi_\epsilon$, in some sense converge towards the solution to

$$\begin{aligned} &A_{ij}\frac{\partial}{\partial x_i \partial x_j}\phi = 0 \quad \text{in } \Omega, \\ &\phi = \overline{\chi_j}\frac{\partial v^{(0)}}{\partial x_j} \quad \text{on } \partial\Omega. \end{aligned} \tag{2.13}$$

$A_{ij}\frac{\partial}{\partial x_i \partial x_j}$ denotes the homogenized, constant coefficient operator associated to $\nabla \cdot (a(\frac{x}{\epsilon})\nabla)$, (*cf.* [1]). If related results in [2,3] give any indication then it would be natural to expect that $\phi_\epsilon \to \phi$ strongly in $L^2(\Omega)$. If this conjecture turned out to be true, then it would mean that there is only one possible first order correction in the case where $\partial\Omega$ has no flat parts (or at least no flat parts with rational normals). Furthermore this first order correction would be zero. To see this we note that the solution to (2.13) has the form $\phi = \overline{\chi_j}\frac{\partial v^{(0)}}{\partial x_j} + \Phi$, where $\Phi$ satisfies

$$\begin{aligned} A_{ij}\frac{\partial}{\partial x_i \partial x_j}\Phi &= \lambda^{(0)}\overline{\rho}\,\overline{\chi_j}\frac{\partial v^{(0)}}{\partial x_j} \text{ in } \Omega, \\ \Phi &= 0 \text{ on } \partial\Omega. \end{aligned}$$

A simple calculation now gives

$$\begin{aligned} \int_\Omega \Phi v^{(0)}\, dx &= -(\lambda^{(0)}\overline{\rho})^{-1}\int_\Omega \Phi A_{ij}\frac{\partial}{\partial x_i \partial x_j} v^{(0)}\, dx \\ &= -(\lambda^{(0)}\overline{\rho})^{-1}\int_\Omega A_{ij}\frac{\partial}{\partial x_i \partial x_j}\Phi\, v^{(0)}\, dx \\ &= -\int_\Omega \overline{\chi_j}\frac{\partial v^{(0)}}{\partial x_j} v^{(0)}\, dx \\ &= 0, \end{aligned}$$

and, therefore

$$\int_\Omega \phi v^{(0)}\, dx = \int_\Omega \overline{\chi_j}\frac{\partial v^{(0)}}{\partial x_j} v^{(0)}\, dx + \int_\Omega \Phi v^{(0)}\, dx = 0.$$

Under the assumption that $\phi_\epsilon \to \phi$ strongly in $L^2(\Omega)$ it therefore follows from (2.11) that the only possible first order correction would be

$$\lambda_*^{(1)} = -\lambda^{(0)} \lim_{\epsilon_k \to 0} \frac{\int_\Omega \rho(\frac{x}{\epsilon_k})\phi_{\epsilon_k} v^{(0)}\, dx}{\int_\Omega \rho(\frac{x}{\epsilon_k})[v^{(0)}]^2\, dx} = -\lambda^{(0)} \int_\Omega \phi v^{(0)}\, dx = 0.$$

If, on the other hand, the domain is a convex polygon, with sides that have normals with rational components, then the functions $\chi_j(\frac{x}{\epsilon})|_{\partial\Omega}$ have more than one weak limit point and it is shown in [2,3] that one should in general expect a continuum of first order corrections. The same phenomenon is clearly expected to hold for domains whose boundary has a flat part with a corresponding rational normal. If it is known that the composite is layered then it is very likely necessary that part of the boundary be flat and parallel to the layers in order to get a continuum of first order corrections (see example B in section 5 of [3]).

The set of polygons with rational normals are clearly of significant practical importance. Furthermore, the fact that there may be a continuum of first order corrections associated to any polygon with rational normals in a sense of closure (stability) makes the continuum of corrections an important phenomenon. This in spite of the fact that the set of polygons with rational normals may be regarded as a set of "measure zero". Based on these observations we feel that a continuum of first order corrections will "for many practical purposes" be the phenomenon that one will observe, even though strictly speaking it is far from being the "generic" event. For the same reason we feel that in many practical situations one will encounter the relatively large phase shift phenomenon we are about to illustrate.

**3. Vibration of the layered slab.** In this section we illustrate the implications of the first order correction $\lambda_*^{(1)}$ for long term vibration. We begin by posing the initial value problem which describes the low frequency vibration of a layered slab of thickness $L$. The fine layering (microstructure) is of thickness $h$. The governing equation of motion is

$$\rho\left(\frac{z}{h}\right)\frac{\partial^2 u}{\partial t^2} = \frac{\partial}{\partial z}a\left(\frac{z}{h}\right)\frac{\partial u}{\partial z}, \quad z \in [0, L]. \tag{3.1}$$

with boundary condition

$$u(0,t) = u(L,t) = 0.$$

For initial conditions, we choose

$$u(z,0) = \sum_{m=1}^{M} b_m \sin\frac{m\pi z}{L}, \quad \frac{\partial u}{\partial t}(z,0) = 0,$$

where $b_m$ is an arbitrary sequence, and $M$ is some integer. By introduction of the variables $\tilde{z} = z/L$ and $\tilde{t} = t/L$ this may be transformed to a problem on the interval [0,1] with the dimensionless small scale

$$\epsilon \equiv h/L$$

and coefficients $a(\tilde{z}/\epsilon)$, $\rho(\tilde{z}/\epsilon)$ (as considered in the previous sections of this note). Here we shall always express formulas in terms of the physical variables $z$ and $t$.

According to our asymptotics, an $O(\epsilon)$ approximation to the solution of this initial value problem is given by

$$u(z,t) = \sum_{m=1}^{M} b_m \sin\frac{m\pi z}{L}\cos\left(\frac{\sqrt{\lambda_m^\epsilon}\, t}{L}\right) + O(\epsilon). \tag{3.2}$$

The homogenized eigenvalues $\lambda_m^{(0)}$ are given by

$$\lambda_m^{(0)} = \frac{m^2\pi^2 A}{\overline{\rho}}, \quad m = 1, 2, \ldots.$$

From (2.9) and (2.10), we can calculate the first order correction $\lambda_{*m}^{(1)}$. Provided $\chi(L/h) = \chi(1/\epsilon) \to \chi(\delta_0)$ it is given by

$$\lambda_{*m}^{(1)} = 2\lambda_m^{(0)}[\chi(\delta_0) - \chi(0)].$$

The fraction $0 \le \delta_0 \le 1$ is determined by the nature of the layering. The last result implies that

$$\lambda_m^{(0)} + \epsilon\lambda_{*m}^{(1)} = \lambda_m^{(0)}\left\{1 + 2\epsilon[\chi(\delta_0) - \chi(0)]\right\}.$$

We asserted earlier that for $t/L \approx 1/\epsilon$ (*i.e.* $t \approx L^2/h$) we may substitute the appropriate $\lambda_m^{(0)} + \epsilon\lambda_{*m}^{(1)}$ into the formula (3.2), to obtain an acceptable approximation for $u$. If we think of the solution in (3.2) as a wave packet, then the nature of the correction $\epsilon\lambda_{*m}^{(1)}$ can be interpreted as a phase shift. The packet travels at the approximate speed

$$c = \sqrt{\frac{A}{\overline{\rho}}\{1 + 2\epsilon[\chi(\delta_0) - \chi(0)]\}}.$$

Note that, to first order, the propagation is dispersionless. The inherent dispersion, which was studied in [4], is contained in the higher order corrections to the eigenvalues.

To better illustrate the nature of the first order correction $\lambda_{*m}^{(1)}$, we consider a two-phase medium given by

$$a(y) = \begin{cases} a_1 & \text{for } 0 \le y < y_0 \\ a_2 & \text{for } y_0 \le y < y_1 \\ a_1 & \text{for } y_1 \le y < 1 \end{cases},$$

and

$$\rho(y) = \begin{cases} \rho_1 & \text{for } 0 \le y < y_0 \\ \rho_2 & \text{for } y_0 \le y < y_1 \\ \rho_1 & \text{for } y_1 \le y < 1 \end{cases},$$

with $0 < a_1 \le a_2$. The volume fraction of component 1 is $\theta = y_0 + (1 - y_1)$. Using these values, we find the effective properties to be

$$A = \left(\frac{\theta}{a_1} + \frac{(1-\theta)}{a_2}\right)^{-1}, \qquad \overline{\rho} = \theta\rho_1 + (1-\theta)\rho_2.$$

The auxilliary function $\chi(y)$ (normalized by $\chi(0) = 0$) is,

$$\chi(y) = \begin{cases} \left(1 - \frac{A}{a_1}\right) y \quad \text{for } 0 \le y < y_0 \\ \left(1 - \frac{A}{a_1}\right) y_0 + \left(1 - \frac{A}{a_2}\right)(y - y_0) \quad \text{for } y_0 \le y < y_1 \\ \left(1 - \frac{A}{a_1}\right) y_0 + \left(1 - \frac{A}{a_2}\right)(y_1 - y_0) + \left(1 - \frac{A}{a_1}\right)(y - y_1) \\ \qquad \text{for } y_1 \le y < 1 \end{cases}$$

To arrive a formula for a $\lambda_{*m}^{(1)}$ for a given composite, we need to look at the limit of $\chi(1/\epsilon)$. Since $\chi(y)$ is 1-periodic, the value $\chi(1/\epsilon)$ is determined by the fractional part of the ratio $1/\epsilon = L/h$. For convenience, define

$$\delta(x) = x - [x],$$

where $[x]$ denotes the integer part of $x \geq 0$. By continuity of $\chi$ it now follows that $\chi(1/\epsilon) \to \chi(\delta_0)$ if $\delta(1/\epsilon) \to \delta_0$. The simplest way to find a sequence $\epsilon_k$ which realizes a particular value $\chi(\delta_0)$ is therefore to pick $\epsilon_k = h_k/L_k$ such that

$$\delta\left(\frac{1}{\epsilon_k}\right) = \delta_0.$$

Here we choose to keep $h$ fixed and change the length $L$ to adjust $\epsilon$. This choice of a sequence allows us to attain any one among the continuum of corrections $\lambda_{*m}^{(1)}$. Since $a_1 \leq A \leq a_2$, the global minimum of $\chi$ is $\chi_- = (1 - A/a_1)y_0 \leq 0$, and the global maximum is $\chi_+ = (1 - A/a_1)y_0 + (1 - A/a_2)(y_1 - y_0) \geq 0$. The interval $[\chi_-, \chi_+]$ depends on $y_0$ and $y_1$. This interval is of length $(\chi_+ - \chi_-) = (1 - A/a_2)(1-\theta)$, which only depends on the volume fraction $\theta$. By choosing $y_0$ and $y_1$ arbitrarily, but corresponding to a fixed volume fraction $\theta$, the intervals $[\chi_-, \chi_+]$ will cover the entire symmetric interval

$$\left[-\left(1 - \frac{A}{a_2}\right)(1-\theta), \left(1 - \frac{A}{a_2}\right)(1-\theta)\right].$$

In summary this implies that the first order correction $\lambda_{*m}^{(1)}$ for a two component composite (with a fixed volume fraction) ranges over the interval

$$\left[-2\lambda_m^{(0)}\left(1 - \frac{A}{a_2}\right)(1-\theta), 2\lambda_m^{(0)}\left(1 - \frac{A}{a_2}\right)(1-\theta)\right].$$

A special case is when the composite contains an integer number of layers, *i.e.*, $\delta(L/h) = 0$. Then, the correction is $\lambda_{*m}^{(1)} \equiv 0$. It is important to note that the first order correction is nonzero otherwise. This means that the vibration of, and wave propagation in, a slab with integer number of layers is markedly different from that with noninteger number of layers.

In order illustrate the differences in approximation associated with long term vibration, we need (for comparison) to obtain very accurate numerical solutions of the initial value problem (3.1). One particularly accurate method for time integration of the partial differential equation in (3.1) is the method of characteristics. The characteristics associated with (3.1) are not straight lines, but they can be "straightened" by either using a coordinate transformation (to travel time) or considering the special case where $\rho \equiv a$. We choose the latter option to keep the exposition simple.

The characteristics are now the straight lines

$$\xi = z - t \text{ and } \eta = z + t.$$

Letting $P(z,t) = a(z/h)u_z(z,t)$ and $V(z,t) = u_t(z,t)$, we find that along the $\xi$- and $\eta$-characteristics, we have

$$adV = dP \text{ and } -adV = dP,$$

respectively. We specialize to the present case of $a(z/h)$ piecewise constant, and arrange the discretization of the interval $[0, L]$ so that the jumps in $a$ occur only at node points. Under this construction, we can integrate the transport equations along the characteristics *exactly*. This leads to an exact finite difference scheme; with the only source of errors being the computer roundoff errors.

For our numerical example, we choose a high-contrast 2-phase medium,

$$a(y) = \begin{cases} 1 & \text{for } 0 \le y < \frac{1}{6} \\ 10 & \text{for } \frac{1}{6} \le y < 1 \end{cases}.$$

This choice gives a volume fraction of $\theta = 1/6$. The layer thickness is $h = 0.6$. Recall that we have set $\rho \equiv a$. The effective constants are found to be

$$\overline{\rho} = \frac{51}{6} \text{ and } A = 4.$$

We will consider two slabs; Slab 1 is of thickness $L = 120.6$ and Slab 2 is of thickness $L = 120.7$.

**Properties of Slab 1** Since $L/h = 201$, we have an integer number of layers and $\lambda_{*m}^{(1)} \equiv 0$. Therefore, if we follow a travelling pulse, it will up to $t \approx L^2/h$, be well approximated by a pulse with an associated phase speed

$$c = \sqrt{\frac{A}{\overline{\rho}}} = \sqrt{\frac{24}{51}} \approx 0.6860$$

and which is perfectly reflected at the ends $z = 0$ and $z = L$. To be more precise: up to $t \approx L^2/h$ an $O(\epsilon)$ approximate solution to the initial value problem is given by (3.2) with $\lambda_m^\epsilon$ replaced by $\lambda_m^{(0)}$. □

**Properties of Slab 2** Since the number of layers is not an integer, the first order correction $\lambda_{*m}^{(1)}$ is nonzero. In fact, we have constructed the composite so that the correction is as large as possible. For this slab, $\delta(L/h) = \delta(1/\epsilon) = 1/6$, hence

$$\lambda_{*m}^{(1)} = 2\lambda_m^{(0)}\chi(1/6) = -\lambda_m^{(0)}.$$

For this medium, the phase speed is, up to $t \approx L^2/h$, well approximated by

$$c = \sqrt{\frac{A}{\overline{\rho}}(1-\epsilon)} \approx 0.6843,$$

using $\epsilon = h/L = 0.6/120.7$. This speed is slower, by an order of $h/L$, from the phase speed of Slab 1. □

The initial condition for $u(z,t)$ is chosen to be a Fourier series approximation of

$$k(z) = \begin{cases} 0 & \text{for} \quad 0 \le z \le 20 \\ \frac{z-20}{2} & \text{for} \quad 20 < z \le 40 \\ 10 - \frac{z-40}{2} & \text{for} \quad 40 < z \le 60 \\ 0 & \text{for} \quad z > 60 \end{cases}.$$

The Fourier coefficients of the expansion for the function are

$$b_m = \frac{L}{(m\pi)^2}\left[2\sin\frac{40m\pi}{L} - \sin\frac{20m\pi}{L} - \sin\frac{60m\pi}{L}\right].$$

In both experiments, we took a total of 10 Fourier coefficients, giving an initial displacement $u(z,0) = f(z)$ for Slab 1 as shown in Figure 1 (the initial displacement in Slab 2 is nearly indistinguishable from that in Slab 1).

The solutions to the initial value problem for Slab 1 and Slab 2 are computed, and the displacements $u(z,t)$ at $z = 40$ are recorded for both slabs. In Figure 2, we have superposed the response at $z = 40$ for Slab 1 and Slab 2 in the time interval $t \in [10200, 10600]$. There is a difference of 0.1 units in the thicknesses of the slabs; amounting to a relative difference of about 0.082%. The difference in the response is certainly not accounted for by the difference in length. What does account for this difference is the difference in effective phase speed.

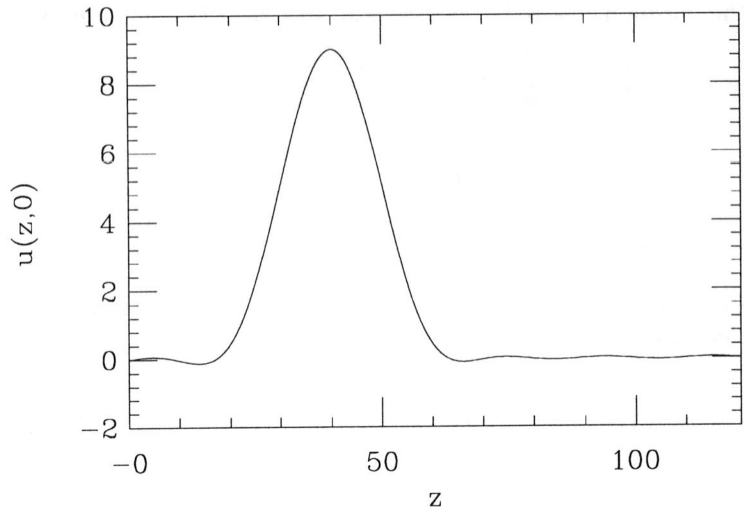


FIGURE 1. *The initial displacement for the problem is an 10 term Fourier approximation of the function* $k(z)$.

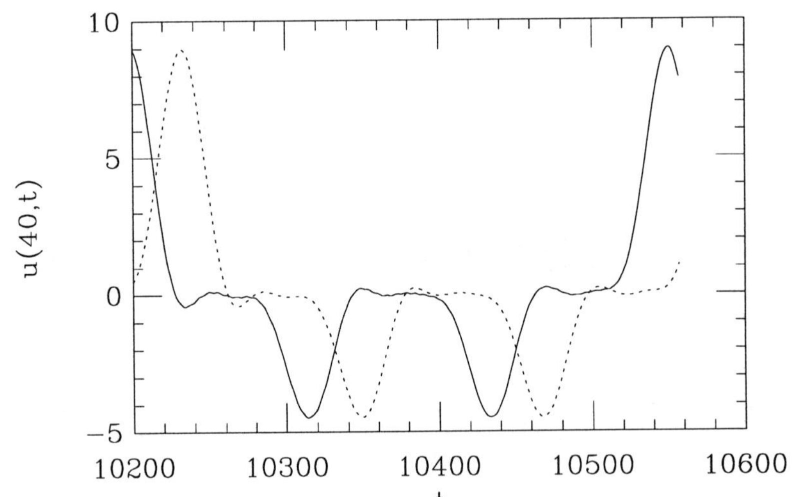


FIGURE 2. *The comparison of the displacement records at* $z = 40$ *for Slab 1 and Slab 2. The response for Slab 1 is shown in solid. Observe the large phase shift difference in the two graphs.*

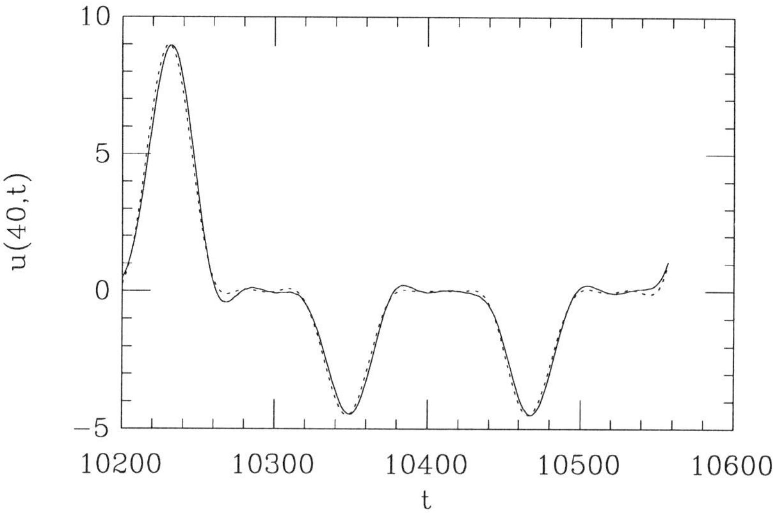


FIGURE 3. *Comparison of the exact solution (Slab 2) with the approximate solution from (3.2) with the "true" eigenvalues, $\lambda_m^\epsilon$, replaced by $\lambda_m^{(0)} + \epsilon\lambda_{*m}^{(1)}$. The approximate solution is shown in dashes. Note that the agreement is very good except for the dispersion at the tail of the main signals.*

To give a feeling for the time scale associated with Figure 2, consider that in Slab 1, it takes 351.60 units of time for a pulse to traverse the thickness of the slab twice. Therefore, in this time window, we are capturing the response after the initial pulse has travelled 60 thicknesses of the slab, or roughly, 12000 layer thicknesses of the microstructure. For these two slabs $L^2/h$ is roughly 24000 so $10200 \leq t \leq 10600$ is below $L^2/h$, but of the same order of magnitude. To see that the correction $\lambda_{*m}^{(1)}$ indeed predicts the phenomenon above we consider the approximate solution given by the sum in the right hand side of (3.2). We compute an approximate solution for the initial value problem associated with Slab 2 by replacing $\lambda_m^\epsilon$ with $\lambda_m^{(0)} + \epsilon\lambda_{*m}^{(1)}$ in this sum. A comparison with the exact solution is made in Figure 3. Note that the agreement between these two solutions is very good, except for the dispersion tail at the end of the main signals. This inherent dispersion is a higher order (in $\epsilon$) effect that is not captured by the approximate solution (see [4] for a discussion of dispersion). The speed of the main signal is indeed very well predicted by the approximate solution using the homogenized eigenvalues with a first order correction in place of the "true" eigenvalues. If we remove the contribution of the correction $\lambda_{*m}^{(1)}$ in the approximate solution, we find that the response we get has a large disagreement with the exact solution. This is shown in Figure 4. These last figures support our claim that the phase shift is largely due to the first order corrections of the eigenvalues, which again are strongly influenced by the interaction of the microstructure with the boundary.

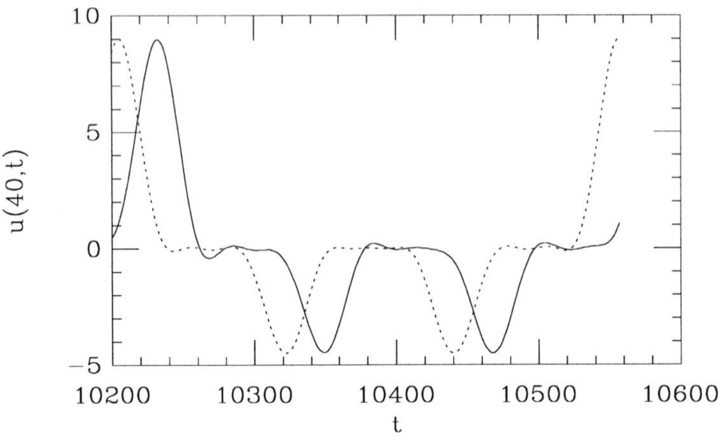


FIGURE 4. *Comparison of the exact solution (Slab 2) with the approximate solution from (3.2), with the "true" eigenvalues, $\lambda_m^\epsilon$, replaced by $\lambda_m^{(0)}$. The approximate solution is shown in dashes. This figure gives clear evidence that the correct phase shift is predicted by the first order correction $\lambda_{*m}^{(1)}$.*

## REFERENCES

[1] A. BENSOUSSAN, J.L. LIONS, AND G. PAPANICOLAOU, *Asymptotic Analysis for Periodic Structures*, North-Holland, Amsterdam 1980.

[2] S. MOSKOW AND M. VOGELIUS, *First order corrections to the homogenized eigenvalues of a periodic composite medium: a convergence proof for convex polygons* (to appear).

[3] F. SANTOSA AND M. VOGELIUS, *First order corrections to the homogenized eigenvalues of a periodic composite medium*, SIAM J. Appl. Math. **53**(6) (1993), pp. 1636–1668.

[4] F. SANTOSA AND W.W. SYMES, *A dispersive effective medium for wave propagation in periodic composites*, SIAM J. Appl. Math. **51**(4) (1991), pp. 984–1005.

# ON THE LINEARISED DYNAMICS OF LINKED MECHANICAL STRUCTURES*

E.J.P. GEORG SCHMIDT†

**Abstract.** In this paper we take a bird's eye view of the linear models which have recently been used by various authors in the area of boundary control of partial differential equations to describe the dynamics of mechanical structures which link together elastic and rigid obects. We will illustrate the modeling procedure with various models, old and new. We provide a general existence theorem for the very complicated systems involved in this modeling by using a familiar variational approach in a functional analytic setting.

**1. Introduction.** This paper is concerned with linear models describing the dynamics of systems of elastic or rigid mechanical elements jointed together in some fashion. Typical elements include elastic strings, beams, point masses, membranes or plates, rigid or elastic bodies. The structure may consist of many elements which are all of the same type, such as networks of strings ([20,9]) or beams ([2,7,9,10]) and structures built out of plates or membranes ([5,6]). It can also combine elements of various types; for example strings and point masses ([21]), strings and beams ([9]), beam and rigid body ([11] and [12]), plate and body ([15]), membrane and string ([17]), plate and Timoshenko beam ([4]).

The aims of this paper are

- to describe a general procedure for the linear modeling of such systems;
- to provide some examples;
- to prove a general existence theorem for the model equations, using familiar variational methods in an appropriate abstract setting.

Our approach, which has by now been repeatedly used in various papers, largely in collaborative work with J.Lagnese, G. Leugering and more recently with Wei Ming, is conceptually simple but does lead to complicated systems of equations and side conditions; it seems rather remarkable that the existence theory can be treated so simply and in such generality!

The work on modelling and existence of solutions has been done for its own sake but also pro vides the framework for results on controllability and stabilizability. We do not treat the latter questions here, referring only to the cited literature.

**2. General model for linked structures.** We suppose that the system contains $n$ elements indexed by $i \in \mathcal{I} = \{1 \leq i \leq n\}$. The state of the $i$–th element is given by a pair $r_i$ and $\dot{r}_i$ of elements belonging to

---

* Research supported by the Natural Sciences and Engineering Research Council of Canada Grant A7271.

† Mailing address: Department of Mathematics and Statistics, McGill University, 805 Sherbrooke Street West, Montreal, Quebec, Canada, H3A2K6.

Hilbert spaces $V_i$ and $H_i$ respectively. These elements respectively describe displacement and displacement velocities of the configuration variables. We let the inner products on these spaces be denoted by $< \cdot, \cdot >_{V_i}$ and $< \cdot, \cdot >_{H_i}$. The kinetic energy corresponding to the above state is given by $\| \dot{r}_i \|_{H_i}^2 = < \dot{r}_i, \dot{r}_i >_{H_i}$. We suppose that the potential energy is given in terms of a continuous, symmetric and positive semidefinite bilinear form $b_i(\cdot, \cdot)$ by $b_i(r_i, r_i)$; sometimes, but not always, this form may coincide with $< \cdot, \cdot >_{V_i}$. Concerning these spaces and forms we make the following standard assumptions:

**I** The space $V_i$ can be compactly and densely imbedded in $H_i$.
**II** For $r_i \in V_i$ one has $\| r_i \|_{V_i} \leq C \| r_i \|_{H_i}$.
**III** For $\lambda > 0$ the bilinear form

$$b_{i,\lambda}(\cdot, \cdot) = b_i(\cdot, \cdot) + \lambda < \cdot, \cdot >_{H_i} \tag{2.1}$$

is an inner product on $V_i$ equivalent to $< \cdot, \cdot >_{V_i}$ .

The appropriate expressions for the kinetic and potential energies are usually obtained for specific physical systems by linearization about an equilibrium; the models we describe therefore apply to small displacements away from equilibrium.

The JUNCTIONS in the linked system, by which we understand both the joints between different elements and the loci of interaction with the exterior of the system (often, but not always, subsets of the boundaries of the elements), are indexed by $j \in \mathcal{J} = \{1 \leq j \leq m\}$. With each j we associate the set $\mathcal{I}_j$ of indices of elements involved in the junction. We distinguish between MULTIPLE JUNCTIONS indexed by $j \in \mathcal{J}_M$ and involving more than one element, and SIMPLE JUNCTIONS indexed by $j \in \mathcal{J}_S$ which involve only one element. The nature of multiple junctions is expressed by continuity conditions on the displacements of the adjoining elements; these can be expressed abstractly in the form

$$\sum_{i \in \mathcal{I}_j} C_{ji} r_i = \mathbf{0}, \tag{2.2}$$

where $U_j$ is some Hilbert space associated with the $j$-th junction and $C_{ji} : V_i \mapsto U_j$ is a bounded linear operator. Such a relation may also be imposed at a simple junction of the system where, for example, an element may be clamped. The indices for which simple junctions are subject to such a condition will be denoted by $\mathcal{J}_S^C$. Conditions of this nature, formulated as bounded, linear constraints are called GEOMETRIC CONDITIONS. The homogeneous condition (2.2) can be replaced by an inhomogeneous condition with $u_j \in U_j$ replacing $\mathbf{0}$; this condition we refer to as a GEOMETRIC CONTROL and in familiar examples these are simply Dirichlet conditions.

The state of the linked system is then described by $r = (r_1, \dots, r_n) \in V$ and $r = (r_1, \dots, r_n) \in H$ where

$$V = \{r \in \prod_{i=1}^{n} V_i : (2.2) \text{ holds for } j \in \mathcal{J}_M \cup \mathcal{J}_S^C\}, \tag{2.3}$$

with corresponding inner product $< \cdot, \cdot >_V$, and

$$H = \text{ closure of } V \text{ in } \prod_{i=1}^{n} H_i, \tag{2.4}$$

with inner product $< \cdot, \cdot >_H$. On $V$ we define

$$b_\lambda(r, \tilde{r}) = \sum_{i=1}^{n} b_{i,\lambda}(r, \tilde{r}), \tag{2.5}$$

and $b(\cdot, \cdot) = b_0(\cdot, \cdot)$. The assumptions I, II and III (formulated without the subscript $i$) then hold for $V$, $H$ and $b_\lambda(\cdot, \cdot)$.

In addition to geometric controls, we allow also DYNAMIC CONTROLS such as distributed controls and controls involving Neumann conditions. In variational formulations of elliptic problems these appear as a linear functionals $u^* \in V^*$. Later we shall be more precise about the distinction between distributed and "Neumann" controls.

The equations which describe the dynamics of the linked system are obtained by applying Hamilton's principle to the Lagrangian function

$$\mathcal{L}(r(t)) = \int_0^T [< \dot{r}(t), \dot{r}(t) >_H - b(r(t), r(t)) + u^*(r(t))] \, dt, \tag{2.6}$$

where now $\dot{r}(t)$ indicates the derivative of $r(t)$ with respect to $t$. The condition that $\mathcal{L}(r(t))$ should be stationary is

$$\int_0^T [< \dot{r}(t), \delta\dot{r}(t) >_H - b(r(t), \delta r(t)) + u^*(\delta r(t))] \, dt \; = \; 0, \tag{2.7}$$

for all variations $\delta r(t)$ of $r(t)$ such that $r(t) + \delta r(t)$ satisfies the geometric conditions. This variational formulation translates into various dynamic equations and free junction conditions. In the next section we illustrate this procedure for a variety of specific linked structures and derive the equations describing their dynamics, while in Section 4 we provide an abstract existence theorem for these equations in the general context outlined above.

## 3. Examples, old and new

**3.1. String networks.** These are the subject of [20] (see also [9] for a general, streamlined approach). Elastic strings, indexed by $i \in \mathcal{I}$ are joined together at certain endpoints. We refer to the locations of endpoints as nodes rather than junctions; these are indexed by $j \in \mathcal{J}$. The model describes the small displacements of the string about an equilibrium in which the $i$-th string is stretched in the direction $v_i$ by a factor $s_i > 1$ and is parametrised by the natural arc length $x$ in $(0, l_i)$. The displacement of the $i$-th is given by a function $r_i \in V_i = W_1(0, l_i; R^3)$, the Sobolev space of $R^3$-valued functions having first order distribution derivatives representable by square integrable functions. The corresponding velocities are $\dot{r}_i \in H_i = L_2(0, l_i; R^3)$. In [20] we linearize a non-linear problem about the equilibrium described above to obtain the expression $b_i(r_i, r_i)$ for the potential energy of the i-th string, where

$$b_i(r_i, \tilde{r}_i) = \frac{1}{2}\int_0^{l_i} Q_i \partial r_i(x) \cdot \partial \tilde{r}_i(x)\, dx.$$

Here $Q_i$ is a $3 \times 3$ matrix having eigenvalues $h_i$ and $h_i(1 - s_i^{-1})$, whose corresponding eigenspaces are the space spanned by $v_i$ and the orthogonal complement of that space; $h_i$ is the constant in Hooke's law for the i-th string; $\partial$ denotes differentiation with respect to $x$. The kinetic energy of the i-th string is defined in terms of the following inner product on $H_i$:

$$< \dot{r}_i, \dot{\tilde{r}}_i >_{H_i} = \frac{1}{2}\int_0^{l_i} \rho_i \dot{r}_i \cdot \dot{\tilde{r}}_i \, dx.$$

where $\rho_i$ is the density of the $i$-th string.

At the multiple nodes the continuity of the displacements of the adjoining strings is expressed by

$$\text{(3.1)} \qquad \text{for } j \in \mathcal{J}_M, \quad r_i(x_{ij}, t) \text{ is the same for all } i \in \mathcal{I}_j,$$

where $x_{ij}$ is the parameter value 0 or $l_i$ of the endpoint of the $i$-th string where it meets the j-th node. This latter condition is not expressed in the form (2.2) but it can certainly be rewritten to conform to it.

For simple nodes we consider four alternatives. For $j \in \mathcal{J}_S$ let $\mathcal{I}_j = \{i\}$. The simple nodes indexed by $j \in \mathcal{J}_S^C$ are clamped (a geometric condition) so that

$$\text{(3.2)} \qquad r_i(x_{ij}, t) = 0.$$

Those indexed by $j \in \mathcal{J}_S^G$ will be subject to a Dirichlet control, geometric in nature,

$$\text{(3.3)} \qquad r_i(x_{ij}, t) = u_j(t).$$

Those indexed by $j \in \mathcal{J}_S^F$ will be free; this corresponds to a homogeneous Neumann condition

$$Q_i \partial r_i(x_{ij}, t) = 0. \tag{3.4}$$

Finally for $j \in \mathcal{J}_S^D$ we impose a Neumann control, dynamic in nature,

$$Q_i \partial r_i(x_{ij}, t) = u_j(t). \tag{3.5}$$

Now following the outline of Section 2 we define

$$V = \{r \in \prod_i^n V_i : (3.1) \text{ and } (3.2) \text{ hold.}\} \tag{3.6}$$

We apply Hamilton's principle to the Lagrangian

$$\mathcal{L}(r(t)) = \sum_{i=1}^n \frac{1}{2} \int_0^T \int_0^{l_i} [\rho_i |\dot{r}_i|^2 - Q_i \partial r_i \cdot \partial r_i]\, dx\, dt + \sum_{j \in \mathcal{J}_S^N} \epsilon_{ij} u_j(t) \cdot r_i(x_{ij}, t)$$

where $\epsilon_{ij} = 1$ if $x_{ij} = l_i$, or $-1$ if $x_{ij} = 0$. Using variations $\delta r_i$ which respect the geometric constraints (3.1), (3.2) and (3.3) the stationary condition (2.7) becomes

$$\sum_{i=1}^n \int_0^T \int_0^{l_i} [\rho_i \dot{r}_i \cdot \delta \dot{r}_i - Q_i \partial r_i \cdot \partial \delta r_i]\, dx\, dt + \sum_{j \in \mathcal{J}_S^D} \epsilon_{ij} u_j(t) \cdot \delta r_i(x_{ij}, t) = 0.$$

First, using variations $\delta r_i$ with support in $(0, l_i) \times (0, T)$, one obtains in a standard way involving integration by parts the equations

$$\rho_i \ddot{r}_i = \partial [Q_i \partial r_i] \quad \text{for} \quad i \in \mathcal{I}. \tag{3.7}$$

Subsequent variations with supports concentrated around multiple nodes yield the following dynamic (balance of forces) conditions at the multiple nodes,

$$\sum_{i \in \mathcal{I}_j} \epsilon_{ij} Q_i \partial r_i(x_{ij}, t) = 0 \quad \text{for } j \in \mathcal{J}_M. \tag{3.8}$$

At the simple nodes indexed by $j \in \mathcal{J}_S^F$ and $j \in \mathcal{J}_S^D$ the variational argument leads to the free condition (3.4) and the inhomogeneous condition (3.5) respectively.

The equations (3.7), the geometric conditions (3.1), (3.2) and (3.3) and the dynamic conditions (3.4), (3.5) and (3.8) have to be supplemented by initial w following the outline of Section 2 we define

$$V = \{r \in \prod_i^n V_i : (3.1) \text{ and } (3.2) \text{ hold.}\} \tag{3.9}$$

conditions prescribing $r(0)$ and $\dot{r}(0)$.

While each equation in (2.4) decomposes into transverse and longitudinal wave equations having different wave speeds, the components corresponding to different strings are "scrambled" at the multiple nodes by the geometric and dynamic node conditions.

In the above model it is natural for the matrices to have no spatial dependence. When the network is subject to gravity the equilibrium configuration will "sag" and then the matrices entering the linearized equations do depend on $x$ (see [21]).

**3.2. Timoshenko beam networks.** We now turn to planar Timoshenko beam networks as developed in [7]; see [9] for a detailed derivation and generalization to non-planar networks. In the planar beam model the configuration of each beam is given by the location of its centre line and the shear angle between the cross section and the normal to the centre line. We suppose that in the equilibrium configuration the center line is straight and the shear angle vanishes. Denoting the direction of the centre line by the unit vector $v_i$ and letting $v_i^{\perp}$ denote an orthogonal unit vector in the plane of the network, we denote by $u_i v_i + w_i v_i^{\perp}$ the displacement of the centre line and by $\psi_i$ the total rotation angle of the cross sections along the center line of the $i$-th beam. Now we take $p_i = 3$ and set $r_i = (u_i, w_i, \psi_i)$, requiring it to belong to the space $V_i = W_1(0, l_i; R^3)$. The corresponding velocity field $\dot{r}_i$ is required to lie in $H_i = L_2(0, l_i; R^3)$. The standard expressions for the potential and kinetic energies motivate the following definitions:

$$b_i(r, \tilde{r}) = \frac{1}{2} \int_0^{l_i} [E_i A_i \partial u_i \partial \tilde{u}_i + K_i(\partial w_i + \psi_i)(\partial \tilde{w}_i + \tilde{\psi}_i) + E_i I_i \partial \psi_i \partial \tilde{\psi}_i] \, dx,$$

and

$$< \dot{r}, \dot{\tilde{r}} >_{H_i} = \frac{1}{2} \int_0^{l_i} [\rho_i \dot{u}_i \dot{\tilde{u}}_i + \rho_i \dot{w}_i \dot{\tilde{w}}_i + I_{\rho_i} \dot{\psi}_i \dot{\tilde{\psi}}_i] \, dx.$$

These expressions involve the following physical constants associated with the $i$–th beam: $\rho_i$ is the density, $A_i$ is the cross-sectional length, $E_i$ is the Young's modulus, $I_i$ is the second moment of inertia, $I_{\rho_i}$ is the polar moment of inertia of a cross section and $K_i$ is the shear modulus.

One here requires the continuity of the displacements at multiple nodes, and for "rigid joints" one also demands continuity of the rotation angles $\psi_i$. Using the same index notation as for string networks, one has the following geometric multiple node conditions:

for $j \in \mathcal{J}_M$, $u_i(x_{ij}, t)v_i + w_i(x_{ij}, t)v_i^{\perp}$ and $\psi_i$ are the same for all $i \in \mathcal{I}_j$.
(3.10)

or simple nodes we consider the same alternatives as for string networks. For $j \in \mathcal{J}_S^C$ the nodes are clamped so that

$$r_i(x_{ij}, t) = \mathbf{0}, \tag{3.11}$$

which of course means that each of the variables $u_i$, $w_i$ and $\psi_i$ is set equal to zero Those indexed by $j \in \mathcal{J}_S^G$ will be subject to a Dirichlet control, in which these variables are prescribed. Those indexed by $j \in \mathcal{J}_S^F$ will be free, which again corresponds to homogeneous Neumann conditions

$$E_i I_i \partial u_i(x_{ij}, t) = 0, K_i[\partial w_i + \psi_i](x_{ij}, t) = 0, E_i I_i \partial \psi_i(x_{ij}, t) = 0. \tag{3.12}$$

Finally for $j \in \mathcal{J}_S^D$ we impose a dynamic Neumann control, in which the above quantities are prescribed; the three prescribed functions can be written as the components of a control function $u_j$. We remark that there are other possibilities. For example at some simple nodes the displacement components $u_i$ and $w_i$ might be prescribed while the angle $\psi$ could be free and therefore subject to a homogeneous Neumann condition .

Define

$$V = \{r \in \prod_i^n V_i : (3.9) \text{ and } (3.10) \text{ hold}\}.$$

Then one applies Hamilton's principle to the Lagrangian (2.6) with

$$u^*(r(t)) = \sum_{j \in \mathcal{J}_S^D} \epsilon_{ij} u_j(t) \cdot r_i(x_{ij}, t).$$

We obtain the following system of equations for each beam:

$$\left\{ \begin{array}{lcl} \rho_i \ddot{u}_i & = & E_i A_i \partial^2 u_i, \\ \rho_i \ddot{w}_i & = & K_i \partial(\partial w_i + \psi_i), \\ I_{\rho_i} \ddot{\psi}_i & = & E_i I_i \partial^2 \psi - K_i(\partial w_i + \psi_i). \end{array} \right. \tag{3.13}$$

In this model the longitudinal displacements are governed by a wave equation, while the transverse displacements interact with the rotation angle in the Timoshenko equations. For $j \in \mathcal{J}_M$ one also obtains dynamic conditions coupling the beams:

$$\left\{ \begin{array}{l} \sum_{i \in \mathcal{I}_j} \epsilon_{ij} E_i I_i \partial \psi_i(x_{ij}, t) = \mathbf{0}, \\ \sum_{i \in \mathcal{I}_j} \epsilon_{ij} [E_i I_i \partial u_i v_i + K_i(\partial w_i + \psi_i)) v_i^\perp](x_{ij}, t) = \mathbf{0}. \end{array} \right. \tag{3.14}$$

The free node conditions (3.10) will be satisfied along with the Neumann control conditions.

**3.3. Beams and strings.** As an example of a structure involving elements of different nature we consider some ways in which strings and beams can be joined together. It is useful to note that the Timoshenko model the beam is thought of as occupying a narrow rectangle $\{xv + yv^\perp : x \in (0, l), y \in (-h, h)\}$. The displacement at the point corresponding to $(x, y)$ is given by $r(x, y, t) = [u(x, t) + y\psi(x, t)]v + w(x, t)v^\perp$ (see the

Appendix to [9] for details). In setting up the expressions for potential and kinetic energies the variable $y$ is integrated away and enters the constants occurring in the Timoshenko equations. However in modeling the joints between Timoshenko beams and other objects it is helpful to think of the displacements in terms of both variables $x$ and $y$.

We shall now consider a planar network consisting of two strings indexed by 1 and 2 attached in various ways to one beam indexed by 3. For $i = 1, 2$ the spaces $V_i$ and $H_i$ as well as the forms $b_i(\cdot,\cdot)$ and inner products $< \cdot,\cdot >_{H_i}$ are chosen as in section 3.1 while for $i = 3$ the choice is made as in section 3.2 In the absence of dynamic controls this leads to the Lagrangian function

$$\begin{aligned} \mathcal{L}r &= \int_0^T \frac{1}{2}\Big[\sum_{i=1}^{2}\int_0^{l_i} [\rho_i|\dot{r}_i|^2 - Q_i\partial r_i.\partial r_i]\,dx \\ &+ \int_0^{l_3} \{[\rho_3\dot{u}_3^2 + \rho_3\dot{w}_3^2 + I_{\rho_3}\dot{\psi}_3^2] \\ &- [E_3A_3\partial u_3{}^2 + K_3(\partial w_3 + \psi_3)^2 + E_3I_3\partial\psi_3{}^2]\}\,dx\Big]\,dt. \end{aligned} \tag{3.15}$$

Now one needs to specify the nature of the joints, which will involve the endpoints of the string and endfaces of the beam. We note first that in all cases variations of $r_i$ inside $(0, l_i)$ will lead to the equations (3.7) for the two strings and the Timoshenko equations (3.12) for $i = 3$. General variations, for the moment not subject to any geometric constraints, then lead one to the variational identity

$$\begin{aligned} &\sum_{i=1}^{2}\Big[[Q_i\partial r_i\cdot\delta r_i](x,t)\Big]_{x=0}^{x=l_i} \\ &+\Big[[E_3A_3\partial u_3\delta u_3 + K_3(\partial w_3+\psi_3)\delta w_3 + E_3I_3\partial\psi_3\delta\psi_3](x,t)\Big]_{x=0}^{x=l_3} = 0. \end{aligned} \tag{3.16}$$

We do not write down the various controls or free conditions at the endpoints which are not joined to another element, where one has all the possibilities listed in the previous sections.

Consider first the situation where *the two strings are attached respectively to the two endpoints of the centerline of the beam.* Here one has the geometric constraints

$$\left\{\begin{array}{lllll} r_1(0,t) &=& r_3(0,0,t) &=& u_3(0,t)v_3 + w_3(0,t)v_3^\perp, \\ r_2(0,t) &=& r_3(l_3,0,t) &=& u_3(l_3,t)v_3 + w_3(l_3,t)v_3^\perp. \end{array}\right. \tag{3.17}$$

We consider variations $\delta r_1$, $\delta u_3$ and $\delta w_3$ which respect the first of the above conditions and vanish at the other endpoints and set $\delta r_2 = \underline{0}$ and

$\delta\psi_3 = 0$. Then, because

$$\delta r_1(0,t) = \delta u_3(0,t)v_3 + \delta w_3(0,t)v_3^\perp,$$

(3.15) reduces to

$$[-Q_1\partial r_1.v_3 - E_3A_3\partial u_3](0,t)\delta u_3(0,t) + [-Q_1\partial r_1.v_3^\perp - K_3(\partial w_3 + \psi_3)](0,t) \\ \delta w_3(0,t) = 0.$$

One deduces the following two dynamic node conditions associated with the junction between the first string and the beam:

$$\left\{ \begin{array}{l} Q_1\partial r_1(0,t)\cdot v_3 + E_3A_3\partial u_3(0,t) = 0, \\ Q_1\partial r_1(0,t)\cdot v_3^\perp + K_3[\partial w_3 + \psi_3](0,t) = 0. \end{array} \right. \tag{3.18}$$

Next we consider the situation where *the two strings are both attached to the centerline of the beam at one end*, which corresponds to the geometric constraint

$$r_1(0,t) = r_2(0,t) = r_3(0,0,t) = u_3(0,t)v_3 + w_3(0,t)v_3^\perp. \tag{3.19}$$

Here we take variations with $\delta\psi_3 = 0$ and $\delta r_1$, $\delta r_2$ and $\delta r_3$ having support near the junction. Then, using

$$\delta r_1(0,t) = \delta r_2(0,t) = \delta u_3(0,t)v_3 + \delta w_3(0,t)v_3^\perp,$$

we get from (3.15) that

$$\left\{ \begin{array}{l} Q_1\partial r_1(0,t)\cdot v_3 + Q_2\partial r_2(0,t)\cdot v_3 + E_3A_3\partial u_3(0,t) = 0, \\ Q_1\partial r_1(0,t)\cdot v_3^\perp + Q_2\partial r_2(0,t)\cdot v_3^\perp + K_3[\partial w_3 + \psi_3](0,t) = 0. \end{array} \right. \tag{3.20}$$

Finally we consider the situation where *the two strings are attached to the two extreme points of one endface of the beam*, leading to the constraints

$$\left\{ \begin{array}{lll} r_1(0,t) = & r_3(0,h,t) & = [u_3(0,t) + h\psi_3(0,t)]v_3 + w_3(0,t)v_3^\perp, \\ r_2(0,t) = & r_3(0,-h,t) & = [u_3(0,t) - h\psi_3(0,t)]v_3 + w_3(0,t)v_3^\perp. \end{array} \right. \tag{3.21}$$

For the variations we therefore must require

$$\left\{ \begin{array}{lll} \delta r_1(0,t) & = & [\delta u_3(0,t) + h\delta\psi_3(0,t)]v_3 + \delta w_3(0,t)v_3^\perp, \\ \delta r_2(0,t) & = & [\delta u_3(0,t) - h\delta\psi_3(0,t)]v_3 + \delta w_3(0,t)v_3^\perp. \end{array} \right.$$

This gives rise to the dynamic conditions (3.20) as before, together with the additional condition

$$h[Q_1\partial r_1(0,t) - Q_2\partial r_2(0,t)]\cdot v_3 + E_3I_3\partial\psi_3(0,t) \;=\; 0. \tag{3.22}$$

The three different arrangements of beam and strings lead to the same equations with different geometric and coupling conditions where they are joined together. Supplementing these by boundary conditions and controls together with initial conditions one is led to systems for which existence of solutions follows from Section 4. The systems behave in different ways. For example it can be shown that in the last situation, where the strings are joined to the extreme points of one endface of the beam and the other endface is clamped, the motion of the whole system can be exactly controlled from the remaining endpoints of the strings. In the other cases the beam cannot be controlled by control of the strings.

**3.4. String, rod and membrane.** We consider a system comprised of a string attached to a membrane by means of a rigid rod of negligible cross-section which is imbedded in the boundary of the membrane and to which the string is attached at the centrepoint. We suppose that the equilibrium configuration lies in a plane with the membrane occupying a domain $\Omega$ lying in the half plane $x_1 < 0$ while the string lies along the interval $(0, l)$ of the $x_1$-axis with the rod along the interval $[-h, h]$ of the $x_2$-axis. We wish to consider small transverse displacements $y_1(x_1, x_2, t)$ of the membrane and $y_2(x_1, t)$ of the string. Here we have used the notation $y_i$ in place of $r_i$. The displacements of the rod are conveniently described by the displacement $y_3(t)$ of its centrepoint, which coincides with $y_2(0, t)$, as well as by the angle $\psi(t)$ which the displaced rod makes with the equilibrium plane. We suppose that the unattached end of the string is clamped and that the displacement of the membrane will be controlled by Dirichlet data on a subset $\Gamma_0$ of the boundary $\partial\Omega$ of the membrane which should of course not meet the segment $\Gamma_2$ of the boundary along which the rod is imbedded. We let $\Gamma_1$ denote the remainder of $\partial\Omega$ which is free. We choose the following spaces:

$$V_1 = W(\Omega),\ H_1 = L_2(\Omega),\ V_2 = \{y \in W^1(0, l) : y(l) = 0\},$$

$$H_2 = L_2(0, l),\ V_3 = H_3 = R^2.$$

We shall define the bilinear forms and inner products to correspond to the potential and kinetic energies of the three elements with all physical constants set to 1 except for the mass of the rod which we shall suppose to be $m$. We set

$$(3.23)\quad b_1(y_1, \tilde{y}_1) = \int_\Omega \nabla y_1 \cdot \nabla \tilde{y}_1 \, dx_1 \, dx_2, \ <\dot{y}_1, \dot{\tilde{y}}_1>_{H_1} = \frac{1}{2}\int_\Omega \dot{y}_1 \dot{\tilde{y}}_1 \, dx_1 \, dx_2,$$

$$(3.24)\quad b_2(y_2, \tilde{y}_2) = \int_0^l \partial_1 y_2 \partial_1 \tilde{y}_2 \, dx_1, \ <\dot{y}_2, \dot{\tilde{y}}_2>_{H_2} = \frac{1}{2}\int_0^l \dot{y}_2 \dot{\tilde{y}}_2 \, dx_1,$$

$$(3.25)\quad \begin{aligned} b_3((y_3, \psi_3), (\tilde{y}_3, \tilde{\psi}_3)) &= 0, <(\dot{y}_3, \dot{\psi}_3), (\dot{\tilde{y}}_3, \dot{\tilde{\psi}}_3)>_{H_3} \\ &= \frac{m}{2}[\dot{y}_3 \dot{\tilde{y}}_3 + \frac{h^2}{3} \dot{\psi}_3 \dot{\tilde{\psi}}_3]. \end{aligned}$$

Here $\partial_1$ denotes the partial derivative with respect to $x_1$.

The fashion in which the membrane and the string are joined to the rod gives rise to the following linearized geometrical constraints (using the approximations $\cos\psi_3 \approx 1$ and $\sin\psi_3 \approx \psi_3$):

$$y_3(t) = y_2(0,t) \text{ and } y_1(0,x_2) = y_2(0,t) + x_2\psi_3(t) \text{ for } x_2 \in [-h,h]. \tag{3.26}$$

Since the string is clamped at its other endpoint we also have the condition

$$y_2(l,t) = 0. \tag{3.27}$$

The Dirichlet control on $\Gamma_0$ gives the condition

$$y_2(\cdot,t) = u(\cdot,t) \quad \text{on} \quad \Gamma_0. \tag{3.28}$$

We now apply Hamilton's principle to the Lagrangian

$$\begin{aligned}
\mathcal{L}(y_1,y_2,y_3,\psi_3) &= \frac{1}{2}\int_0^T \Big[\int_\Omega [|\dot{y}_1(x_1,x_2,t)|^2 - |\nabla y_1(x_1,x_2,t)|^2]\,dx_1\,dx_2 \\
&+ \int_0^l [|\dot{y}_2(x_1,t)|^2 - |\partial_1 y_2(x_1,t)|^2]\,dx_1 \\
&+ m[|\dot{y}_3(t)|^2 + \frac{h^2}{3}|\dot{\psi}_3(t)|^2]\Big]\,dt
\end{aligned}$$

with variations in the configuration functions respecting the geometric conditions (3.26), (3.27) and (3.28).

Variations in $y_1$ and $y_2$ having their supports in $\Omega$ and $(0,l)$ respectively lead to the two wave equations

$$\ddot{y}_1(x_1,x_2,t) = \Delta y_1(x_1,x_2,t) \text{ for } (x_1,x_2,t) \in \Omega \times [0,T] \tag{3.29}$$

and

$$\ddot{y}_2(x_1,t) = \partial_1^2 y_1(x_1,t) \text{ for } (x_1,t) \in (0,l) \times [0,T]. \tag{3.30}$$

Variations in $y_1$ which vanish on $\Gamma_0$ and $\Gamma_2$ lead to the Neumann condition

$$\frac{\partial}{\partial\nu} y_1 = 0 \quad \text{on} \quad \Gamma_1. \tag{3.31}$$

What happens along the rod is more interesting. We can vary the functions in such a way that the variations $\delta\psi_3(t)$ and $\delta y_2(0,t)$ are essentially arbitrary; they then determine

$$\begin{aligned}
\delta y_1(0,x_2,t) &= \delta y_2(0,t) + x_2\delta\psi_3(t) \\
\text{for } x_2 \in [-h,h],\ \delta y_3(t) &= \delta y_2(0,t).
\end{aligned} \tag{3.32}$$

Taking variations of the Lagrangian, integrating by parts over $\Omega$ and $(0,l)$ and discarding the terms involving the two wave equations one is left with

$$\int_0^T \Big[\int_{-h}^{h} \partial_1 y_1(0,x_2,t)[\delta y_2(0,t) + x_2\delta\psi_3(t)]\,dx_2 - \partial_1 y_2(0,t)\delta y_2(0,t) + m[\dot{y}_3(t)\delta\dot{y}_3(t) + \frac{h^2}{3}\dot{\psi}_3(t)\delta\dot{\psi}_3(t)]\Big]\,dt = 0.$$

Using (3.26) and (3.32) and integrating the latter terms by parts with respect to. This can be rewritten as

$$\int_0^T \Big[[\int_{-h}^{h} \partial_1 y_1(0,x_2,t)\,dx_2 - \partial_1 y_2(0,t) - m\ddot{y}_2(t)]\delta y_2(0,t) + [\int_{-h}^{h} \partial_1 y_1(0,x_2,t)x_2\,dx_2 - m\frac{h^2}{3}\ddot{\psi}_3(t)]\delta\psi_3(t)\Big]\,dt = 0.$$

This leads to the following non-local, dynamic, multiple node conditions:

$$\int_{-h}^{h} \partial_1 y_1(0,x_2,t)\,dx_2 - \partial_1 y_2(0,t) = m\ddot{y}_2(t)$$

and

$$\int_{-h}^{h} \partial_1 y_1(0,x_2,t)x_2\,dx_2 = m\frac{h^2}{3}\ddot{\psi}_3(t). \tag{3.33}$$

The complete problem now becomes to solve the equations (3.29) and (3.30), subject to the geometric conditions (3.26), (3.27) and (3.28), the dynamic conditions (3.31) and (3.32), as well as initial conditions on $y_1$, $\dot{y}_1$, $y_2$, $\dot{y}_2$, \$$y_3$, $\dot{y}_3$, $\psi_3$, $\dot{\psi}_3$. The existence of solutions for this apparently complicated system involving in particular boundary conditions which are non local and contain second order time derivatives can be treated quite simply as a special case of the result in the next section.

We point out that in [17] Puel and Zuazua have treated a system linking a membrane to a string, without the intervention of a rod. Their model also involves a non-local junction condition. While their modelling of the junction does not seem to be physically realizable, it does have the virtue of leading to a result on exact controllability, which this author has not been able to achieve for the model described here.

**4. The existence theorem.** The general approach is quite familiar from treatments of the wave equation. The equations of the previous section can all be rewritten in the form

$$\frac{d}{dt} r(t) \times \dot{r}(t) = \dot{r}(t) \times \mathcal{A}r(t), \tag{4.1}$$

where $\mathcal{A}$ is a linear map associated with the stationary problem and defined in terms of the bilinear form $b(\cdot,\cdot)$. The solution $r(t) \times \dot{r}(t)$ defines a trajectory in $V \times H$.

We want to introduce controls into the problem. These can be of two kinds: geometric controls (such as controls entering as Dirichlet data) and dynamic controls (such as controls in the form of Neumann data or forcing terms). The geometric controls enter very simply as a linear constraint

$$\mathcal{G}r(t) = u^G(t), \tag{4.2}$$

where the geometric control function $u^G(t)$ takes its values in a space $U^G$ and $\mathcal{G} : V \mapsto U^G$ is a bounded, linear map which is assumed to be surjective. The dynamic controls enter as linear functionals in the weak formulation of the problem. We distinguish between distributed controls defined by linear functionals on $V$ having the form $< f, \phi >_H$ with $f$ in some subspace of $H$, and "Neumann controls" $u^N$ belonging to a space $U^N$ which can be imbedded in the dual space $V^*$ of $V$ and satisfies the following "independence" assumption over the space $V_0 = \ker(\mathcal{G})$:

$$\left\{ \begin{array}{l} f \in H,\ u^N \in U^N, < f, \phi >_H + u^N(\phi) = 0 \\ \text{for all } \ \phi \in V_0 \ \text{ implies } f = 0 \text{ and } u^N = 0. \end{array} \right. \tag{4.3}$$

**Remark**: From this assumption it follows in particular that $V_0$ is dense in $H$. It also follows that if one lets $\mathcal{R}_0 : V^* \mapsto V_0^*$ be the restriction of linear functionals on $V$ to $V_0$ the composition $\mathcal{R}_0 \circ \mathcal{N} : U^N \mapsto V_0^*$ is injective.

We shall next give a weak sense to the problem

$$[\lambda - \mathcal{A}]r = f, \ \ \mathcal{N}r = u^N, \ \ \mathcal{G}r = u^G, \tag{4.4}$$

where $\mathcal{A} : V \mapsto H$ and $\mathcal{N} : V \mapsto U^N$ will be defined below on a suitable domain dense in $V$ and will both turn out to be surjective.

The stationary problem (4.4) takes the following weak form involving test functions in $V_0 = \ker(\mathcal{G})$: given $f \in H$, $u^G \in U^G$ and $u^N \in U^N$ find $r \in V$ such that

$$\left\{ \begin{array}{rcl} b_\lambda(r, \phi) & = & < f, \phi >_H + u^N(\phi), \ \text{ for any } \phi \in V_0, \\ \mathcal{G}r & = & u^G. \end{array} \right. \tag{4.5}$$

**Remark**: The homogeneous geometric conditions (such as the geometric conditions on junctions and the clamped endpoint conditions in string and beam networks) are built into the space $V$ while the homogeneous dynamic conditions (such as the dynamic multiple node conditions and the free endpoint conditions on string and beam networks) are implicit in the variational formulation.

THEOREM 4.1. *Let $f \in H$, $u^G \in U^G$ and $u^N \in U^N$. For $\lambda > 0$ the weak equation (4.5) has a unique solution in $V$, which satisfies the estimate*

$$\| r \|_V \leq C[\| f \|_H + \| u^N \|_{U^N} + \nu^G \|_{U^G}]. \tag{4.6}$$

*For $\lambda = 0$ the same is true if $b(\cdot,\cdot)$ is positive definite on $V_0$. Otherwise there exist solutions if and only if for some $r^G \in V$ with $\mathcal{G}r^G = u^G$ one has $< f, \phi >_H + u^N(\phi) - b(r^G, \phi) = 0$ for any $\phi$ in*

$$V_{00} = \{r \in V_0 : b(r, \phi) = 0 \text{ for all } \phi \in V_0\}. \tag{4.7}$$

*The solutions are not unique. The solution set is of the form $r_p + V_{00}$, where $r_p$ is any particular solution; this may be chosen to satisfy the estimate (4.6).*

*Proof.* Set $l(\phi) =< f, \phi >_H + u^N(\phi) \in V^*$. Pick any $r^G \in V$ such that $\mathcal{G}r^G = u^G$ and seek a solution of (4.5) having the form $r = r^G + r$ with $r \in V_0$. This must satisfy

$$b_\lambda(r, \phi) = l(\phi) - b(r^G, \phi) \text{ for all } \phi \in V_0.$$

Since, for $\lambda > 0$, $b_\lambda(\cdot,\cdot)$ is an inner product on $V_0$ it follows from the Riesz representation theorem that the latter equation indeed has a unique solution $r$ so that $r$ is a solution of (4.5). It is easy to check that this solution is unique. Moreover one has the estimate

$$\begin{aligned} \| r \| &\leq \| l(\cdot) - b_\lambda(r^G, \cdot) \|_{V_0^*} \\ &\leq C[\| f \|_H + \| u^N \|_{V_0^*} + \| r^G \|_V]. \end{aligned}$$

The estimate (4.6) now follows from this if one remarks that as a consequence of the boundedness and surjectivity of $\mathcal{G} : V \mapsto U^G$

$$|||u^G||| = \inf\{\| r^G \|_V \in V : \mathcal{G}r^G = u^G\}$$

defines a norm on $U^G$ equivalent to $\| u^G \|_{U^G}$.

Now we consider the case $\lambda = 0$. In case $b(\cdot,\cdot)$ is positive definite on $V_0$ it is an inner product equivalent to $< \cdot, \cdot >_V$ on $V_0$ and the previous argument goes through. Otherwise one checks easily that the given condition for solvability is necessary, and also that it is independent of the choice of $r^G$. The sufficiency of the condition is obtained by using it to factor the functional $l(\phi)$, defined as above, through the quotient space $V_0/V_{00}$ and by applying the Riesz representation theorem on that space to the quotient functional. □

THEOREM 4.2. *Let $\mathcal{S}_\lambda : H \times U^N \times U^G \mapsto V$ be defined for $\lambda > 0$ and in case $b(\cdot,\cdot)$ is positive definite on $V_0$ also for $\lambda = 0$ by*

$$\mathcal{S}_\lambda(f, u^N, u^G) = r, \tag{4.8}$$

*where $r$ is the solution of the weak problem (4.5). Then*

**a.** *$\mathcal{S}_\lambda$ is injective.*

**b.** *$\mathcal{S}_\lambda$ has an image $P$ which is independent of $\lambda$ and dense in $V$; $P_0 = P \cap V_0$ is dense in $H$.*

c. *One can define a surjective map $\mathcal{A}_\lambda : P \mapsto H$ by $\mathcal{A}_\lambda \mathcal{S}_\lambda(f, u^N, u^G) = f$; then $\mathcal{A}_\lambda - \lambda$ is independent of $\lambda$.*

**d.** *One can define a surjective map $\mathcal{N} : P \mapsto U^N$ by $\mathcal{N}\mathcal{S}_\lambda(f, u^N, u^G) = u^N$; $\mathcal{N}$ does not depend on $\lambda$.*

*Proof.* To prove (a) note that if $\mathcal{S}_\lambda(f, u^N, u^G) = \mathcal{S}_\lambda(\tilde{f}, \tilde{u}^N, \tilde{u}^G) = r$ then certainly $u^G = \mathcal{G}r = \tilde{u}^G$. Moreover $< f - \tilde{f}, \phi >_H + [u^N - \tilde{u}^N](\phi) = 0$ for all $\phi \in V_0$; by our independence assumption (4.3) $\tilde{f} = f$ and $\tilde{u}^N = u^N$.

Next we note that (4.5) is equivalent to

$$(4.9)\begin{cases} b_\mu(r, \phi) &= < f + [\mu - \lambda]r, \phi >_H + u^N(\phi), \quad \text{for any } \phi \in V_0, \\ \mathcal{G}r &= u^G. \end{cases}$$

From this it follows that

$$\mathcal{S}_\mu(f + [\mu - \lambda]\mathcal{S}_\lambda(f, u^N, u^G), u^N, u^G) = \mathcal{S}_\lambda(f, u^N, u^G) \tag{4.10}$$

Therefore $P = \operatorname{im}(\mathcal{S}_\lambda)$ is independent of $\lambda$.

To show that $P$ is dense in $V$ we note first that any $r \in V$ can be written as $r = r_0 + r_1$ with $r_1 = \mathcal{S}_\lambda(0, 0, \mathcal{G}r) \in P$ and $r_0 = r - r_1 \in V_0$, so that it is enough to show that $P_0$ is dense in $V_0$. Suppose that $\tilde{r} \in V_0$ is orthogonal to $P_0$ with respect to the inner product $b_\lambda(\cdot, \cdot)$. Then, for any $f \in H$,

$$0 = b_\lambda(\mathcal{S}(f, 0, 0), \tilde{r}) = < f, \tilde{r} >_H,$$

so that $\tilde{r} = 0$ and the density follows.

To show that $P_0$ is dense in $H$ suppose that $\tilde{f}$ is orthogonal to $P_0$ in $H$ so that in particular (4.5) with $r = \phi = \tilde{r} = \mathcal{S}_\lambda(\tilde{f}, 0, 0) \in V_0$ becomes

$$0 = < \tilde{f}, \tilde{r} >_H = b_\lambda(\tilde{r}, \tilde{r}).$$

It follows that $\tilde{r} = 0$ so that $\tilde{f}$ also vanishes.

Because $\mathcal{S}_\lambda$ is injective $\mathcal{A}_\lambda$ is well defined. It follows from (4.9) with $r = \mathcal{S}_\lambda(f, u^N, u^G)$ that $\mathcal{A}_\mu r = \mathcal{A}_\lambda r + [\mu - \lambda]r$. Hence $\mathcal{A}_\lambda - \lambda$ is independent of $\lambda$. From (4.9) it also follows at once that $\mathcal{N}$ is well defined and independent of $\lambda$. □

By assertion **c** of the last theorem we can define $\mathcal{A} = \mathcal{A}_\lambda - \lambda$. We also define $\mathcal{A}_0$ as the restriction of $\mathcal{A}$ to

$$\operatorname{dom}(\mathcal{A}_0) = \{r \in P : \mathcal{G}r = 0, \ \mathcal{N}r = 0\}.$$

Then

THEOREM 4.3. *$\mathcal{A}_0$ is a self adjoint operator on $H$. Its spectrum consists of a monotone decreasing sequence $\{-\lambda_j\}_{k=1}^{\infty}$ of eigenvalues each* *having finite multiplicity. If $b(\cdot,\cdot)$ is positive definite on $V_0$ the eigenvalues are all negative; otherwise if it is only non-negative the first eigenvalue $-\lambda_1$ is 0. The operator $\mathcal{B}_0$ on $V_0 \times H$ defined by*

$$\mathcal{B}_0[r \times \dot{r}] = \dot{r} \times \mathcal{A}_0 r, \ \textit{for}\ r \times \dot{r} \in dom(\mathcal{B}_0) = dom(\mathcal{A}_0) \times V_0$$

*is skew adjoint and hence generates a group of unitary operators $\{\mathcal{U}_0(t)\}_{t\in R}$ on $V_0 \times H$.*

The proof of this theorem is entirely standard and exploits the symmetry of $b(\cdot,\cdot)$ as well as the compactness of the imbedding of $V_0$ in $H$ which together with previous arguments and definitions imply that, for $\lambda > 0$, $\mathcal{A}_0 - \lambda$ is the inverse of the compact, bounded and symmetric operator $f \mapsto \mathcal{S}_\lambda(f, 0, 0)$ on $H$.

For initial data $r^0 \times \dot{r}^0 \in \text{dom}(\mathcal{B}_0)$ and $f(t) \in C^1([0,T]; H)$ the $V_0$-component of

$$\mathcal{U}_0(t)[r^0 \times \dot{r}^0] + \int_0^t \mathcal{U}_0(t-s)[0 \times f(s)]ds$$

provides a solution of the hyperbolic system

$$(4.11) \quad \ddot{r} = \mathcal{A}r + f, \quad \mathcal{N}r = 0, \quad \mathcal{G}r = 0, \quad r(0) = r^0, \quad \dot{r}(0) = \dot{r}^0,$$

such that $r(t) \in C_0([0,T]; P_0) \cap C^1([0,T]; V_0) \cap C^2([0,T]; H)$. For details see [Pazy [16], Chapter 4, Theorem 2.4 and Corollary 2.5] .

We also wish to consider the inhomogeneous hyperbolic problem

$$(4.12)\ \ddot{r} = \mathcal{A}r + f, \quad \mathcal{N}r = u^N, \quad \mathcal{G}r = u^G, \quad r(0) = r^0, \quad \dot{r}(0) = \dot{r}^0.$$

We can prove the following existence theorem.

THEOREM 4.4. *Let $f \in C^1([0,T]; H)$, $u^D \in C^3([0,T]; U^N)$, $u^G \in C^3([0,T]; U^G)$ and $r^0 \times \dot{r}^0 \in P \times V$ satisfy the following compatibility conditions:*

$$(4.13) \qquad u^N(0) = \mathcal{N}r^0, \quad u^G(0) = \mathcal{G}r^0, \quad \dot{u}^G(0) = \mathcal{G}\dot{r}^0.$$

*Then (4.11) has a solution $r(t)$ in $C([0,T]; P) \cap C^1([0,T]; V) \cap C^2([0,T]; H)$.*

*Proof.* The solution can be obtained as the sum $r(t) = \tilde{r}(t) + s(t)$ where, for each t, $\tilde{r}(t)$ is the solution of

$$(4.14) \qquad \mathcal{A}_\lambda \tilde{r}(t) = 0, \quad \mathcal{N}\tilde{r}(t) = u^N(t), \quad \mathcal{G}\tilde{r}(t) = u^G(t),$$

while $s(t)$ is the solution of

$$\ddot{s} = \mathcal{A}s + \tilde{f}, \quad \mathcal{N}s = 0, \quad \mathcal{G}s = 0, \quad s(0) = s^0 = r^0 - \tilde{r}(0), \quad \dot{s}(0) = \dot{s}^0 = \dot{r}^0 - \dot{\tilde{r}}(0), \tag{4.15}$$

with

$$\tilde{f}(t) = [f - \lambda\tilde{r} - \ddot{\tilde{r}}](t).$$

One can verify that $\tilde{f} \in C^1([0,T];H)$ and also easily check $s^0 \times \dot{s}^0 \in \text{dom}(\mathcal{B}_0)$. The solution of problem (4.15) is then obtained with the aid of $\mathcal{U}_0(t)$, as described above. □

One would, of course, like to weaken the the regularity assumption on the data, possibly allowing for weaker solutions. This can often be done with the help of a priori inequalities. Such inequalities, which are also crucial to the proof of results on controllability, are frequently obtained by continuity starting with smoother data which allow enough regularity to justify extensive integrations by parts. These regularity questions cannot be treated in abstract generality. However a useful first step towards establishing additional regularity, at least with respect to $t$, can be taken using the following result which can be inferred from [Pazy [16], Chapter 4].

LEMMA 4.1. *Let $\mathcal{B}$ be the generator of a strongly continuous semigroup $\mathcal{U}(t)$ on a Banach space $X$. Suppose that $f(t) \in C_0^k((0,T];X)$ and that $x^0 \in dom(\mathcal{B}^k)$. Then the equation*

$$\frac{d}{dt}x(t) = \mathcal{B}x(t) + f(t), \quad x(0) = x^0$$

*has a solution*

$$x(t) \in C^k([0,T];X) \cap C^{k-1}([0,T]; dom(\mathcal{B})),$$

*where $dom(\mathcal{B})$ is normed by the graph norm $\| x \|_{\mathcal{B}}^2 = \| \mathcal{B}x \|_X^2 + \| x \|_X^2$. Moreover, for $i = 1, \ldots, k$, the i-th order derivatives $x^{(i)}(t)$ satisfy the equation*

$$\frac{d}{dt}x^{(i)}(t) = \mathcal{B}x^{(i)}(t) +^{(i)} (t), \quad x^{(i)}(0) = \mathcal{B}^{(i)}x^0.$$

Applying this to the solution $s(t)$ of (4.15) and taking into account the regularity of $\tilde{r}(t)$ which follows from the hypotheses one readily deduces the assertion of the next regularity theorem.

THEOREM 4.5. *Let $r^0 \times \dot{r}^0 \in dom(\mathcal{B}_0^k)$ and suppose that $f(t) \in C_0^k((0,T];H)$, $u^N(t) \in C_0^{k+2}((0,T];U^N)$ and $u^G(t) \in C_0^k((0,T];U^G)$. Then the solution $r(t)$ of (4.12) belongs to $C^{k+1}([0,T]:H) \cap C^k([0,T]:V) \cap C^{k-1}([0,T]:P)$, where $P$ is topologised by the norm*

$$\| r \|_P^2 = \| \mathcal{A}r \|_H^2 + \| \mathcal{N}r \|_{U^N}^2 + \| \mathcal{G}r \|_{U^G}^2 .$$

## REFERENCES


[1] J.A.C. Bresse, *Cours de Mécanique Appliqué*, Mallet Bachelier (1859).

[2] G. Chen, M. Delfour, M. Krall and G. Payre, *Modeling, stabilization and control of serially connected beams*, SIAM J. Cont. and Opt. **25** (1987), pp. 526-546.

[3] P.R. Garabedian, *Partial Differential Equations*, John Wiley and Sons (1964).

[4] J.E. Lagnese, *Modelling and Controllability of Plate-Beam Networks*, manuscript, March (1992).

[5] J.E. Lagnese, *Boundary Controllability in Transmission Problems for Thin Plates*, to appear in "Differential Equations, Dynamical Systems, and Control Science: A Festschrift in Honor of Lawrence Markus".

[6] J.E. Lagnese and J.E. Leugering, *Modelling of Dynamic Networks of Thin Elastic Plates*, to appear in Math. Meth. Appl. Sci.

[7] J.E. Lagnese, G. Leugering and E.J.P.G. Schmidt, *Control of Planar Networks of Timoshenko Beams*, to appear in SIAM J. Cont. and Opt.

[8] J.E. Lagnese, G. Leugering and E.J.P.G. Schmidt, *Modelling of dynamic networks of thin thermoelastic beams*, to appear in Math. Meth. Appl. Sci.

[9] J.E. Lagnese, G. Leugering and E.J.P.G. Schmidt, *On the analysis and control of hyyperbolic systems associated with vibrating networks*, to appear in Proceedings of the Royal Society of Edinburgh.

[10] G. Leugering and E.J.P.G. Schmidt, *On the control of networks of vibrating strings andd beams*, Proc of the 28th IEEE CDC **3** (1989), pp. 2287-2290.

[11] W. Littman and L. Markus, *Exact boundary controllability of a hybrid system of elasticity*, Arch. Rat. Mech. Anal. **103** (1988), pp. 193-236.

[12] W. Littman and L. Markus, *Stabilization of a hybrid system of elasticity*, Ann. di mat. pura ed appl., Serie IV **CLII** (1988), pp. 281-330.

[13] W. Littman and L. Markus, *Remarks on exact controllability and stabilization of a hybrid system in elasticity through boundary damping*, Proceedings of a Symposium in Santiago, Spain (1987).

[14] W. Littman and L. Markus, *Some recent results on control and stabilization of flexible structures*, Proc. of COMCON Workshop on Stabilization of Flexible Structures, pp. 151-161, Montpellier (1987).

[15] S. Nicaise, *Contrôlabilité exacte d'un problème couplépluri -dimensionel*, C.R. Acad Sci Paris, Série I, Math. **311** (1990), pp. 19-22.

[16] A. Pazy, *Semigroups of Linear Operators and Applications to Partial Differential Equations*, Springer–Verlag (1983).

[17] J.P. Puel and E. Zuazua, *Exact controllability for a model of a multidimensional flexible structure*, manuscript (October 1991).

[18] J. Rauch and M. Taylor, *Exponential Decay of Solutions to Hyperbolic Equations in Bounded Domains*, Indiana Univ. Math. Jour. **24** (1974), pp. 79-86.

[19] D.L. Russell, *Controllability and stabilizability theory for linear partial differential equations: recent progress and open questions*, SIAM Rev. **20**(1978), pp. 639-739.

[20] E.J.P.G. Schmidt, *On the modelling and exact controllability of networks of vibrating strings*, SIAM J. Cont. and Opt., **30**(1992), pp. 229-245.

[21] E.J.P.G. Schmidt and Wei Ming, *On the modelling and analysis of networks of vibrating strings and masses*, Report #91-13 from the Department of Mathematics and Statistics, McGill University (1991).